External Magnetic Field Effects on Hydrothermal Treatment of Nanofluid

Numerical and Analytical Studies

External Magnetic Field Effects on Hydrothermal Treatment of Nanofluid

Numerical and Analytical Studies

Mohsen Sheikholeslami Kandelousi
(Mohsen Sheikholeslami)
Department of Mechanical Engineering,
Babol University of Technology

Davood Domairry Ganji
Department of Mechanical Engineering,
Babol University of Technology

ELSEVIER

AMSTERDAM • BOSTON • HEIDELBERG • LONDON
NEW YORK • OXFORD • PARIS • SAN DIEGO
SAN FRANCISCO • SINGAPORE • SYDNEY • TOKYO
William Andrew is an imprint of Elsevier

William Andrew is an imprint of Elsevier
The Boulevard, Langford Lane, Kidlington, Oxford, OX5 1GB, UK
50 Hampshire Street, 5th Floor, Cambridge, MA 02139, USA

Notices

Knowledge and best practice in this field are constantly changing. As new research and experience broaden our understanding, changes in research methods, professional practices, or medical treatment may become necessary.

Practitioners and researchers must always rely on their own experience and knowledge in evaluating and using any information, methods, compounds, or experiments described herein. In using such information or methods they should be mindful of their own safety and the safety of others, including parties for whom they have a professional responsibility.

To the fullest extent of the law, neither the Publisher nor the authors, contributors, or editors, assume any liability for any injury and/or damage to persons or property as a matter of products liability, negligence or otherwise, or from any use or operation of any methods, products, instructions, or ideas contained in the material herein.

British Library Cataloguing-in-Publication Data
A catalogue record for this book is available from the British Library

Library of Congress Cataloging-in-Publication Data
A catalog record for this book is available from the Library of Congress

ISBN: 978-0-323-43138-5

Contents

List of Figures

List of Tables

Preface

In this book, we provide readers with the fundamentals of the effect of magnetic field on nanofluid hydrothermal behavior. As a magnetic field is used, two important effects should be considered: Magnetohydrodynamic (MHD) and Ferrohydrodynami (FHD). Several methods exist to investigate MHD and FHD. This book introduces the applications of new numerical and semianalytical methods for such problems. The book also discusses different models for simulation of nanofluid. There are two models for simulating nanofluid flow and heat transfer: single-phase and two-phase model. In the single-phase model, nanoparticles are in thermal equilibrium, and there are not any slip velocities between the nanoparticles and fluid molecules; thus, they have a uniform mixture of nanoparticles. In the two-phase model, the nanoparticles cannot accompany fluid molecules because of some slip mechanisms such as Brownian motion and thermophoresis, so the volume fraction of nanofluid may not be uniform anymore, and there would be a variable concentration of nanoparticles in a mixture. This text is suitable for senior undergraduate students, postgraduate students, engineers, and scientists.

Chapter 1 of this book deals with the essential fundamentals of MHD and FHD. The different models for simulation of nanofluid are discussed. Furthermore, the governing equations for natural convection and mixed convection of nanofluid in the presence of a magnetic field are presented. Chapter 2 deals with the control volume finite element method (CVFEM). This method combines interesting characteristics from both the finite-volume and finite-element methods. CVFEM combines the flexibility of the finite-element methods to discretize complex geometry with the conservative formulation of the finite-volume methods in which the variables have easy physical interpretation in terms of fluxes, forces, and sources. After introducing this method, several examples are presented in which MHD and FHD effects are considered.

Chapter 3 gives a complete account of the new semianalytical methods. The names of our selected methods are homotopy perturbation method, Galerkin optimal homotopy asymptotic method, differential transformation method, Adomian decomposition method, and homotopy analysis method. At first we present the basic idea and application of these methods in a simple example. Then we provide different examples about MHD nanofluid flow and heat transfer. In Chapter 4, the lattice Boltzmann method (LBM) is presented. In traditional computational fluid dynamics methods, Navier–Stokes equations are solved on discrete nodes, elements, or volumes. In other words, the nonlinear partial differential equations convert into a set of nonlinear algebraic equations, which are solved iteratively. In LBM, the fluid is replaced by fractious particles. These particles stream along given directions (lattice links) and collide at the lattice sites. Another advantage of LBM is that it can handle complex phenomena such as moving boundaries (multiphase, solidification, and melting problems) naturally without the need for a face-tracing method as it is in the traditional computational fluid dynamics. Several examples included in Chapter 4 give readers a full account of the theory and practice associated with the LBM. Several sample codes of the aforementioned numerical and semianalytical methods are presented in the Appendix. Readers will be able to extend this code and solve all the examples that exist in this book.

Mohsen Sheikholeslami Kandelousi (Mohsen Sheikholeslami)
Davood Domairry Ganji

Nomenclature

A	Amplitude
B	Magnetic induction
C_p	Specific heat at constant pressure
C_f, $\tilde{C}_f$	Skin friction coefficients
c_s	Speed of sound in lattice scale
D_B	Brownian diffusion coefficient
D_T	Thermophoretic diffusion coefficient
Ec	Eckert number
En	Heat transfer enhancement
e_α	Discrete lattice velocity in direction
f_k^{eq}	Equilibrium distribution
g	Internal energy distribution functions
g^{eq}	Equilibrium internal energy distribution functions
g_z	Acceleration due to gravity
Gr_f	Grashof number
H_x, H_y	Components of the magnetic field intensity
H	The magnetic field strength
Ha	Hartmann number
J	Electric current
Le	Lewis number
M	Magnetization
N	Number of undulations
Nb	Brownian motion parameter
Nt	Thermophoretic parameter
Nu	Nusselt number
Nr	Buoyancy ratio number
Pr	Prandtl number
S	Squeeze number
Sc	Lewis number
T	Fluid temperature
T_c'	Curie temperature
u,v	Velocity components in the x-direction and y-direction
U,V	Dimensionless velocity components in the x-direction and y-direction
$u_w(x)$	velocity of the stretching surface
x,y	Space coordinates
X,Y	Dimensionless space coordinates
r	Nondimensional radial distance
k	Thermal conductivity
L	Gap between inner and outer boundary of the enclosure
Ra	Rayleigh number

q''	Heat flux
Rd	Radiation parameter

Greek symbols

γ	Angle measured from right plane
ζ	Inclination angle
γ	Angle of turn of the semiannulus enclosure
ε	Eccentricity
α	Thermal diffusivity
σ	Electrical conductivity
ϕ	Volume fraction
μ	Dynamic viscosity
ω,Ω	Vorticity and dimensionless vorticity
υ	Kinematic viscosity
ψ and Ψ	Stream function and dimensionless stream function
Θ	Dimensionless temperature
ρ	Fluid density
β	Thermal expansion coefficient
μ_0	Magnetic permeability of vacuum
σ_e	Stefan–Boltzmann constant
β_R	Mean absorption coefficient
λ	Dimensionless suction/injection parameter

Subscripts

c	Cold
h	Hot
loc	Local
ave	Average
nf	Nanofluid
f	Base fluid
p	Solid particles
in	Inner
out	Outer
eq	Equilibrium distribution function
neq	Nonequilibrium distribution function

CHAPTER

1

Magnetohydrodynamic and ferrohydrodynamic

CHAPTER OUTLINE

1.1 MAGNETOHYDRODYNAMIC

1.1.1 DEFINITION

Magnetohydrodynamic (MHD) (magnetofluid dynamics or hydromagnetics) is the study of the dynamics of electrically conducting fluids. Examples of such fluids include plasmas, liquid metals, and salt water or electrolytes. The word magnetohydrodynamic is derived from *magneto-* meaning magnetic field, *hydro-* meaning liquid, and *-dynamic* meaning movement. The field of MHD was initiated by Hannes Alfvén [1], for which he received the Nobel Prize in physics in 1970. The fundamental concept behind MHD is that magnetic fields can induce currents in a moving conductive fluid, which in turn creates forces on the fluid and changes the magnetic field itself (Fig. 1.1). The set of equations that describe MHD are a combination of the Navier–Stokes equations of fluid dynamics and Maxwell's equations of electromagnetism. These differential equations have to be solved simultaneously, either analytically or numerically.

Fig. 1.2 shows how a magnetic field influences the conductive fluid flow. In Fig. 1.2a, we have a current that runs down the screw and into the magnetic field, traveling at a 90° angle through the magnet and out the wire; this current and magnetic field cause a force that is orthogonal to both forces, causing the magnet to spin while it is magnetically attached to the screw. The same thing happens in Fig. 1.2b; however, instead of the force being exerted on the magnet, the salt water that the

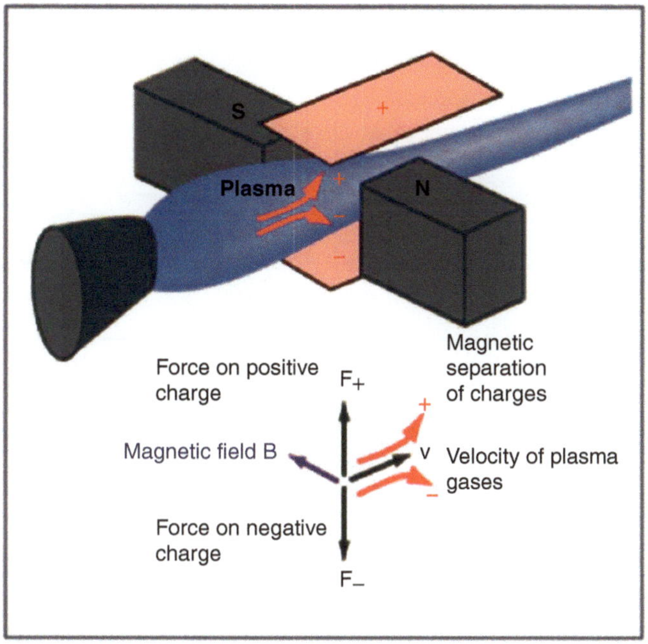

FIGURE 1.1 Induced Current in a Moving Conductive Fluid in the Presence of a Magnetic Field

magnet is submerged in rotates instead because of the resulting Lorentz force that is applied to the water and not the magnet. Fig. 1.2c shows the right-hand side law to determine the direction of the Lorentz forces.

1.1.2 MATHEMATICAL MODEL

The equations of MHD describe the motion of a conducting fluid in a magnetic field. This fluid is usually either a liquid metal or plasma. In both cases, the conductivity ought to be regarded as a tensor if the gyro frequency exceeds the collision frequency. (If there are several collisions per gyro orbit, then the influence of the magnetic field on the transport coefficients will be minimal.) However, to keep the mathematics simple, we shall treat the conductivity as a constant scalar. In fact, it turns out that for many of our applications, it is adequate to take the conductivity as infinite.

Two key physical effects occur in MHD (Fig. 1.3), and understanding them well is the key to developing physical intuition in this subject. The first effect arises when a good conductor moves into a magnetic field. Electric current is induced in the conductor, which, by Lenz's law, creates its own magnetic field. This induced magnetic field tends to cancel the original, externally supported field, thereby, in effect, excluding the magnetic field lines from the conductor. Conversely, when the magnetic field penetrates the conductor and the conductor is moved out of the field, the

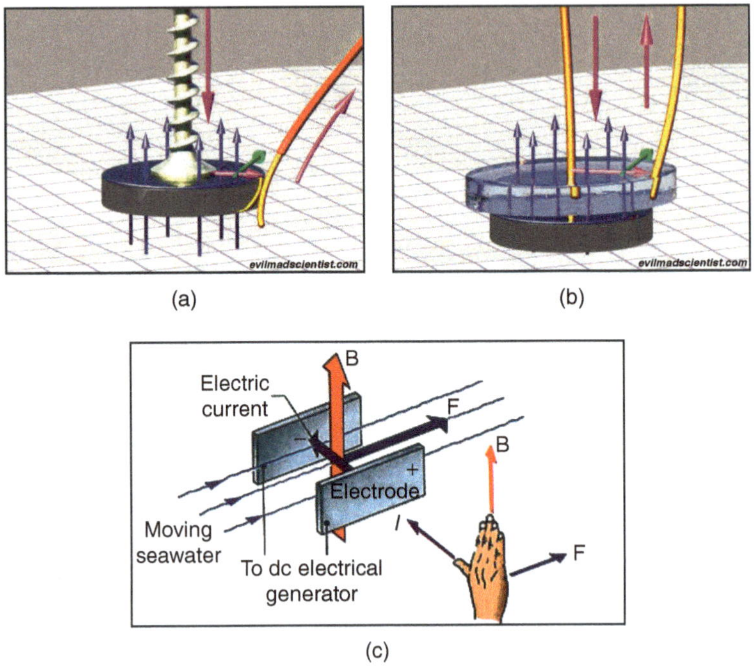

FIGURE 1.2

(a, b) Effect of a magnetic field on conductive fluid flow. (c) Right-hand side law. The blue line indicates the magnetic field, the purple line indicates current, and the green line indicates motion.

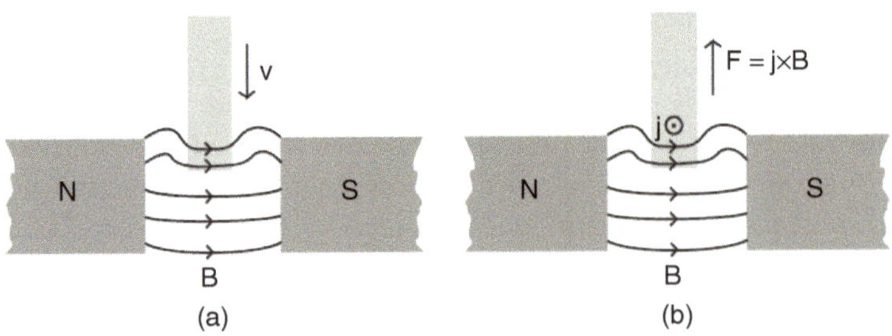

FIGURE 1.3 The Two-Key Physical Effects Occurring in Magnetohydrodynamic

(a) A moving conductor modifies the magnetic field by appearing to drag the field lines with it. When the conductivity is infinite, the field lines appear to be frozen into the moving conductor. (b) When the electric current, owing in the conductor, crosses the magnetic field lines, there will be a Lorentz force, which will accelerate the fluid.

induced field reinforces the applied field. The net result is that the lines of force appear to be dragged along with the conductor – they "go with the flow." Naturally, if the conductor is a fluid with complex motions, the ensuing magnetic field distribution can become quite complex, and the current will build up until its growth is balanced by Ohmic dissipation.

The second key effect is dynamical. When currents are induced by a motion of a conducting fluid through a magnetic field, a Lorentz (or $\vec{j} \times \vec{B}$) force will act on the fluid and modify its motion. In MHD, the motion modifies the field, and the field in turn reacts back and modifies the motion. This makes the theory highly nonlinear.

1.1.2.1 Lorentz force law

The Lorentz force law is the basis for the MHD generator. The Lorentz force law states that the charged particle experiences a force when is moving in the electromagnetic field. This force can be explained as

$$\vec{F} = q\left(\vec{u} \times \vec{B}\right) \tag{1.1}$$

where $\vec{F}$ is the force acting on charged particle, q is the charge of the particle, $\vec{u}$ is the velocity of particle, and $\vec{B}$ is the magnetic induction.

1.1.2.2 Faraday's law

When a charged particle moves in a magnetic field, it experiences the retarding force as well as producing voltage. This is the basis of Faraday's law (Fig. 1.4).

1.1.2.3 Maxwell's equations

To describe an electromagnetic field, we use a number of variables:

$\vec{B}$: formally the magnetic flux density, but we shall refer to it as the magnetic field
$\vec{H}$: the magnetic field strength
$\vec{E}$: the electric field

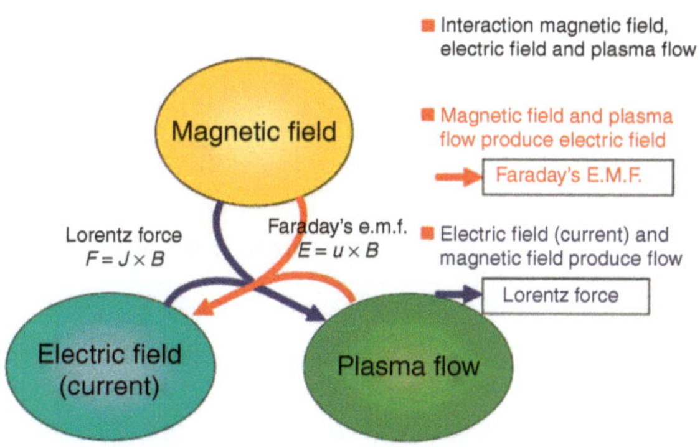

FIGURE 1.4 Relation Between the Lorentz Force and Faraday's Law

$\vec{D}$: the electric displacement
$\vec{j}$: the electric current density
ρ_c: the electric charge density

These are related through Maxwell's equations, the equations governing the evolution of electric and magnetic fields:

$$\vec{\nabla} \cdot \vec{D} = \rho_c \tag{1.2}$$

$$\vec{\nabla} \times \vec{E} = -\frac{\partial \vec{B}}{\partial t} \tag{1.3}$$

$$\vec{\nabla} \cdot \vec{B} = 0 \tag{1.4}$$

$$\vec{\nabla} \times \vec{H} = \vec{j} + \frac{\partial \vec{D}}{\partial t} \tag{1.5}$$

where in an isotropic medium (which we shall assume)

$$\vec{D} = \varepsilon \vec{E}, \quad \vec{B} = \mu \vec{H} \tag{1.6}$$

where ε is the permittivity (or dielectric constant) and μ is the magnetic permeability of the medium. For our purposes, ε and μ can be approximated by their values in a vacuum: $\varepsilon_0 = 8.854 \times 10^{-12}\ \mathrm{Fm}^{-1}$, $\mu_0 = 4\pi \times 10^{-7}\ \mathrm{Hm}^{-1}$. The speed of light is $c = (\mu_0 \varepsilon_0)^{-0.5} = 2.998 \times 10^8\ \mathrm{ms}^{-1}$.

1.1.2.4 The Navier–Stokes equation

The equation governing the flow of a fluid is the Navier–Stokes (or momentum) equation:

$$\rho\left(\frac{\partial \vec{u}}{\partial t} + \left(\vec{u} \cdot \vec{\nabla}\right)\vec{u}\right) = -\vec{\nabla} p + \rho \upsilon \nabla^2 \vec{u} + \text{other forces} \tag{1.7}$$

where ρ is the fluid density, $\vec{u}$ is its velocity, p is the pressure, and υ is the kinematic viscosity.

When the fluid contains electrical charge ρ_c per unit volume, then there is a force per unit volume of $\rho_c \vec{E}$.

When an electric current density $\vec{j}$ flows through the fluid, there is a force per unit volume of $\vec{j} \times \vec{B}$.

Then the Navier–Stokes equation becomes

$$\rho\left(\frac{\partial \vec{u}}{\partial t} + \left(\vec{u} \cdot \vec{\nabla}\right)\vec{u}\right) = -\vec{\nabla} p + \rho \upsilon \nabla^2 \vec{u} + \rho_c \vec{E} + \vec{j} \times \vec{B} + \text{other forces} \tag{1.8}$$

1.1.2.5 Ohm's law

Ohm's law asserts that the total electric current owing in a conductor is proportional to the total electric field. In addition to the field $\vec{E}$ acting on a fluid at rest, a fluid moving with velocity u in the presence of a magnetic field $\vec{B}$ is subject to an additional electric field $\vec{u}\times\vec{B}$. Ohm's law then gives

$$\vec{j} = \sigma\left(\vec{E} + \vec{u}\times\vec{B}\right) \tag{1.9}$$

where the constant of proportionality σ is called the electrical conductivity.

1.1.3 MAGNETOHYDRODYNAMIC APPROXIMATION

The equations we have introduced are capable of describing a great range of phenomena. One example, of course, is electromagnetic waves. The wave equation that describes these is derived from Maxwell's equations. In the applications that we are interested in here, speeds are very small compared with the speed of light c. We can make use of this to introduce substantial simplifications to the governing equations. To demonstrate this, let U be a typical fluid speed, T, a typical time scale; L, a typical length scale (with $U = \ L = T$); B, a typical magnetic field strength; and E, a typical strength of the electric field. Eq. (1.3) then requires that

$$\frac{E}{L} \approx \frac{B}{T} \Rightarrow \frac{E}{B} \approx \frac{L}{T} = U \tag{1.10}$$

Let us then consider equation (1.5).

$$\frac{\left|\partial\vec{D}/\partial t\right|}{\left|\vec{\nabla}\times\vec{H}\right|} \approx \frac{D/T}{H/L} = \frac{\varepsilon_0 E/T}{B/\mu_0 L} = \varepsilon_0\mu_0\frac{E}{B}\frac{L}{T} \approx \frac{1}{c^2}\left(\frac{L}{T}\right)^2 = \frac{U^2}{c^2} \tag{1.11}$$

For speeds small compared with that of light (i.e., for $U << c$), the displacement current $\partial\vec{D}/\partial t$ can therefore be neglected in (1.5), which becomes

$$\vec{\nabla}\times\vec{H} = \vec{j} \tag{1.12}$$

Let us now consider the forces in (1.8). From (1.2)

$$|\rho_c| \approx \frac{D}{L} \tag{1.13}$$

and from (1.11)

$$\left|\vec{j}\right| \approx \frac{H}{L} \tag{1.14}$$

Then making use of (1.10), (1.12), and (1.13)

$$\frac{\left|\rho_c \vec{E}\right|}{\left|\vec{j}\times\vec{B}\right|} \approx \frac{(D/L)E}{(H/L)B} = \frac{\varepsilon_0 E^2}{B^2/\mu_0} = \varepsilon_0\mu_0 \frac{E^2}{B^2} \approx \left(\frac{U}{c}\right)^2 \tag{1.15}$$

Hence the electric force $\rho_c \vec{E}$ can be neglected compared with the magnetic force (usually called the Lorentz force) $\vec{j}\times\vec{B}$ in (1.8), giving

$$\rho\left(\frac{\partial\vec{u}}{\partial t} + \left(\vec{u}\cdot\vec{\nabla}\right)\vec{u}\right) = -\vec{\nabla} p + \rho\upsilon\nabla^2\vec{u} + \rho_c\vec{E} + \vec{j}\times\vec{B} + \text{other forces} \tag{1.16}$$

1.1.4 THE MAGNETIC INDUCTION EQUATION

Having made the aforementioned simplifications, we can eliminate $\vec{E}$ completely from the problem by combining (1.3), (1.4), (1.9), and (1.11).

Using $\vec{B} = \mu_0 \vec{H}$ and Ohm's law (1.9), (1.11) becomes

$$\eta\vec{\nabla}\times\vec{B} = \left(\vec{E} + \vec{u}\times\vec{B}\right) \tag{1.17}$$

where $\eta = \dfrac{1}{\mu_0\sigma}$.

Taking $\vec{\nabla}\times$ (1.15):

$$\vec{\nabla}\times\left(\eta\vec{\nabla}\times\vec{B}\right) = \vec{\nabla}\times\left(\vec{E} + \vec{u}\times\vec{B}\right) \tag{1.18}$$

Assuming η to be constant and using a standard vector identity, this becomes

$$\eta\left(\vec{\nabla}\left(\vec{\nabla}\cdot\vec{B}\right) - \nabla^2\vec{B}\right) = \vec{\nabla}\times\vec{E} + \vec{\nabla}\times\left(\vec{u}\times\vec{B}\right) \tag{1.19}$$

Using (1.3) and (1.4), this becomes

$$-\eta\nabla^2\vec{B} = -\frac{\partial\vec{B}}{\partial t} + \vec{\nabla}\times\left(\vec{u}\times\vec{B}\right) \tag{1.20}$$

or

$$\frac{\partial\vec{B}}{\partial t} = \vec{\nabla}\times\left(\vec{u}\times\vec{B}\right) + \eta\nabla^2\vec{B} \tag{1.21}$$

where η is called the magnetic diffusivity.

Eq. (1.17) is known as the magnetic induction equation. It describes the evolution of the magnetic field $\vec{B}$. The first term on the right-hand side is the induction term that describes the interaction of the field with the flow $\vec{u}$. It is the only term that can generate field. The second term on the right-hand side is a diffusive term. In the absence of a flow $\vec{u}$, we will show later that the diffusive term leads to a decay of the field.

1.1.5 MASS CONTINUITY

Mass can neither be created nor destroyed, so the density ρ can only be changed by the redistribution of matter. Consider a volume V (not moving with the fluid) enclosed by a surface S with unit normal vector n.

The rate of decrease of mass in V must equal the rate of flow of mass out of V, that is,

$$\frac{\partial \vec{B}}{\partial t} = \vec{\nabla} \times \left(\vec{u} \times \vec{B} \right) + \eta \nabla^2 \vec{B} \tag{1.22}$$

Using the divergence theorem, this becomes

$$-\frac{\partial}{\partial t} \int_V \rho \, dV = \int_V \left(\vec{\nabla} \cdot \rho \vec{u} \right) dV \tag{1.23}$$

Because the volume V is fixed, this implies that

$$\int_V \left(\frac{\partial \rho}{\partial t} + \vec{\nabla} \cdot \rho \vec{u} \right) dV = 0 \tag{1.24}$$

Furthermore, because the volume V is arbitrary, (1.18) can only be true if

$$\frac{\partial \rho}{\partial t} + \vec{\nabla} \cdot \rho \vec{u} = 0 \text{ holds everywhere.} \tag{1.25}$$

1.1.6 SUMMARY FOR INCOMPRESSIBLE FLUID

Our system of an electrically conducting incompressible fluid moving with velocity $\vec{u}(x,t)$ through a magnetic field $\vec{B}(x,t)$ is described (in the limit $\left|\vec{u}\right| \ll c$) by

$$\rho \left(\frac{\partial \vec{u}}{\partial t} + \left(\vec{u} \cdot \vec{\nabla} \right) \vec{u} \right) = -\vec{\nabla} p + \rho \upsilon \nabla^2 \vec{u} + \frac{1}{\mu_0} \left(\vec{\nabla} \times \vec{B} \right) \times \vec{B} + \text{other forces} \tag{1.26}$$

$$\frac{\partial \vec{B}}{\partial t} = \vec{\nabla} \times \left(\vec{u} \times \vec{B} \right) + \eta \nabla^2 \vec{B} \tag{1.27}$$

$$\vec{\nabla} \cdot \vec{u} = 0 \tag{1.28}$$

A list of important MHD flow nondimensional parameters is presented in Table 1.1.

Table 1.1 List of Nondimensional Parameters

Nondimensional Parameter	Definition
Reynolds number	$\mathrm{Re} = \dfrac{\text{Inertia forces}}{\text{Viscous forces}} = \dfrac{U_0 L}{\nu}$
Magnetic Reynolds number	$\mathrm{Re}_\mathrm{m} = \dfrac{\text{Convection of B}}{\text{Diffusion of B}} = \dfrac{\text{Induced field}}{\text{Applied field}} = \dfrac{U_0 L}{\upsilon_\mathrm{m}} = \mu_0 \sigma U_0 L$
Hartmann number	$Ha = \left(\dfrac{\text{Electromagnetic forces}}{\text{Viscous forces}}\right)^{1/2} = B_0 L \sqrt{\dfrac{\sigma}{\upsilon\rho}}$
Alfvén number	$\mathrm{Al} = \dfrac{\text{Magnetic field energy}}{\text{Kinetic energy}} = \dfrac{N}{\mathrm{Re}_\mathrm{m}} = \dfrac{B_0^2}{\mu_0 \rho U_0^2}$
Stuart number (interaction parameter)	$N \equiv St = \dfrac{\text{Electromagnetic forces}}{\text{Inertia forces}} = \dfrac{Ha^2}{\mathrm{Re}} = \dfrac{\sigma B_0^2 L}{\rho U_0}$
Batchelor number (magnetic Prandtl number)	$Bt \equiv \mathrm{Pr}_\mathrm{m} = \dfrac{\mathrm{Re}_\mathrm{m}}{\mathrm{Re}} = \mu_0 \sigma \upsilon = \dfrac{\upsilon}{\upsilon_\mathrm{m}}$

1.2 FERROHYDRODYNAMIC

1.2.1 DEFINITION

Magnetic nanofluid (ferrofluid) is a magnetic colloidal suspension consisting of a base liquid and magnetic nanoparticles with a size range of 5 to 15 nm in diameter coated with a surfactant layer. The most often used magnetic material is single-domain particles of magnetite, iron, or cobalt and the base liquids such as water or kerosene. A ferrofluid acts as a fluid that is affected by an external magnetic field, and externally applied magnetic fields can be used to control and direct the flow of ferrofluids because of which it is applicable in numerous fields such as electronic packing, mechanical engineering, thermal engineering, aerospace, and bioengineering [2–5]. Numerical analysis of the heat transfer enhancement and fluid flow characteristics of a rotating cylinder under the influence of magnetic dipole in the backward-facing step geometry was studied by Selimefendigil and Oztop [6]. They found that the effect of cylinder rotation on the local Nusselt number distribution is more pronounced at a low Reynolds number. Wrobel et al. [7] studied the thermomagnetic convective flow of paramagnetic fluid in an annular enclosure with a round rod core and a cylindrical outer wall, numerically and experimentally. Their results show that magnetizing force affects the heat transfer rate, and a strong magnetic field can control the magnetic convection of paramagnetic fluid. Tzirtzilakis et al. [8] used the same model to numerically study the turbulent and two-dimensional (2D) flow of electrical conductive blood. They reported that the influence of the magnetic field decreases in the presence of turbulence. Tzirtzilakis et al. [9] investigated biomagnetic fluid flow in a three-dimensional (3D) rectangular duct. Their investigations showed that the flow is noticeably influenced in the presence of magnetic field. Aminfar et al. [10] investigated the effect of positive and negative magnetic field gradients on the hydrodynamic and thermal behavior of a ferrofluid mixed convection flow in a vertical tube. They showed that negative gradients enhance the Nusselt number, and positive gradients decrease it. Free convection of ferrofluid in a cavity heated from below in the presence of an external magnetic field was studied by Sheikholeslami and Gorji-Bandpy [11]. They found that particles with a smaller size have a better ability to dissipate heat, and a larger volume fraction provides a stronger driving force, which leads to an increase in the temperature profile.

1.2.2 MATHEMATICAL MODEL

The laminar incompressible flow of a homogeneous electrically conducting fluid is considered. As the fluid flows under the influence of a magnetic field, two major forces will act on it. The first one is the magnetization force caused by the tendency of the erythrocytes to orient with the magnetic field. The second one is the Lorentz force that arises because of the electric current generating from the moving ions in the fluid. The governing equations of flow for an incompressible, homogeneous electrically fluid are as follows [12].

Continuity equation:

$$\vec{\nabla}\cdot\vec{V}=0 \tag{1.29}$$

Momentum equations:

$$\rho\frac{D\vec{V}}{Dt}=-\vec{\nabla}p+\rho\vec{F}+\mu\nabla^2\vec{V}+\vec{J}\times\vec{B}+\mu_0\left(\vec{M}\cdot\vec{\nabla}\right)\vec{H} \tag{1.30}$$

Magnetic field equations:

$$\vec{\nabla}\times\vec{H}=\vec{J}=\sigma\left(\vec{V}\times\vec{B}\right) \tag{1.31}$$

$$\vec{\nabla}\cdot\vec{B}=\vec{\nabla}\cdot\left(\vec{H}+\vec{M}\right)=0 \tag{1.32}$$

Energy equation:

In the non-isothermal case, it is also necessary to consider in the mathematical model the energy equation containing the temperature T of the fluid. This equation can be written as

$$\rho C_{\mathrm{p}}\frac{DT}{Dt}=-\mu_0 T\frac{\partial M}{\partial T}\frac{DH}{Dt}+\frac{\vec{J}\cdot\vec{J}}{\sigma}+k\nabla^2 T+\mu\Phi \tag{1.33}$$

In the aforementioned equations, $\vec{V}=(u,v,w)$ is the velocity field. Also, we have $\frac{D\phi}{Dt}=\frac{\partial\phi}{\partial t}+\vec{V}\vec{\nabla}\phi$ and $\vec{\nabla}\phi=\left(\frac{\partial\phi}{\partial x},\frac{\partial\phi}{\partial y},\frac{\partial\phi}{\partial z}\right)$ with x, y, and z the axes of a 3D system.

ρ is the fluid density; p is the pressure; $\vec{F}$ is the body force per unit volume; μ is the dynamical viscosity; $\mu_0 = 4\pi \times 10^{-7}\left(\mathrm{A\,m^2}\right)$ is the magnetic permeability of the vacuum; $\vec{M}$ is the magnetization; $\vec{H}$ is the magnetic field intensity; $\vec{B}$ is the magnetic induction; σ is the electrical conductivity of the fluid; $\vec{J}$ is the density of the electric current; T is the temperature; k is the coefficient of thermal conductivity of the fluid; C_p is the specific heat at constant pressure; and Φ is the dissipation function, which for the 3D case has the form:

$$\Phi = 2\left[\left(\frac{\partial u}{\partial x}\right)^2 + \left(\frac{\partial v}{\partial y}\right)^2 + \left(\frac{\partial w}{\partial z}\right)^2\right] + \left(\frac{\partial v}{\partial x} + \frac{\partial u}{\partial y}\right)^2 + \left(\frac{\partial w}{\partial y} + \frac{\partial v}{\partial z}\right)^2 \\ + \left(\frac{\partial u}{\partial z} + \frac{\partial w}{\partial x}\right)^2 - \frac{2}{3}\left(\frac{\partial u}{\partial x} + \frac{\partial v}{\partial y} + \frac{\partial w}{\partial z}\right)^2 \tag{1.34}$$

The magnetic force attributable to magnetization per unit volume is generally $\mu_0\left(\vec{M}\cdot\vec{\nabla}\right)\vec{H}$ [13,14]. Under the assumption of equilibrium magnetization ($\vec{M}$ and $\vec{H}$ parallel) and keeping in mind that the used magnetic field is solenoid $\vec{\nabla}\cdot\vec{B} = 0$, $\vec{J} = \vec{\nabla}\times\vec{H}$ is the induced current and $\vec{B} = \mu_0\vec{H}$, it is valid, as mentioned in Ref. [14], that

$$\begin{aligned} \mu_0\left(\vec{M}\cdot\vec{\nabla}\right)\vec{H} &= \mu_0\frac{M}{H}\left(\vec{H}\cdot\vec{\nabla}\right)\vec{H} \\ &= \mu_0\frac{M}{H}\left(\frac{1}{2}\vec{\nabla}\left(\vec{H}\cdot\vec{H}\right) - \vec{H}\times\left(\vec{\nabla}\times\vec{H}\right)\right) \\ &= \mu_0\frac{M}{H}\frac{1}{2}\vec{\nabla}\left(\vec{H}^2\right) - \mu_0\frac{M}{H}\left(\vec{H}\times\vec{J}\right) \\ &= \mu_0 M\vec{\nabla}H - \frac{M}{H}\left(\vec{B}\times\vec{J}\right) \\ &= \mu_0 M\vec{\nabla}H + \frac{M}{H}\left(\vec{J}\times\vec{B}\right) \end{aligned} \tag{1.35}$$

Thus, the magnetization force along with the Lorenz force (superposition) is

$$\mu_0 M\vec{\nabla}H + \frac{M}{H}\left(\vec{J}\times\vec{B}\right) + \vec{J}\times\vec{B} = \mu_0 M\vec{\nabla}H + \left(1 + \frac{M}{H}\right)\left(\vec{J}\times\vec{B}\right) \tag{1.36}$$

It also holds, as mentioned, for example, in Ref. [14], that $M \ll H$ thus $\frac{M}{H} \ll 1$ and the force attributable to the imposed magnetic field is finally $\mu_0 M\vec{\nabla}H + \vec{J}\times\vec{B}$.

The term $\mu_0 M\vec{\nabla}H$ in Eq. (1.35) represents the component of magnetic force per unit volume and depends on the existence of the magnetic gradient. This term is well known from FHD, which is the so-called Kelvin force. The term $\vec{J}\times\vec{B}$ appearing in Eq. (1.30) represents the Lorentz force per unit volume and arises because of the electrical conductivity. This term is well known from MHD. The term $\mu_0 T\frac{\partial M}{\partial T}\frac{DH}{Dt}$ in Eq. (1.33) represents the thermal power per unit volume caused by the magneto caloric effect. This term arise from FHD. Also, the term $\frac{\vec{J}\cdot\vec{J}}{\sigma}$ in (1.33) represents the Joule heating.

1.2.3 MAGNETIZATION EQUATIONS

The behavior of a conductive fluid when it is exposed to magnetic field magnetized is described by the magnetization property *M*. Magnetization is the measure of how much the magnetic field affects the magnetic fluid. In an equilibrium situation, the magnetization property is generally determined by the fluid temperature, density, and magnetic field intensity. Various equations, describing the dependence of *M* on these quantities, are given in Ref. [14]. The simplest is the linear equation for isothermal cases:

$$M = \chi H \tag{1.37}$$

where χ is a constant called magnetic susceptibility. The following relation is the linear equation of state, given in Ref. [15]:

$$M = K\left(T_c - T\right) \tag{1.38}$$

where K is a constant called pyromagnetic coefficient and T_c is the Curie temperature.

An equation involving the magnetic intensity H and the temperature T is given in Ref. [15]:

$$M = K'\left(T_c - T\right)H \tag{1.39}$$

where K' is a constant.

1.2.4 MAGNETIZATION EQUATIONS (SATURATION MODEL, EQUILIBRIUM MODEL, MAGNETIC VISCOSITY MODEL)

The general form of continuity and momentum equations are as follows:

Continuity equation:

$$\vec{\nabla}\cdot\vec{V} = 0 \tag{1.40}$$

Linear momentum equation:

$$\rho\frac{D\vec{V}}{Dt} = -\vec{\nabla}p + \rho\vec{F} + \vec{J}\times\vec{B} + \mu_0\left(\vec{M}\cdot\vec{\nabla}\right)\vec{H} + \mu\nabla^2\vec{V} + \zeta\left(\nabla^2\vec{V} + 2\vec{\nabla}\times\vec{\omega}\right) \tag{1.41}$$

$\mu\nabla^2\vec{V}$ is the force acting on the element due to the viscous shearing of the fluid. Here, μ is the shearing viscosity.

$+\zeta\left(\nabla^2\vec{V} + 2\vec{\nabla}\times\vec{\omega}\right)$ is the force arising because of the relative difference in the velocity of the fluid element and its particulate matter caused by the applied magnetic field. This term models the antisymmetric part of the stress tensor.

Here, ζ is the rotational viscosity caused by the magnetic field, and $\vec{\omega}$ is the spinning rate for the particulate matter.

Angular momentum equation

$$\rho I \frac{D\vec{\omega}}{Dt} = \mu_0 \vec{M} \times \vec{H} + \eta' \nabla^2 \vec{\omega} + 2\xi\left(\vec{\nabla} \times \vec{V} - 2\vec{\omega}\right) \tag{1.42}$$

where I is the moment of inertia of the element under consideration. Eq. (1.3) states that the internal angular momentum for the fluid element is balanced by the following right-hand side terms:

$\mu_0 \vec{M} \times \vec{H}$ is the body couple term and is analogous to a body couple on a small magnetic body in the presence of a local magnetic field.

$\eta' \nabla^2 \vec{\omega}$ is the surface couple term and is dependent on the internal strain within the element, analogous to the viscous stress force that is a function of the strain rate of the fluid element. η' is known as the spin diffusion viscosity.

$2\xi\left(\vec{\nabla} \times \vec{V} - 2\vec{\omega}\right)$ is the angular momentum exchange term and arises because of the lack of synchronization between the rate of rotation of the fluid element and the rate of internal spin of the matter making up the fluid element.

The complex mathematical model can be simplified based on the experimental observations and order of magnitude analysis. According to Khashan and Haik [16], it is clear that the angular momentum equation (1.9) with the time rate change of the angular momentum and the diffusion term are much smaller than that of the magnetic torque.

$$\vec{\omega} = \frac{\mu_0}{4\xi}\left(\vec{M} \times \vec{H}\right) + \frac{1}{2}\left(\vec{\nabla} \times \vec{V}\right) \tag{1.43}$$

Three different models were derived to describe the physical phenomena. These models are the saturation model, the equilibrium model, and magnetic viscosity model.

1.2.4.1 Saturation model

When the saturation magnetization is implemented, the magnetization vector becomes parallel to the applied field. Accordingly, the algebraic angular momentum Eq. (1.42) can be further simplified as

$$\vec{\omega} = \frac{1}{2}\left(\vec{\nabla} \times \vec{V}\right) \tag{1.44}$$

With $\vec{\nabla} \times \vec{\omega} = -\frac{1}{2}\left(\nabla^2 \vec{V}\right)$, the linear momentum and angular momentum Eqs (1.8) and (1.10), respectively, can be reduced into a single equation as

$$\rho \frac{D\vec{V}}{Dt} = -\vec{\nabla} p + \rho \vec{F} + \vec{J} \times \vec{B} + \mu_0 \left(\vec{M} \cdot \vec{\nabla}\right)\vec{H} + \mu \nabla^2 \vec{V} \tag{1.45}$$

The magnetization $\vec{M}$ will be effectively equal to the saturation magnetization M_s acting along the field line, that is,

$$\vec{M} = M_s \frac{\vec{H}}{H} \tag{1.46}$$

At high fields, the alignment of the cells with the applied field occurs in less than 1 s. Thus, the saturation model provides a simplified model that describes the conductive behavior at high magnetic fields (5 T and higher).

1.2.4.2 Equilibrium model

For magnetic fields with less than 5 T, the saturation magnetization for the conductive fluid is not reached; therefore, an equation for the magnetization is needed. In this regard, the theoretical calculations are in agreement with the experimental observations of Higashi et al. [17]. Experiments [18] also showed that the blood magnetization can be modeled as

$$\vec{M} = \chi \vec{H} \tag{1.47}$$

where χ is the average magnetic susceptibility of the conductive fluid, which can be determined experimentally.

When the cells are coupled with super paramagnetic particles, the magnetization of these particles will dominate. The magnetic susceptibility of these particles has a linear magnetization relation at low fields (equilibrium model–dominant behavior).

The equilibrium model retains the angular momentum equation to the algebraic expression of the equality of the symmetry of the stress tensor, and the equation reduces to a modified Navier–Stokes equation. This modification is considered valid especially when the cellular components are subjected to a sufficiently high magnetic field or are attached to super paramagnetic particles. Thus, the momentum equation using the equilibrium model is

$$\rho \frac{D\vec{V}}{Dt} = -\vec{\nabla} p + \rho \vec{F} + \vec{J} \times \vec{B} + \mu_0 \vec{\nabla} \vec{H}^2 + \mu \nabla^2 \vec{V} \tag{1.48}$$

1.2.4.3 Magnetic viscosity model

In this model, the orientational effects are coupled to the shear stress via an apparent viscosity term. To account for the orientational effect, the viscosity term needs to be modified to describe the effect of the magnetic field on the conductive flow. In the magnetic viscosity model, the linear momentum equation is reduced to

$$\rho \frac{D\vec{v}}{Dt} = -\vec{\nabla} p + \rho \vec{F} + \vec{J} \times \vec{B} + \mu_0 \vec{\nabla} \vec{H}^2 + \eta^* \nabla^2 \vec{V} \tag{1.49}$$

where η^* is the effective viscosity, which is determined experimentally, Because the orientational effects are functions of the applied field, the effective viscosity is also a function of the applied field ($\eta^* = \eta^*(\vec{H})$).Thus, the solution of Eq. (1.49) is coupled with the solution of the field strength from Maxwell equations.

1.3 NANOFLUID

1.3.1 DEFINITION

Low thermal conductivity of conventional heat transfer fluids such as water, oil, and ethylene glycol mixture is a serious limitation in improving the performance and compactness of many types of engineering equipment such as heat exchangers and electronic devices. To overcome this disadvantage, there is a strong motivation to develop advanced heat transfer fluids with substantially higher conductivity. An innovative way to improve the thermal conductivities of fluids is to suspend small solid particles in the fluid. Various types of powders such as metallic, nonmetallic, and polymeric particles can be added into fluids to form slurries. The thermal conductivities of fluids with suspended particles are expected to be higher than those of common fluids. Nanofluids are a new kind of heat transfer fluid containing a small quantity of nanosized particles (usually smaller than 100 nm) that are uniformly and stably suspended in a liquid. The dispersion of a small amount of solid nanoparticles in conventional fluids changes their thermal conductivity remarkably. Compared with the existing techniques for enhancing heat transfer, the nanofluids show a superior potential for increasing heat transfer rates in a variety of cases [19–21].

1.3.2 MODEL DESCRIPTION

In the literature, convective heat transfer with nanofluids can be modeled using mainly the two-phase or single approach. In the two-phase approach, the velocity between the fluid and particles might not be zero because of several factors such as gravity, friction between the fluid and solid particles, Brownian forces, Brownian diffusion, sedimentation, and dispersion. In the second approach, the nanoparticles can be easily fluidized; therefore, one may assume that the motion slip between the phases, if any, would be considered negligible [22]. The latter approach is simpler and more computationally efficient.

1.3.2.1 Single-phase model

Although nanofluids are solid–liquid mixtures, the approach conventionally used in most studies of natural convection handles the nanofluid as a single-phase (homogenous) fluid. In fact, because of the extreme size and low concentration of the suspended nanoparticles, the particles are assumed to move with same velocity as the fluid. Also, by considering the local thermal equilibrium, the solid particle–liquid mixture may then be approximately considered to behave as a conventional single-phase fluid

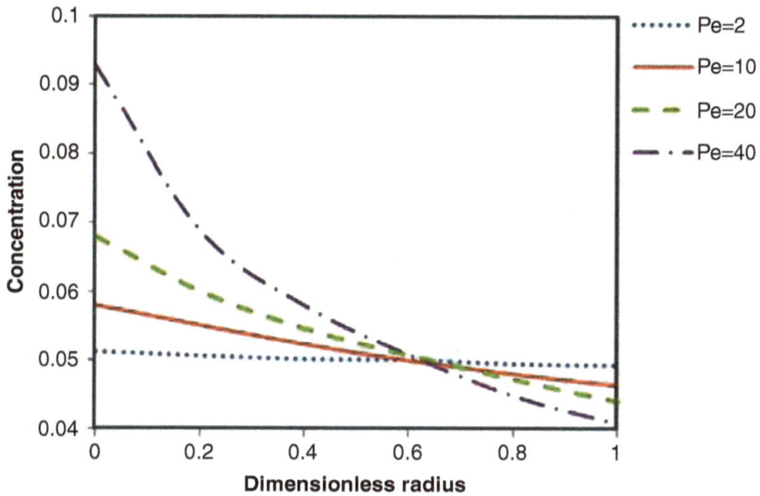

FIGURE 1.5 Relation Between the Particle Concentration Distribution and the Peclet Number [27]

with properties that are to be evaluated as functions of those of the constituents. The governing equations for a homogenous analysis of natural convection are continuity, momentum, and energy equations with their density, specific heat, thermal conductivity, and viscosity modified for nanofluid application. The specific governing equations for various studied enclosures are not shown here; they can be found in different references [23–26]. It should be mentioned that sometimes this assumption is not correct. For example, Ding and Wen [27] show that this assumption may not always remain true for a nanofluid. They investigated the particle migration in a nanofluid for a pipe flow and stated that at Peclet numbers exceeding 10, the particle distribution is significantly nonuniform. Fig. 1.5 shows this relation between concentration and Peclet number for a pipe for 0.05 particle volume fraction showing that homogenous flow assumption is conditional.

Nevertheless, many studies have performed the numerical simulation using single-phase assumption and reported acceptable results for the heat transfer and hydrodynamic properties of the flow.

1.3.2.2 Two-phase model

Several authors have tried to establish convective transport models for nanofluids [28–30]. Nanofluid is a two-phase mixture in which the solid phase consists of nanosized particles. In view of the nanoscale size of the particles, it may be questionable whether the theory of conventional two-phase flow can be applied in describing the flow characteristics of nanofluid. On the other hand, several factors such as gravity; friction between the fluid and solid particles and Brownian forces; and the phenomena of Brownian diffusion, sedimentation, and dispersion may affect a nanofluid flow. Consequently, the slip velocity between the fluid and particles cannot be neglected for simulating nanofluid flows. Because the two-phase approach considers the movement between the solid and fluid molecule, it may have better prediction in nanofluid study. To fully describe and predict the flow and behavior of complex flows, different multiphase theories have been proposed and used. The large number of published articles concerning multiphase flows typically use mixture theory to predict the behavior of nanofluids [28–30]. A comprehensive survey of convective transport in nanofluids was made by Buongiorno [31] using a model in which Brownian motion and thermophoresis are accounted for. Buongiorno developed a two-component, four-equation, nonhomogeneous equilibrium model for mass, momentum, and heat transfer in nanofluids. The nanofluid is treated as a two-component mixture (base fluid and nanoparticles) with the following assumptions: no chemical reactions, negligible external forces, dilute mixture ($\phi = 1$), negligible viscous dissipation, negligible radiative heat transfer, and nanoparticle and base fluid locally in thermal equilibrium. Invoking these assumptions, the following equations represent the mathematical formulation of the nonhomogenous single-phase model for the governing equations as formulated by Buongiorno [31].

Continuity equation:

$$\nabla \cdot v = 0 \tag{1.50}$$

where v is the velocity.

Nanoparticle continuity equation:

$$\frac{\partial \phi}{\partial t} + v \cdot \nabla \phi = \nabla \cdot \left(D_{\mathrm{B}} \nabla \phi + D_{\mathrm{T}} \frac{\nabla T}{T} \right) \tag{1.51}$$

where ϕ is nanoparticle volume fraction and D_{B} is the Brownian diffusion coefficient given by the Einstein–Stokes's equation:

$$D_{\mathrm{B}} = \frac{k_{\mathrm{B}} T}{3\pi \mu d_{\mathrm{p}}} \tag{1.52}$$

where μ is the viscosity of the fluid; d_{p} is the nanoparticle diameter; $k_{\mathrm{B}} = 1.385 \times 10^{-23}$ is Boltzmann constant; and D_{T} is the thermophoretic diffusion coefficient, which is defined as

$$D_{\mathrm{T}} = \left(\frac{\mu}{\rho} \right) \left(0.26 \frac{k}{k + k_{\mathrm{p}}} \right) \tag{1.53}$$

In Eq. (1.53), k and k_{p} are the thermal conductivity of the fluid and particle materials, respectively.

Momentum equation:

$$v \cdot \nabla v = -\frac{1}{\rho_{\mathrm{nf}}} \nabla p + \nabla \cdot \tau + g \tag{1.54}$$

where

$$\tau = -\mu_{\text{nf}}\left(\nabla v + (\nabla v)^{t}\right) \tag{1.55}$$

where the superscript t indicates the transpose of ∇v. Also, p is pressure.

Energy equation:

$$v \cdot \nabla T = \nabla(\alpha_{\text{nf}} \nabla T) + \frac{\rho_{\text{p}} c_{\text{p}}}{\rho_{\text{nf}} c_{\text{nf}}}\left(D_{\text{B}} \nabla \phi \cdot \nabla T + D_{\text{T}} \frac{\nabla T \cdot \nabla T}{T}\right) \tag{1.56}$$

where ϕ and T are the nanoparticle concentration and temperature of nanofluid, respectively.

This nanofluid model can be characterized as a "two-fluid" (nanoparticles and base fluid), four-equation (mass, momentum, energy), nonhomogeneous (nanoparticle and fluid slip velocity allowed) equilibrium (nanoparticle and fluid temperature differences not allowed) model. Note that the conservation equations are strongly coupled. That is, v depends on ϕ via viscosity; ϕ depends on T mostly because of thermophoresis; T depends on ϕ via thermal conductivity and via the Brownian and thermophoretic terms in the energy equation; ϕ and T obviously depends on v because of the convection terms in the nanoparticle continuity and energy equations, respectively.

In a numerical study by Behzadmehr et al. [32], for the first time, a two-phase mixture model was implemented to investigate the behavior of Cu–water nanofluid in a tube, and the results were compared with previous works using a single-phase approach. The authors claimed that the simulation, which was done by assuming that base fluid and particles behave separately, possessed results that are more precise compared with the previous computational modeling. They implemented the mixture theory for their work. It was suggested that the continuity, momentum, and energy equations be written for a mixture of fluid and a solid phase. Some assumptions were also stated for the model such as a strong coupling between two phases and the fluid being closely followed by the particles with each phase owning a different velocity, leading to a term called *slip velocity of nanoparticles*:

$$V_{\text{pf}} = V_{\text{p}} - V_{\text{f}} = \frac{\rho_{\text{p}} d_{\text{p}}^{2}}{18\mu_{\text{f}} f_{\text{drag}}} \frac{(\rho_{\text{p}} - \rho_{\text{m}})}{\rho_{\text{p}}} a, \quad a = g - (V_{\text{m}} \cdot \nabla) V_{\text{m}}, \quad f_{\text{drag}} = \begin{cases} 1 + 0.15\,\text{Re}_{\text{p}}^{0.687} \\ 0.0183\,\text{Re}_{\text{p}} \end{cases} \tag{1.57}$$

The conservation equations (continuity, momentum, and energy, respectively) will be written for the mixture as follows:

$$\nabla \cdot (\rho_{\text{m}} V_{\text{m}}) = 0 \tag{1.58}$$

$$\nabla \cdot (\rho_{\text{m}} V_{\text{m}} V_{\text{m}}) = -\nabla P_{\text{m}} + \nabla \cdot [\tau - \tau_{t}] + \rho_{\text{m}} g + \nabla \cdot \left(\sum_{k=1}^{n} \phi_{k} \rho_{k} V_{\text{dr},k} V_{\text{dr},k}\right) \tag{1.59}$$

$$\nabla \cdot (\phi_{\text{p}} V_{k} (\rho_{k} h_{k} + p)) = \nabla \cdot (k_{\text{eff}} \nabla T - C_{\text{p}} \rho_{\text{m}} vt) \tag{1.60}$$

where $V_{dr,p}$ is the particle draft velocity that is related to the slip velocity and is defined as:

$$V_{dr,p} = V_P - V_f = V_{pf} - \sum_{k=1}^{n} \frac{\phi_k \rho_k}{\rho_m} V_{fk} \tag{1.61}$$

1.3.3 PHYSICAL PROPERTIES OF THE NANOFLUID FOR THE SINGLE-PHASE MODEL

Base nanofluid properties have been published over the past few years. However, only recently have some data on temperature-dependent properties been provided even though they are only for nanofluid effective thermal conductivity and effective absolute viscosity.

1.3.3.1 Density

In the absence of experimental data for nanofluid densities, constant-value temperature independent values, based on nanoparticle volume fraction, are used [33]:

$$\rho_{nf} = \rho_f(1-\phi) + \rho_p\phi \tag{1.62}$$

1.3.3.2 Specific heat capacity

It has been suggested that the effective specific heat can be calculated using the following equation as reported by Pak and Cho [34] as

$$\left(C_p\right)_{nf} = \left(C_p\right)_f (1-\phi) + \left(C_p\right)_p \phi \tag{1.63}$$

Other authors suggest an alternative approach based on the heat capacity concept [35]:

$$\left(\rho C_p\right)_{nf} = \left(\rho C_p\right)_f (1-\phi) + \left(\rho C_p\right)_p \phi \tag{1.64}$$

These two formulations may, of course, lead to different results for specific heat. Because of the lack of experimental data, both formulations are considered equivalent in estimating nanofluid-specific heat capacity [36].

1.3.3.3 Thermal expansion coefficient

The thermal expansion coefficient of nanofluid can be obtained as follows [33]:

$$(\rho\beta)_{nf} = (\rho\beta)_f (1-\phi) + (\rho\beta)_p \phi \tag{1.65}$$

1.3.3.4 Electrical conductivity

The effective electrical conductivity of nanofluid can be calculated as [37]:

$$\sigma_{nf}/\sigma_f = 1 + 3(\sigma_p/\sigma_f - 1)\phi/((\sigma_p/\sigma_f + 2) - (\sigma_p/\sigma_f - 1)\phi) \tag{1.66}$$

1.3.3.5 Dynamic viscosity

Various models have been suggested to model the viscosity of a nanofluid mixture that take into account the percentage of nanoparticles suspended in the base fluid. The classic Brinkman model [38] seems to be a proper one that has been extensively used in the studies on numerical simulation concerning nanofluids. Eq. (1.67) shows the relation between the nanofluid viscosity, base fluid viscosity, and nanoparticle concentration in this model.

$$\mu_{\mathrm{nf}} = \mu_{\mathrm{f}} / (1-\phi)^{2.5} \tag{1.67}$$

However, in some recent computational studies, other models have been selected to be used in the numerical process, such as the work done by Abu-Nada and Chamkha [39] to investigate the convection of CuO–ethylene glycol–water nanofluid in a an enclosure in which the Namburu correlation for viscosity [40] was applied:

$$\log(\mu_{\mathrm{nf}}) = Ae^{-BT} \tag{1.68}$$

where

$$\begin{aligned} A &= 1.837\phi^2 - 29.643\phi + 165.65 \\ B &= 4\times10^{-6}\phi^2 - 0.001\phi + 0.0186 \end{aligned} \tag{1.69}$$

In their study, the results were compared with those of the viscosity modeled by Brinkman. It was outlined that as far as a value for normalized average Nusselt number for the fluid is concerned, for various values of Rayleigh number, Brinkman model owns a prediction of higher value compared with that for the Namburu model showing the notable role of the viscosity model used in the calculations. The authors also state that a combination of different models might as well be implemented that will show different dependence on volume concentration as well as the geometry aspect ratio yet along with the limitation that the models include only the ones mentioned in the study. Other studies have also shown that different models might lead to different results, such as that by Arefmanesh and Mahmoodi [41] on a numerical simulation using Maïga et al. [42] and Brinkman models showing the difference in the results for the heat transfer of Al_2O_3–water nanofluid in a square cavity. The difference was said to be dependent on Richardson number and the volume concentration of particles. A number of suggested relations for viscosity models used in numerical studies are also presented in Table 1.2.

1.3.3.6 Thermal conductivity

Different nanofluid models based on a combination of the different formulas for the thermal conductivity adopted in the studies of natural convection are summarized in Table 1.3. Also, Table 1.4 demonstrates values of thermophysical properties for different materials used as suspended particles in nanofluids [62,63].

Table 1.2 Different Models for Viscosity of Nanofluids Used in Simulation

Model	Equation
Brinkman's model [38]	$\mu_{nf} = \dfrac{\mu_f}{(1-\phi)^{2.5}}$
Einstein's model [43]	$\mu_{nf} = (2.5\phi+1)\mu_f,\ \phi<0.05$
Pak and Cho's correlation [44]	$\mu_{nf} = \mu_f(1+39.11\phi+533.9\phi^2)$
Nguyen et al.'s [45] model (fitted by Abu-Nada [46])	$\mu_{CuO} = -0.6967+15.937/T+1.238\phi+1356.14/T^2 - 0.259\phi^2 - 30.88\phi/T - 19652.74/T^3 + 0.01593\phi^3 + 4.38206\phi^2/T + 147.573\phi/T^2$ $\mu_{Al_2O_3} = \exp(3.003 - 0.04203T - 0.5445\phi + 0.0002553T^2 + 0.0524\phi^2 - 1.622\phi^{-1})$
Jang et al.'s [47] model	$\mu_{nf} = (2.5\phi+1)\mu_f\left[1+\eta(d_p/H)^{-2\varepsilon}\phi^{2/3}(\varepsilon+1)\right]$
Koo and Kleinstreuer [48]	$\mu_{nf} = 5\times10^4\beta\phi\rho_f\sqrt{\dfrac{k_BT}{d_p\rho_p}}f(T,\phi),\ \begin{cases}\beta = 0.0137(100\phi)^{-0.8229} & for\ \ \phi<1\% \\ \beta = 0.0011(100\phi)^{-0.7272} & for\ \ \phi>1\%\end{cases}$
Maiga et al.'s model [49]	$\mu_{nf} = \mu_f(1+7.3\phi+123\phi^2)$
Brownian model [50]	$\mu_{nf} = \mu_f(1+2.5\phi+6.17\phi^2)$
Nguyen et al.'s model [51]	$\mu_{nf} = \mu_f(1+0.025\phi+0.015\phi^2)$
Masoumi et al. [52]	$\mu_{nf} = \mu_f + \rho_p V_B d_p^2/(72C\delta)$
Gherasim et al. [53]	$\mu_{nf} = \mu_f 0.904e^{14.8\phi}$

Table 1.3 Different Models for Thermal Conductivity of Nanofluids Used in Simulation

Model	Equation
Wasp model [23]	$\dfrac{k_{nf}}{k_f} = \dfrac{k_p+2k_f-2\phi(k_f-k_p)}{k_p+2k_f+\phi(k_f-k_p)}$
Jang and Choi's model [54]	$\dfrac{k_{nf}}{k_f} = (1-\phi)+Bk_p\phi+18\times10^6\dfrac{3d_f}{d_p}k_f \mathrm{Re}_{dp}^2 \mathrm{Pr}\phi$
Bruggeman's model [55]	$k_{nf} = 0.25k_f(3\phi-1)k_p/k_f+[3(1-\phi)-1]+\sqrt{\Delta_B}$ $\Delta_B = [(3\phi-1)k_p/k_f+(3(1-\phi)-1)]^2+8k_p/k_f$
Chon et al.'s [56] model	$k_{nf} = k_f\left[1+64.7\phi^{0.7640}(d_f/d_p)^{0.369}(k_f/k_p)^{0.7476}\mathrm{Pr}_T^{0.9955}\mathrm{Re}^{1.2321}\right]$

Table 1.3 Different Models for Thermal Conductivity of Nanofluids Used in Simulation *(cont.)*

Model	Equation
Koo and Kleinstreuer [48]	$k_{nf}=k_f\left[\frac{k_p+2k_f-2\phi(k_f-k_p)}{k_p+2k_f+\phi(k_f-k_p)}\right]+5\times10^4\beta\phi\rho_f(C_p)_f\sqrt{\frac{k_BT}{d_p\rho_p}}f(T,\phi)$
Charuyakorn et al. [57]	$\frac{k_{nf}}{k_f}=\left[\frac{k_p+2k_f-2\phi(k_f-k_p)}{k_p+2k_f+\phi(k_f-k_p)}\right](1+b\phi Pe_p^m)$
Stationary model [58]	$k_{nf}=k_f\left[1+k_p\phi d_f/\left(k_f(1-\phi)d_p\right)\right]$
Yu and Choi [59]	$\frac{k_{nf}}{k_f}=\left[\frac{k_p+2k_f-2\phi(k_f-k_p)(1+\eta)^3}{k_p+2k_f+\phi(k_f-k_p)(1+\eta)^3}\right]$
Patel et al. [60]	$\frac{k_{nf}}{k_f}=1+\frac{k_pd_f\phi}{k_fd_p(1-\phi)}\left[1+c\frac{2k_BTd_p}{\pi\alpha_f\mu_fd_p^2}\right]$
Mintsa et al. [61]	$k_{nf}=k_f\,(1.72\phi+1.0)$

Table 1.4 Thermophysical Properties of the Nanofluid [62,63]

	ρ (kg/m^{-3})	C_p (j/kgk^{-1})	k(W/m·k^{-1})	β(K^{-1})	σ $(\Omega\cdot m)^{-1}$
Pure water	997.1	4,179	0.613	21×10^{-5}	0.05
Copper (Cu)	8,933	385	401	1.67×10^{-5}	5.96×10^{7}
Silver (Ag)	10,500	235	429	1.89×10^{-5}	3.60×10^{7}
Alumina (Al_2O_3)	3,970	765	40	0.85×10^{-5}	1×10^{-10}
Titanium oxide (TiO_2)	4,250	686.2	8.9538	0.9×10^{-5}	1×10^{-12}

1.4 MAGNETOHYDRODYNAMIC NANOFLUID FLOW AND HEAT TRANSFER

The existence of a magnetic field has a noticeable effect on heat transfer reduction under natural convection, but in many engineering applications, increasing heat transfer from solid surfaces is a goal. Therefore, the use of nanofluids with higher thermal conductivity can be considered a promising solution. It is also assumed that the uniform magnetic field $\left(\vec{B}=B_x\vec{e}_x+B_y\vec{e}_y\right)$ of constant magnitude $B=\sqrt{B_x^2+B_y^2}$ is applied, where $\vec{e}_x$ and $\vec{e}_y$ are unit vectors in the Cartesian coordinate system. The orientation of the magnetic field forms an angle λ with horizontal axis such that $\lambda=B_x/B_y$. The electric current J and the electromagnetic force F are defined by $J=\sigma\left(\vec{V}\times\vec{B}\right)$ and $F=\sigma\left(\vec{V}\times\vec{B}\right)\times\vec{B}$, respectively [64].

1.4.1 MATHEMATICAL MODELING FOR THE SINGLE-PHASE MODEL

1.4.1.1 Natural convection

The flow is 2D, laminar, and incompressible. The radiation, viscous dissipation, induced electric current, and Joule heating are neglected. The magnetic Reynolds number is assumed to be small so that the induced magnetic field can be neglected compared with the applied magnetic field. The flow is considered to be steady, 2D, and laminar. Neglecting displacement currents, induced magnetic field, dissipation, and Joule heating and using the Boussinesq approximation, the governing equations of nanofluid heat transfer and fluid flow can be obtained in dimensional form as follows [65]:

$$\frac{\partial u}{\partial x}+\frac{\partial v}{\partial y}=0 \tag{1.70}$$

$$\begin{aligned} u\frac{\partial u}{\partial x}+v\frac{\partial u}{\partial y}&=-\frac{1}{\rho_{\mathrm{nf}}}\frac{\partial P}{\partial x}+\upsilon_{\mathrm{nf}}\left(\frac{\partial^2 u}{\partial x^2}+\frac{\partial^2 u}{\partial y^2}\right)\\ &+\frac{\sigma_{\mathrm{nf}}B^2}{\rho_{\mathrm{nf}}}\left(v\sin\lambda\cos\lambda-u\sin^2\lambda\right) \end{aligned} \tag{1.71}$$

$$\begin{aligned} u\frac{\partial v}{\partial x}+v\frac{\partial v}{\partial y}&=-\frac{1}{\rho_{\mathrm{nf}}}\frac{\partial P}{\partial y}+\upsilon_{\mathrm{nf}}\left(\frac{\partial^2 v}{\partial x^2}+\frac{\partial^2 v}{\partial y^2}\right)+\beta_{\mathrm{nf}}g\left(T-T_{\mathrm{c}}\right)\\ &+\frac{\sigma_{\mathrm{nf}}B^2}{\rho_{\mathrm{nf}}}\left(u\sin\lambda\cos\lambda-v\cos^2\lambda\right) \end{aligned} \tag{1.72}$$

$$u\frac{\partial T}{\partial x}+v\frac{\partial T}{\partial y}=\alpha_{\mathrm{nf}}\left(\frac{\partial^2 T}{\partial x^2}+\frac{\partial^2 T}{\partial y^2}\right) \tag{1.73}$$

where ϕ is the solid volume fraction of nanoparticles and nf is the nanofluid properties.

The stream function and vorticity are defined as follows:

$$u=\frac{\partial\psi}{\partial y},\quad v=-\frac{\partial\psi}{\partial x},\quad \omega=\frac{\partial v}{\partial x}-\frac{\partial u}{\partial y} \tag{1.74}$$

The stream function satisfies the continuity Eq. (1.70). The vorticity equation is obtained by eliminating the pressure between the two momentum equations, that is, by taking y-derivative of Eq. (1.71) and subtracting from it the x-derivative of Eq. (1.72). This gives:

$$\begin{aligned} \frac{\partial\psi}{\partial y}\frac{\partial\omega}{\partial x}-\frac{\partial\psi}{\partial x}\frac{\partial\omega}{\partial y}&=\upsilon_{\mathrm{nf}}\left(\frac{\partial^2\omega}{\partial x^2}+\frac{\partial^2\omega}{\partial y^2}\right)+\beta_{\mathrm{nf}}g\left(\frac{\partial T}{\partial x}\right)\\ &+\frac{\sigma_{\mathrm{nf}}B^2}{\rho_{\mathrm{nf}}}\left(-\frac{\delta v}{\delta y}\sin\lambda\cos\lambda+\frac{\delta u}{\delta y}\sin^2\lambda\right.\\ &\left.+\frac{\delta u}{\delta x}\sin\lambda\cos\lambda-\frac{\delta v}{\delta x}\cos^2\lambda\right) \end{aligned} \tag{1.75}$$

$$\frac{\partial \psi}{\partial y}\frac{\partial T}{\partial x}-\frac{\partial \psi}{\partial x}\frac{\partial T}{\partial y}=\alpha_{\mathrm{nf}}\left(\frac{\partial^2 T}{\partial x^2}+\frac{\partial^2 T}{\partial y^2}\right) \tag{1.76}$$

$$\frac{\partial^2 \psi}{\partial x^2}+\frac{\partial^2 \psi}{\partial y^2}=-\omega \tag{1.77}$$

By introducing the following nondimensional variables:

$$\begin{aligned} &X=\frac{x}{L},\ Y=\frac{y}{L},\ \Omega=\frac{\omega L^2}{\alpha_{\mathrm{f}}},\ \Psi=\frac{\psi}{\alpha_{\mathrm{f}}},\ U=\frac{uL}{\alpha_{\mathrm{f}}},\ V=\frac{vL}{\alpha_{\mathrm{f}}},\\ &\Theta=\frac{T-T_{\mathrm{c}}}{\Delta T}\quad \begin{cases}\Delta T=T_{\mathrm{h}}-T_{\mathrm{c}}\\ \Delta T=(q''L/k_{\mathrm{f}})\end{cases}\end{aligned} \tag{1.78}$$

where in Eq. (1.78), $L=r_{\mathrm{out}}-r_{\mathrm{in}}=r_{\mathrm{in}}$. Also, $\Delta T=T_{\mathrm{h}}-T_{\mathrm{c}}$ and $\Delta T=(q''L/k_{\mathrm{f}})$ are used for constant temperature and constant heat flux boundary conditions, respectively. Using the dimensionless parameters, the equations now become:

$$\begin{aligned}\frac{\partial \Psi}{\partial Y}\frac{\partial \Omega}{\partial X}-\frac{\partial \Psi}{\partial X}\frac{\partial \Omega}{\partial Y}&=\mathrm{Pr}\frac{(\mu_{\mathrm{nf}}/\mu_{\mathrm{nf}})}{(\rho_{\mathrm{nf}}/\rho_{\mathrm{nf}})}\left(\frac{\partial^2 \Omega}{\partial X^2}+\frac{\partial^2 \Omega}{\partial Y^2}\right)+\mathrm{Ra}\,\mathrm{Pr}\left[\frac{\beta_{\mathrm{nf}}}{\beta_f}\right]\left(\frac{\partial \Theta}{\partial X}\right)\\ &+\mathrm{Ha}^2\,\mathrm{Pr}\frac{(\sigma_{\mathrm{nf}}/\sigma_{\mathrm{nf}})}{(\rho_{\mathrm{nf}}/\rho_{\mathrm{nf}})}\left(-\frac{\delta V}{\delta Y}\tan\lambda+\frac{\delta U}{\delta Y}\tan^2\lambda+\frac{\delta U}{\delta X}\tan\lambda-\frac{\delta V}{\delta X}\right)\end{aligned} \tag{1.79}$$

$$\frac{\partial \Psi}{\partial Y}\frac{\partial \Theta}{\partial X}-\frac{\partial \Psi}{\partial X}\frac{\partial \Theta}{\partial Y}=\left[\frac{\dfrac{k_{\mathrm{nf}}}{k_{\mathrm{f}}}}{\dfrac{(\rho C_{\mathrm{p}})_{\mathrm{nf}}}{(\rho C_{\mathrm{p}})_{\mathrm{f}}}}\right]\left(\frac{\partial^2 \Theta}{\partial X^2}+\frac{\partial^2 \Theta}{\partial Y^2}\right) \tag{1.80}$$

$$\frac{\partial^2 \Psi}{\partial X^2}+\frac{\partial^2 \Psi}{\partial Y^2}=-\Omega \tag{1.81}$$

where the Rayleigh number, Hartmann number, and Prandtl number are defined as follows:

$$Ra=g\beta_{\mathrm{f}}L^3\Delta \mathrm{T}/(\alpha_{\mathrm{f}}\upsilon_{\mathrm{f}}),\ \ Ha=LB_x\sqrt{\sigma_{\mathrm{f}}/\mu_{\mathrm{f}}},\ \ \mathrm{Pr}=\upsilon_{\mathrm{f}}/\alpha_{\mathrm{f}} \tag{1.82}$$

1.4.1.2 Mixed convection

Similar to the previous section, the governing equations of nanofluid heat transfer and fluid flow can be obtained in dimensional form as follows:

$$\frac{\partial u}{\partial x}+\frac{\partial v}{\partial y}=0 \tag{1.83}$$

$$u\frac{\partial u}{\partial x}+v\frac{\partial u}{\partial y}=-\frac{1}{\rho_{\mathrm{nf}}}\frac{\partial P}{\partial x}+\upsilon_{\mathrm{nf}}\left(\frac{\partial^2 u}{\partial x^2}+\frac{\partial^2 u}{\partial y^2}\right)+\beta_{\mathrm{nf}}g\left(\frac{\partial T}{\partial x}\right)+\frac{\sigma_{\mathrm{nf}}B^2}{\rho_{\mathrm{nf}}}\left(v\sin\lambda\cos\lambda-u\sin^2\lambda\right) \tag{1.84}$$

$$u\frac{\partial v}{\partial x}+v\frac{\partial v}{\partial y}=-\frac{1}{\rho_{\mathrm{nf}}}\frac{\partial P}{\partial y}+\upsilon_{\mathrm{nf}}\left(\frac{\partial^2 v}{\partial x^2}+\frac{\partial^2 v}{\partial y^2}\right)+\frac{\sigma_{\mathrm{nf}}B^2}{\rho_{\mathrm{nf}}}\left(u\sin\lambda\cos\lambda-v\cos^2\lambda\right) \tag{1.85}$$

$$u\frac{\partial T}{\partial x}+v\frac{\partial T}{\partial y}=\alpha_{\mathrm{nf}}\left(\frac{\partial^2 T}{\partial x^2}+\frac{\partial^2 T}{\partial y^2}\right) \tag{1.86}$$

By introducing the following nondimensional variables:

$$X=\frac{x}{L},\ Y=\frac{y}{L},\ \Omega=\frac{\omega L}{u_{\mathrm{r}}},\ \Psi=\frac{\psi}{u_{\mathrm{r}}L},\ U=\frac{u}{u_{\mathrm{r}}},\ V=\frac{v}{u_{\mathrm{r}}},\quad \Theta=\frac{T-T_{\mathrm{c}}}{\Delta T}\ \begin{cases}\Delta T=T_{\mathrm{h}}-T_{\mathrm{c}}\\ \Delta T=(q''L/k_{\mathrm{f}})\end{cases} \tag{1.87}$$

where in Eq. (4.18), $L=r_{\mathrm{out}}-r_{\mathrm{in}}=r_{\mathrm{in}}$. Also $\Delta T=T_{\mathrm{h}}-T_{\mathrm{c}}$ and $\Delta T=\left(q''L/k_{\mathrm{f}}\right)$ are used for constant temperature and constant heat flux boundary conditions, respectively. The vorticity equation is obtained by eliminating the pressure between the two momentum equations, that is, by taking y-derivative of Eq. (4.15) and subtracting from it the x-derivative of Eq. (4.16). Using the dimensionless parameters, the equations now become:

$$\frac{\partial\Psi}{\partial Y}\frac{\partial\Omega}{\partial X}-\frac{\partial\Psi}{\partial X}\frac{\partial\Omega}{\partial Y}=\frac{1}{\mathrm{Re}}\frac{(\mu_{\mathrm{nf}}/\mu_{\mathrm{nf}})}{(\rho_{\mathrm{nf}}/\rho_{\mathrm{nf}})}\left(\frac{\partial^2\Omega}{\partial X^2}+\frac{\partial^2\Omega}{\partial Y^2}\right)+\frac{Gr}{\mathrm{Re}^2}\frac{\beta_{\mathrm{nf}}}{\beta_{\mathrm{f}}}\frac{\partial\Theta}{\partial X}+\frac{Ha^2}{\mathrm{Re}}\frac{(\sigma_{\mathrm{nf}}/\sigma_{\mathrm{nf}})}{(\rho_{\mathrm{nf}}/\rho_{\mathrm{nf}})}\left(-\frac{\delta V}{\delta Y}\tan\lambda+\frac{\delta U}{\delta Y}\tan^2\lambda+\frac{\delta U}{\delta X}\tan\lambda-\frac{\delta V}{\delta X}\right) \tag{1.88}$$

$$\frac{\partial\Psi}{\partial Y}\frac{\partial\Theta}{\partial X}-\frac{\partial\Psi}{\partial X}\frac{\partial\Theta}{\partial Y}=\frac{1}{\mathrm{Re}\,\mathrm{Pr}}\left[\frac{\frac{k_{\mathrm{nf}}}{k_{\mathrm{f}}}}{\frac{(\rho C_{\mathrm{p}})_{\mathrm{nf}}}{(\rho C_{\mathrm{p}})_{\mathrm{f}}}}\right]\left(\frac{\partial^2\Theta}{\partial X^2}+\frac{\partial^2\Theta}{\partial Y^2}\right) \tag{1.89}$$

$$\frac{\partial^2\Psi}{\partial X^2}+\frac{\partial^2\Psi}{\partial Y^2}=-\Omega \tag{1.90}$$

where $\mathrm{Re}=\frac{\rho_{\mathrm{f}}u_{\mathrm{r}}L}{\mu_{\mathrm{f}}}$, $Gr=\frac{g\beta\Delta TL^3}{\upsilon^2}$ and $Ri=\frac{Gr}{\mathrm{Re}^2}$ are the Reynolds number, Grashof number, and Richardson number, respectively.

1.4.2 MATHEMATICAL MODELING FOR THE TWO-PHASE MODEL

1.4.2.1 Natural convection

The nanofluid's density, ρ is

$$\begin{aligned}\rho &= \phi\rho_{\mathrm{p}} + (1-\phi)\rho_{\mathrm{f}} \\ &\cong \phi\rho_{\mathrm{p}} + (1-\phi)\left\{\rho_{\mathrm{f}_0}\left(1-\beta(T-T_{\mathrm{c}})\right)\right\}\end{aligned} \tag{1.91}$$

where ρ_{f} is the base fluid's density, T_{c} is a reference temperature, ρ_{f_0} is the base fluid's density at the reference temperature, and β is the volumetric coefficient of expansion. Taking the density of base fluid as that of the nanofluid, the density ρ in Eq. (1.91) thus becomes

$$\rho \cong \phi\rho_{\mathrm{p}} + (1-\phi)\left\{\rho_0\left(1-\beta(T-T_{\mathrm{c}})\right)\right\} \tag{1.92}$$

where ρ_0 is the nanofluid's density at the reference temperature.

The continuity, momentum under Boussinesq approximation, and energy equations for the laminar and steady-state natural convection in a 2D enclosure can be written in dimensional form as follows [66]:

$$\frac{\partial u}{\partial x} + \frac{\partial v}{\partial y} = 0 \tag{1.93}$$

$$\rho_f\left\{u\frac{\partial u}{\partial x} + v\frac{\partial u}{\partial y}\right\} = -\frac{\partial P}{\partial x} + \mu\left(\frac{\partial^2 u}{\partial x^2} + \frac{\partial^2 u}{\partial y^2}\right) + \sigma B^2\left(v\sin\lambda\cos\lambda - u\sin^2\lambda\right) \tag{1.94}$$

$$\begin{aligned}\rho_f\left\{u\frac{\partial v}{\partial x} + v\frac{\partial v}{\partial y}\right\} &= -\frac{\partial P}{\partial y} + \mu\left(\frac{\partial^2 v}{\partial x^2} + \frac{\partial^2 v}{\partial y^2}\right) - (\phi-\phi_{\mathrm{c}})\left(\rho_{\mathrm{p}} - \rho_{\mathrm{f}_0}\right)g \\ &\quad + (1-\phi_{\mathrm{c}})\rho_{\mathrm{f}_0}\left(T-T_{\mathrm{c}}\right)g + \sigma B^2\left(u\sin\lambda\cos\lambda - v\cos^2\lambda\right)\end{aligned} \tag{1.95}$$

$$u\frac{\partial T}{\partial x} + v\frac{\partial T}{\partial y} = \alpha\left(\frac{\partial^2 T}{\partial x^2} + \frac{\partial^2 T}{\partial y^2}\right) + \frac{(\rho c)_{\mathrm{p}}}{(\rho c)_{\mathrm{f}}}\left[\begin{array}{l} D_{\mathrm{B}}\left\{\dfrac{\partial \phi}{\partial x}\cdot\dfrac{\partial T}{\partial x} + \dfrac{\partial \phi}{\partial y}\cdot\dfrac{\partial T}{\partial y}\right\} \\ +(D_{\mathrm{T}}/T_{\mathrm{c}})\left\{\left(\dfrac{\partial T}{\partial x}\right)^2 + \left(\dfrac{\partial T}{\partial y}\right)^2\right\}\end{array}\right] \tag{1.96}$$

$$u\frac{\partial \phi}{\partial x} + v\frac{\partial \phi}{\partial y} = D_{\mathrm{B}}\left\{\frac{\partial^2 \phi}{\partial x^2} + \frac{\partial^2 \phi}{\partial y^2}\right\} + \left(\frac{D_T}{T_{\mathrm{c}}}\right)\left\{\frac{\partial^2 T}{\partial x^2} + \frac{\partial^2 T}{\partial y^2}\right\} \tag{1.97}$$

The stream function and vorticity are defined as follows:

$$u = \frac{\partial \psi}{\partial y},\quad v = -\frac{\partial \psi}{\partial x},\quad \omega = \frac{\partial v}{\partial x} - \frac{\partial u}{\partial y} \tag{1.98}$$

Also, the following nondimensional variables should be introduced [67]:

$$X = \frac{x}{L},\ Y = \frac{y}{L},\ \Omega = \frac{\omega L^2}{\alpha},\ \Psi = \frac{\psi}{\alpha},\ \Theta = \frac{T - T_{\mathrm{c}}}{T_{\mathrm{h}} - T_{\mathrm{c}}},\ \Phi = \frac{\phi - \phi_{\mathrm{c}}}{\phi_{\mathrm{h}} - \phi_{\mathrm{c}}},\ U = \frac{uL}{\alpha},\ V = \frac{vL}{\alpha} \tag{1.99}$$

By using these dimensionless parameters, the equations become:

$$\begin{aligned}\left[\frac{\partial \Psi}{\partial Y}\frac{\partial \Omega}{\partial X}-\frac{\partial \Psi}{\partial X}\frac{\partial \Omega}{\partial Y}\right]&=\Pr\left(\frac{\partial^2 \Omega}{\partial X^2}+\frac{\partial^2 \Omega}{\partial Y^2}\right)+\Pr \mathrm{Ra}\left(\frac{\partial \Theta}{\partial X}-\mathrm{Nr}\frac{\partial \Phi}{\partial X}\right)\\&+\mathrm{Ha}^2\Pr\left(-\frac{\delta V}{\delta Y}\tan\lambda+\frac{\delta U}{\delta Y}\tan^2\lambda+\frac{\delta U}{\delta X}\tan\lambda-\frac{\delta V}{\delta X}\right)\end{aligned} \tag{1.100}$$

$$\begin{aligned}\frac{\partial \Psi}{\partial Y}\frac{\partial \Theta}{\partial X}-\frac{\partial \Psi}{\partial X}\frac{\partial \Theta}{\partial Y}&=\left(\frac{\partial^2 \Theta}{\partial X^2}+\frac{\partial^2 \Theta}{\partial Y^2}\right)+\mathrm{Nb}\left(\frac{\partial \Phi}{\partial X}\frac{\partial \Theta}{\partial X}+\frac{\partial \Phi}{\partial Y}\frac{\partial \Theta}{\partial Y}\right)\\&+\mathrm{Nt}\left(\left(\frac{\partial \Theta}{\partial X}\right)^2+\left(\frac{\partial \Theta}{\partial Y}\right)^2\right)\end{aligned} \tag{1.101}$$

$$\frac{\partial \Psi}{\partial Y}\frac{\partial \Phi}{\partial X}-\frac{\partial \Psi}{\partial X}\frac{\partial \Phi}{\partial Y}=\frac{1}{\mathrm{Le}}\left(\frac{\partial^2 \Phi}{\partial X^2}+\frac{\partial^2 \Phi}{\partial Y^2}\right)+\frac{\mathrm{Nt}}{\mathrm{Nb\,Le}}\left(\frac{\partial^2 \Theta}{\partial X^2}+\frac{\partial^2 \Theta}{\partial Y^2}\right) \tag{1.102}$$

$$\frac{\partial^2 \Psi}{\partial X^2}+\frac{\partial^2 \Psi}{\partial Y^2}=-\Omega \tag{1.103}$$

where thermal Rayleigh number, buoyancy ratio number, Prandtl number, Brownian motion parameter, thermophoretic parameter, Lewis number, and Hartmann number of nanofluid are defined as follows:

$$\begin{aligned}&\mathrm{Ra}=(1-\phi_c)\rho_{f_0} g\beta L^3(T_h-T_c)/(\mu\alpha),\ \mathrm{Nr}=(\rho_p-\rho_0)(\phi_h-\phi_c)/[(1-\phi_c)\rho_{f_0}\beta L(T_h-T_c)],\\&\Pr=\mu/\rho_f\alpha,\ \mathrm{Nb}=(\rho c)_p D_B(\phi_h-\phi_c)/((\rho c)_f\alpha),\ \mathrm{Nt}=(\rho c)_p D_T(T_h-T_c)/[(\rho c)_f\alpha T_c],\\&\mathrm{Le}=\alpha/D_B,\ \mathrm{Ha}=LB_x\sqrt{\sigma/\mu}\end{aligned} \tag{1.104}$$

1.4.2.2 Mixed convection

To obtain the governing equations for force convection, the nondimensional variables should be introduced as:

$$\begin{aligned}&X=\frac{x}{L},\ Y=\frac{y}{L},\ \Omega=\frac{\omega L}{u_r},\ \Psi=\frac{\psi}{u_r L},\ U=\frac{u}{u_r},\ V=\frac{v}{u_r},\\&\Theta=\frac{T-T_c}{T_h-T_c},\ \Phi=\frac{\phi-\phi_c}{\phi_h-\phi_c}\end{aligned} \tag{1.105}$$

So the governing equations are as follows:

$$\begin{aligned}\left[\frac{\partial \Psi}{\partial Y}\frac{\partial \Omega}{\partial X}-\frac{\partial \Psi}{\partial X}\frac{\partial \Omega}{\partial Y}\right]&=\frac{1}{\mathrm{Re}}\left(\frac{\partial^2 \Omega}{\partial X^2}+\frac{\partial^2 \Omega}{\partial Y^2}\right)+\frac{\mathrm{Gr}}{\mathrm{Re}^2}\left(\frac{\partial \Theta}{\partial X}-\mathrm{Nr}\frac{\partial \Phi}{\partial X}\right)\\&+\frac{\mathrm{Ha}^2}{\mathrm{Re}}\left(-\frac{\delta V}{\delta Y}\tan\lambda+\frac{\delta U}{\delta Y}\tan^2\lambda+\frac{\delta U}{\delta X}\tan\lambda-\frac{\delta V}{\delta X}\right)\end{aligned} \tag{1.106}$$

$$\frac{\partial \Psi}{\partial Y}\frac{\partial \Theta}{\partial X}-\frac{\partial \Psi}{\partial X}\frac{\partial \Theta}{\partial Y}=\frac{1}{\mathrm{Re}\,\mathrm{Pr}}\left(\frac{\partial^2 \Theta}{\partial X^2}+\frac{\partial^2 \Theta}{\partial Y^2}\right)+\frac{\mathrm{Nb}}{\mathrm{Re}}\left(\frac{\partial \Phi}{\partial X}\frac{\partial \Theta}{\partial X}+\frac{\partial \Phi}{\partial Y}\frac{\partial \Theta}{\partial Y}\right)+\frac{\mathrm{Nt}}{\mathrm{Re}}\left(\left(\frac{\partial \Theta}{\partial X}\right)^2+\left(\frac{\partial \Theta}{\partial Y}\right)^2\right) \tag{1.107}$$

$$\frac{\partial \Psi}{\partial Y}\frac{\partial \Phi}{\partial X}-\frac{\partial \Psi}{\partial X}\frac{\partial \Phi}{\partial Y}=\frac{1}{\mathrm{Le}\,\mathrm{Re}}\left(\frac{\partial^2 \Phi}{\partial X^2}+\frac{\partial^2 \Phi}{\partial Y^2}\right)+\frac{\mathrm{Nt}}{\mathrm{Nb}\,\mathrm{Le}\,\mathrm{Re}}\left(\frac{\partial^2 \Theta}{\partial X^2}+\frac{\partial^2 \Theta}{\partial Y^2}\right) \tag{1.108}$$

$$\frac{\partial^2 \Psi}{\partial X^2}+\frac{\partial^2 \Psi}{\partial Y^2}=-\Omega \tag{1.109}$$

where the thermal Reynolds number, buoyancy ratio number, Prandtl number, Brownian motion parameter, thermophoretic parameter, Lewis number, Hartmann number, Grashof number, and Richardson number of nanofluid are defined as follows:

$$\begin{gathered}\mathrm{Re}=\frac{\rho_{\mathrm{f}} L u_{\mathrm{r}}}{\mu},\ \mathrm{Nr}=(\rho_{\mathrm{p}}-\rho_0)(\phi_{\mathrm{h}}-\phi_{\mathrm{c}})/[(1-\phi_{\mathrm{c}})\rho_{\mathrm{f}_0}\beta L(T_{\mathrm{h}}-T_{\mathrm{c}})],\ \mathrm{Pr}=\mu/\rho_{\mathrm{f}}\alpha,\\ \mathrm{Nb}=(\rho c)_{\mathrm{p}} D_{\mathrm{B}}(\phi_{\mathrm{h}}-\phi_{\mathrm{c}})/((\rho c)_{\mathrm{f}}\upsilon),\ \mathrm{Nt}=(\rho c)_{\mathrm{p}} D_{\mathrm{T}}(T_{\mathrm{h}}-T_{\mathrm{c}})/[(\rho c)_{\mathrm{f}}\upsilon T_{\mathrm{c}}],\ \mathrm{Le}=\upsilon/D_{\mathrm{B}},\\ \mathrm{Ha}=\mathrm{LB}_x\sqrt{\sigma/\mu},\ \mathrm{Gr}=(1-\phi_{\mathrm{c}})\rho_{\mathrm{f}_0} g\beta L^3\rho_{\mathrm{f}}(T_{\mathrm{h}}-T_{\mathrm{c}})/(\mu^2),\ \mathrm{Ri}=\frac{\mathrm{Gr}}{\mathrm{Re}^2}\end{gathered} \tag{1.110}$$

1.5 FERROHYDRODYNAMIC NANOFLUID FLOW AND HEAT TRANSFER

For the expression of the magnetic field strength, it can be considered that the magnetic source represents a magnetic wire placed vertically to the x-y plane at the point $(\bar{a},\bar{b})$. The components of the magnetic field intensity ($\overline{H}_x,\overline{H}_y$) and the magnetic field strength ($\overline{H}$) can be considered as [68]:

$$\overline{H}_x=\frac{\gamma}{2\pi}\frac{1}{(x-\bar{a})^2+(y-\bar{b})^2}(y-\bar{b}) \tag{1.111}$$

$$\overline{H}_y=-\frac{\gamma}{2\pi}\frac{1}{(x-\bar{a})^2+(y-\bar{b})^2}(x-\bar{a}) \tag{1.112}$$

$$\overline{H}=\sqrt{\overline{H}_x^2+\overline{H}_y^2}=\frac{\gamma}{2\pi}\frac{1}{\sqrt{(x-\bar{a})^2+(y-\bar{b})^2}} \tag{1.113}$$

where γ the magnetic field strength at the source (of the wire) and $(\bar{a},\bar{b})$ is the position where the source is located.

1.5.1 MATHEMATICAL MODELING FOR THE SINGLE-PHASE MODEL

1.5.1.1 Natural convection

The flow is considered to be steady, 2D, incompressible, and laminar. Using the Boussinesq approximation, the governing equations of heat transfer and fluid flow for nanofluid can be obtained as follows [68]:

$$\frac{\partial u}{\partial x}+\frac{\partial v}{\partial y}=0 \tag{1.114}$$

$$\rho_{\mathrm{nf}}\left(u\frac{\partial u}{\partial x}+v\frac{\partial u}{\partial y}\right)=-\frac{\partial P}{\partial x}+\mu_{\mathrm{nf}}\left(\frac{\partial^2 u}{\partial x^2}+\frac{\partial^2 u}{\partial y^2}\right)+\mu_0 M\frac{\partial \overline{H}}{\partial x}-\sigma_{\mathrm{nf}}B_y^2 u+\sigma_{\mathrm{nf}}B_x B_y v \tag{1.115}$$

$$\begin{aligned}\rho_{\mathrm{nf}}\left(u\frac{\partial v}{\partial x}+v\frac{\partial v}{\partial y}\right)=&-\frac{\partial P}{\partial y}+\mu_{\mathrm{nf}}\left(\frac{\partial^2 v}{\partial x^2}+\frac{\partial^2 v}{\partial y^2}\right)+\mu_0 M\frac{\partial \overline{H}}{\partial y}-\sigma_{\mathrm{nf}}B_x^2 v+\sigma_{\mathrm{nf}}B_x B_y u\\&+\rho_{\mathrm{nf}}\beta_{\mathrm{nf}}g(T-T_{\mathrm{c}})\end{aligned} \tag{1.116}$$

$$\begin{aligned}(\rho C_{\mathrm{p}})_{\mathrm{nf}}\left(u\frac{\partial T}{\partial x}+v\frac{\partial T}{\partial y}\right)=&\,k_{\mathrm{nf}}\left(\frac{\partial^2 T}{\partial x^2}+\frac{\partial^2 T}{\partial y^2}\right)+\sigma_{\mathrm{nf}}(uB_y-vB_x)^2\\&-\mu_0 T\frac{\partial M}{\partial T}\left(u\frac{\partial \overline{H}}{\partial x}+v\frac{\partial \overline{H}}{\partial y}\right)+\mu_{\mathrm{nf}}\left\{2\left(\frac{\partial u}{\partial x}\right)^2+2\left(\frac{\partial v}{\partial x}\right)^2+\left(\frac{\partial u}{\partial x}+\frac{\partial v}{\partial y}\right)^2\right\}\end{aligned} \tag{1.117}$$

The terms $\mu_0 M\frac{\partial \overline{H}}{\partial x}$ and $\mu_0 M\frac{\partial \overline{H}}{\partial y}$ in (4.62) and (4.63), respectively, represent the components of magnetic force per unit volume and depend on the existence of the magnetic gradient on the corresponding x and y directions. These two terms are well known from FHD, which is the so-called Kelvin force. The terms $-\sigma_{\mathrm{nf}}B_y^2 u+\sigma_{\mathrm{nf}}B_x B_y v$ and $-\sigma_{\mathrm{nf}}B_x^2 v+\sigma_{\mathrm{nf}}B_x B_y u$ appearing in (4.62) and (4.63), respectively, represent the Lorentz force per unit volume toward the x and y directions and arise because of the electrical conductivity of the fluid. These two terms are known in MHD. The principles of MHD and FHD are combined in the mathematical model presented in Ref. [68], and the aforementioned terms arise together in the governing equations (4.62) and (4.63). The term $\mu_0 T\frac{\partial M}{\partial T}\left(u\frac{\partial \overline{H}}{\partial x}+v\frac{\partial \overline{H}}{\partial y}\right)$ in Eq. (4.64) represents the thermal power per unit volume caused by the magneto caloric effect. Also, the term $\sigma_{\mathrm{nf}}(uB_y-vB_x)^2$ in (4.64) represents the Joule heating. For the variation of the magnetization M, with the magnetic field intensity $\overline{H}$ and temperature T, the following relation derived experimentally in Ref. [68] is considered:

$$M=K'\overline{H}(T_{\mathrm{c}}'-T) \tag{1.118}$$

where K' is a constant and T_{c}' is the Curie temperature.

In the aforementioned equations, μ_0 is the magnetic permeability of the vacuum $\left(4\pi \times 10^{-7}\ \text{Tm/A}\right)$, $\overline{H}$ is the magnetic field strength, $\overline{B}$ is the magnetic induction $\left(\overline{B} = \mu_0 \overline{H}\right)$, and the bar above the quantities denotes that they are dimensional. The effective density (ρ_{nf}) and heat capacitance $\left(\rho C_{\text{p}}\right)_{\text{nf}}$ of the nanofluid are defined as [33]:

$$\rho_{\text{nf}} = \rho_{\text{f}}(1-\phi) + \rho_{\text{s}}\phi \tag{1.119}$$

$$\left(\rho C_{\text{p}}\right)_{\text{nf}} = \left(\rho C_{\text{p}}\right)_{\text{f}}(1-\phi) + \left(\rho C_{\text{p}}\right)_{\text{s}}\phi \tag{1.120}$$

where ϕ is the solid volume fraction of nanoparticles. The thermal diffusivity of the nanofluids is

$$\alpha_{\text{nf}} = \frac{k_{\text{nf}}}{\left(\rho C_{\text{p}}\right)_{\text{nf}}} \tag{1.121}$$

and the thermal expansion coefficient of the nanofluid can be determined as

$$\beta_{\text{nf}} = \beta_{\text{f}}(1-\phi) + \beta_{\text{s}}\phi \tag{1.122}$$

The dynamic viscosity of the nanofluids given by Brinkman [33] is

$$\mu_{\text{nf}} = \frac{\mu_{\text{f}}}{(1-\phi)^{2.5}} \tag{1.123}$$

The effective thermal conductivity of the nanofluid can be approximated by the Maxwell–Garnetts model as [33]:

$$\frac{k_{\text{nf}}}{k_{\text{f}}} = \frac{k_{\text{s}} + 2k_{\text{f}} - 2\phi(k_{\text{f}} - k_{\text{s}})}{k_{\text{s}} + 2k_{\text{f}} + \phi(k_{\text{f}} - k_{\text{s}})} \tag{1.124}$$

and the effective electrical conductivity of nanofluid was presented [68] as follows:

$$\frac{\sigma_{\text{nf}}}{\sigma_{\text{f}}} = 1 + \frac{3\left(\dfrac{\sigma_{\text{s}}}{\sigma_{\text{f}}} - 1\right)\phi}{\left(\dfrac{\sigma_{\text{s}}}{\sigma_{\text{f}}} + 2\right) - \left(\dfrac{\sigma_{\text{s}}}{\sigma_{\text{f}}} - 1\right)\phi} \tag{1.125}$$

The stream function and vorticity are defined as:

$$u = \frac{\partial \psi}{\partial y},\ v = -\frac{\partial \psi}{\partial x},\ \omega = \frac{\partial v}{\partial x} - \frac{\partial u}{\partial y} \tag{1.126}$$

The stream function satisfies the continuity Eq. (1.114). The vorticity equation is obtained by eliminating the pressure between the two momentum equations, that is, by taking y-derivative of Eq. (1.115) and subtracting from it the x-derivative of Eq. (1.116). By introducing the following nondimensional variables:

$$X=\frac{x}{L},\ Y=\frac{y}{L},\ \Omega=\frac{\omega L^2}{\alpha_f},\ \Psi=\frac{\psi}{\alpha_f},\ \Theta=\frac{T-T_c}{T_h-T_c},\ U=\frac{uL}{\alpha_f},\ V=\frac{vL}{\alpha_f},$$
$$H=\frac{\overline{H}}{\overline{H_0}},\ H_x=\frac{\overline{H_x}}{\overline{H_0}},\ H_y=\frac{\overline{H_y}}{\overline{H_0}} \tag{1.127}$$

where in Eq. (1.127) $\overline{H_0}=\overline{H}(\bar{a},0)=\dfrac{\gamma}{2\pi|b|}$ and $L=r_{out}-r_{in}=r_{in}$. Using the dimensionless parameters, the equations now become:

$$\begin{aligned}\frac{\partial\Psi}{\partial Y}\frac{\partial\Omega}{\partial X}-\frac{\partial\Psi}{\partial X}\frac{\partial\Omega}{\partial Y}&=\Pr\left[\frac{\mu_{nf}/\mu_f}{\rho_{nf}/\rho_f}\right]\left(\frac{\partial^2\Omega}{\partial X^2}+\frac{\partial^2\Omega}{\partial Y^2}\right)+\mathrm{Ra}\Pr\left[\frac{\beta_{nf}}{\beta_f}\right]\left(\frac{\partial\Theta}{\partial X}\right)\\&+\mathrm{Mn_F}\Pr\left(\frac{\rho_f}{\rho_{nf}}\right)\left\{\frac{\partial H}{\partial X}\frac{\partial\Theta}{\partial Y}-\frac{\partial H}{\partial Y}\frac{\partial\Theta}{\partial X}\right\}H-\mathrm{Ha}^2\Pr\left[\frac{\sigma_{nf}/\sigma_f}{\rho_{nf}/\rho_f}\right]\\&\times\left\{\frac{\partial V}{\partial X}H_x^2+V\left(2H_x\frac{\partial H_x}{\partial X}\right)-\frac{\partial U}{\partial X}H_xH_y-U\frac{\partial H_x}{\partial X}H_y-U\frac{\partial H_y}{\partial X}H_x\right.\\&\left.-\frac{\partial U}{\partial Y}H_y^2-U\left(2H_y\frac{\partial H_y}{\partial Y}\right)+\frac{\partial V}{\partial Y}H_xH_y+V\frac{\partial H_x}{\partial Y}H_y+V\frac{\partial H_y}{\partial Y}H_x\right\}\end{aligned} \tag{1.128}$$

$$\begin{aligned}\frac{\partial\Psi}{\partial Y}\frac{\partial\Theta}{\partial X}-\frac{\partial\Psi}{\partial X}\frac{\partial\Theta}{\partial Y}&=\left(\frac{k_{nf}}{k_f}\right)\Big/\left(\frac{(\rho C_P)_{nf}}{(\rho C_P)_f}\right)\left(\frac{\partial^2\Theta}{\partial X^2}+\frac{\partial^2\Theta}{\partial Y^2}\right)\\&+\mathrm{Ha}^2\,Ec\left(\frac{\sigma_{nf}}{\sigma_f}\right)\Big/\left(\frac{(\rho C_P)_{nf}}{(\rho C_P)_f}\right)\{UH_y-VH_x\}^2\\&+\mathrm{Mn_F}\mathrm{Ec}\frac{(\rho C_P)_f}{(\rho C_P)_{nf}}\left\{U\frac{\partial H}{\partial X}+V\frac{\partial H}{\partial Y}\right\}H(\varepsilon_1+\Theta)\\&+\mathrm{Ec}\left(\frac{\mu_{nf}}{\mu_f}\right)\Big/\left(\frac{(\rho C_P)_{nf}}{(\rho C_P)_f}\right)\left\{2\left(\frac{\partial U}{\partial X}\right)^2+2\left(\frac{\partial V}{\partial X}\right)^2+\left(\frac{\partial U}{\partial X}+\frac{\partial V}{\partial Y}\right)^2\right\}\end{aligned} \tag{1.129}$$

$$\frac{\partial^2\Psi}{\partial X^2}+\frac{\partial^2\Psi}{\partial Y^2}=-\Omega \tag{1.130}$$

where $\mathrm{Ra_f}=g\beta_f L^3(T_h-T_c)/(\alpha_f\upsilon_f),\mathrm{Pr_f}=\upsilon_f/\alpha_f,\mathrm{Ha}=L\mu_0H_0\sqrt{\sigma_f/\mu_f},\varepsilon_1=T_1/\Delta T$, $\mathrm{Ec}=(\mu_f\alpha_f)/\left[(\rho C_P)_f\,\Delta T\,L^2\right]$ and $\mathrm{Mn_F}=\mu_0H_0^2K'(T_h-T_c)L^2/(\mu_f\alpha_f)$ are the Rayleigh number, Prandtl number, Hartmann number arising from MHD, temperature number, Eckert number, and magnetic number arising from FHD the for the base fluid, respectively. The thermophysical properties of the nanofluid are given in Table 1.5 [68].

Table 1.5 Thermophysical Properties of Water and Nanoparticles [68]

	ρ (kg/m^{-3})	C_p (j/kgk^{-1})	k(W/m·k^{-1})	$\beta\times10^5$(K^{-1})	d_p(nm)	σ $(\Omega\cdot m)^{-1}$
Pure water	997.1	4,179	0.613	21	–	0.05
Fe_3O_4	5,200	670	6	1.3	47	25,000

1.5.1.2 Mixed convection

To obtain the governing equations for force convection, the nondimensional variables should be introduced as:

$$X=\frac{x}{L},\ Y=\frac{y}{L},\ \Omega=\frac{\omega L}{u_r},\ \Psi=\frac{\psi}{Lu_r},\ \Theta=\frac{T-T_c}{T_h-T_c},\ U=\frac{u}{u_r},\ V=\frac{v}{u_r},$$
$$H=\frac{\overline{H}}{\overline{H_0}},\ H_x=\frac{\overline{H_x}}{\overline{H_0}},\ H_y=\frac{\overline{H_y}}{\overline{H_0}} \tag{1.131}$$

where in Eq. (1.131) $\overline{H_0}=\overline{H}(\overline{a},0)=\frac{\gamma}{2\pi|b|}$. Using the dimensionless parameters, the equations now become:

$$\begin{aligned}\frac{\partial\Psi}{\partial Y}\frac{\partial\Omega}{\partial X}-\frac{\partial\Psi}{\partial X}\frac{\partial\Omega}{\partial Y}=&\frac{1}{\mathrm{Re}}\left[\frac{\mu_{nf}/\mu_f}{\rho_{nf}/\rho_f}\right]\left(\frac{\partial^2\Omega}{\partial X^2}+\frac{\partial^2\Omega}{\partial Y^2}\right)\\&+\mathrm{Mn_F}\left(\frac{\rho_f}{\rho_{nf}}\right)H\left\{\frac{\partial H}{\partial X}\frac{\partial\Theta}{\partial Y}-\frac{\partial H}{\partial Y}\frac{\partial\Theta}{\partial X}\right\}-\frac{\mathrm{Ha}^2}{\mathrm{Re}}\left[\frac{\sigma_{nf}/\sigma_f}{\rho_{nf}/\rho_f}\right]\\&\times\left\{\frac{\partial V}{\partial X}H_x^2+V\left(2H_x\frac{\partial H_x}{\partial X}\right)-\frac{\partial U}{\partial X}H_xH_y-U\frac{\partial H_x}{\partial X}H_y-U\frac{\partial H_y}{\partial X}H_x\right.\\&\left.-\frac{\partial U}{\partial Y}H_y^2-U\left(2H_y\frac{\partial H_y}{\partial Y}\right)+\frac{\partial V}{\partial Y}H_xH_y+V\frac{\partial H_x}{\partial Y}H_y+V\frac{\partial H_y}{\partial Y}H_x\right\}\\&+\frac{Gr}{\mathrm{Re}^2}\left[\frac{\beta_{nf}}{\beta_f}\right]\left(\frac{\partial\Theta}{\partial X}\right)\end{aligned} \tag{1.132}$$

$$\begin{aligned}\frac{\partial\Psi}{\partial Y}\frac{\partial\Theta}{\partial X}-\frac{\partial\Psi}{\partial X}\frac{\partial\Theta}{\partial Y}=&\frac{1}{\mathrm{Pr\,Re}}\left(\frac{k_{nf}}{k_f}\right)\Big/\left(\frac{(\rho C_P)_{nf}}{(\rho C_P)_f}\right)\left(\frac{\partial^2\Theta}{\partial X^2}+\frac{\partial^2\Theta}{\partial Y^2}\right)\\&+\mathrm{Ha}^2\frac{\mathrm{Ec}}{\mathrm{Re}}\left(\frac{\sigma_{nf}}{\sigma_f}\right)\Big/\left(\frac{(\rho C_P)_{nf}}{(\rho C_P)_f}\right)\{UH_y-VH_x\}^2\\&+\mathrm{Mn_F Ec}\frac{(\rho C_P)_f}{(\rho C_P)_{nf}}\left\{U\frac{\partial H}{\partial X}+V\frac{\partial H}{\partial Y}\right\}H(\varepsilon_1+\Theta)\\&+\frac{\mathrm{Ec}}{\mathrm{Re}}\left(\frac{\mu_{nf}}{\mu_f}\right)\Big/\left(\frac{(\rho C_P)_{nf}}{(\rho C_P)_f}\right)/\left\{2\left(\frac{\partial U}{\partial X}\right)^2+2\left(\frac{\partial V}{\partial X}\right)^2+\left(\frac{\partial U}{\partial X}+\frac{\partial V}{\partial Y}\right)^2\right\}\end{aligned} \tag{1.133}$$

$$\frac{\partial^2\Psi}{\partial X^2}+\frac{\partial^2\Psi}{\partial Y^2}=-\Omega \tag{1.134}$$

where $\mathrm{Re}=\frac{\rho_f L u_r}{\mu_f}$, $\mathrm{Gr}=g\beta\Delta T L^3/\upsilon^2$, $\mathrm{Ha}=LH_0\mu_0\sqrt{\sigma_f/\mu_f}$, $\varepsilon_1=T_1/\Delta T$, $\mathrm{Ec}=\left(\rho_f u_r^2\right)/\left[\left(\rho C_P\right)_f \Delta T\right]$, and $\mathrm{Mn}_F=\mu_0 H_0^2 K'(T_h-T_c)/\left(\rho_f u_r^2\right)$ are the Reynolds number, Grashof number, Hartmann number arising from MHD, temperature number, Eckert number, and magnetic number arising from FHD the for the base fluid, respectively.

1.5.2 MATHEMATICAL MODELING FOR TWO-PHASE MODEL

1.5.2.1 Natural convection

The nanofluid's density ρ is

$$\begin{aligned}\rho&=\phi\rho_p+(1-\phi)\rho_f\\&\cong\phi\rho_p+(1-\phi)\left\{\rho_{f_0}\left(1-\beta(T-T_c)\right)\right\}\end{aligned} \tag{1.135}$$

where ρ_f is the base fluid's density, T_c is a reference temperature, ρ_{f_0} is the base fluid's density at the reference temperature, and β is the volumetric coefficient of expansion. Taking the density of base fluid as that of the nanofluid, the density ρ in equation (1.135) thus becomes

$$\rho\cong\phi\rho_p+(1-\phi)\left\{\rho_0\left(1-\beta(T-T_c)\right)\right\} \tag{1.136}$$

where ρ_0 is the nanofluid's density at the reference temperature.

The continuity, momentum under Boussinesq approximation, and energy equations for the laminar and steady-state natural convection in a 2D enclosure can be written in dimensional form as follows:

$$\frac{\partial u}{\partial x}+\frac{\partial v}{\partial y}=0 \tag{1.137}$$

$$\begin{aligned}\rho_f\left\{u\frac{\partial u}{\partial x}+v\frac{\partial u}{\partial y}\right\}=-\frac{\partial P}{\partial x}+\mu\left(\frac{\partial^2 u}{\partial x^2}+\frac{\partial^2 u}{\partial y^2}\right)+\mu_0 M\frac{\partial \overline{H}}{\partial x}\\-\sigma B_y^2 u+\sigma B_x B_y v\end{aligned} \tag{1.138}$$

$$\begin{aligned}\rho_f\left\{u\frac{\partial v}{\partial x}+v\frac{\partial v}{\partial y}\right\}=-\frac{\partial P}{\partial y}+\mu\left(\frac{\partial^2 v}{\partial x^2}+\frac{\partial^2 v}{\partial y^2}\right)-(\phi-\phi_c)\left(\rho_p-\rho_{f_0}\right)g\\+(1-\phi_c)\rho_{f_0}\left(T-T_c\right)g+\mu_0 M\frac{\partial\overline{H}}{\partial y}-\sigma B_x^2 v+\sigma B_x B_y u\end{aligned} \tag{1.139}$$

$$u\frac{\partial T}{\partial x}+v\frac{\partial T}{\partial y}=\alpha\left(\frac{\partial^2 T}{\partial x^2}+\frac{\partial^2 T}{\partial y^2}\right)+\frac{(\rho c)_\mathrm{p}}{(\rho c)_\mathrm{f}}\left[\begin{array}{l}D_\mathrm{B}\left\{\frac{\partial \phi}{\partial x}.\frac{\partial T}{\partial x}+\frac{\partial \phi}{\partial y}.\frac{\partial T}{\partial y}\right\}\\ +(D_\mathrm{T}/T_\mathrm{c})\left\{\left(\frac{\partial T}{\partial x}\right)^2+\left(\frac{\partial T}{\partial y}\right)^2\right\}\end{array}\right]$$
$$+\left(uB_y-vB_x\right)^2-\mu_0 T\frac{\partial M}{\partial T}\left(u\frac{\partial \overline{H}}{\partial x}+v\frac{\partial \overline{H}}{\partial y}\right)$$
$$+\mu\left\{2\left(\frac{\partial u}{\partial x}\right)^2+2\left(\frac{\partial v}{\partial x}\right)^2+\left(\frac{\partial u}{\partial x}+\frac{\partial v}{\partial y}\right)^2\right\} \tag{1.140}$$

$$u\frac{\partial \phi}{\partial x}+v\frac{\partial \phi}{\partial y}=D_\mathrm{B}\left\{\frac{\partial^2 \phi}{\partial x^2}+\frac{\partial^2 \phi}{\partial y^2}\right\}+\left(\frac{D_\mathrm{T}}{T_\mathrm{c}}\right)\left\{\frac{\partial^2 T}{\partial x^2}+\frac{\partial^2 T}{\partial y^2}\right\} \tag{1.141}$$

The terms $\mu_0 M\frac{\partial \overline{H}}{\partial x}$ and $\mu_0 M\frac{\partial \overline{H}}{\partial y}$ in (1.138) and (1.139), respectively, represent the components of magnetic force per unit volume and depend on the existence of the magnetic gradient on the corresponding x and y directions. These two terms are well known from FHD, which is the so-called Kelvin force. The terms $-\sigma_\mathrm{nf}B_y^2u+\sigma_\mathrm{nf}B_xB_yv$ and $-\sigma_\mathrm{nf}B_x^2v+\sigma_\mathrm{nf}B_xB_yu$ appearing in (1.138) and (1.139), respectively, represent the Lorentz force per unit volume toward the x and y directions and arise because of the electrical conductivity of the fluid. These two terms are known in MHD. The principles of MHD and FHD are combined in the mathematical model presented in Ref. [68], and the aforementioned terms arise together in the governing equations (1.138) and (1.139). The term $\mu_0 T\frac{\partial M}{\partial T}\left(u\frac{\partial \overline{H}}{\partial x}+v\frac{\partial \overline{H}}{\partial y}\right)$ in Eq. (1.140) represents the thermal power per unit volume caused by the magnetocaloric effect. Also, the term $\sigma_\mathrm{nf}\left(uB_y-vB_x\right)^2$ in (1.140) represents the Joule heating. For the variation of the magnetization M, with the magnetic field intensity $\overline{H}$ and temperature T, the following relation derived experimentally in Ref. [68] is considered:

$$M=K'\overline{H}\left(T_\mathrm{c}'-T\right) \tag{1.142}$$

where K' is a constant and T_c' is the Curie temperature.

In the aforementioned equations, μ_0 is the magnetic permeability of the vacuum $\left(4\pi\times10^{-7}\ \mathrm{Tm/A}\right)$, $\overline{H}$ is the magnetic field strength, $\overline{B}$ is the magnetic induction $\left(\overline{B}=\mu_0\overline{H}\right)$, and the bar above the quantities denotes that they are dimensional.

The stream function and vorticity are defined as follows:

$$u=\frac{\partial \psi}{\partial y},\ v=-\frac{\partial \psi}{\partial x},\ \omega=\frac{\partial v}{\partial x}-\frac{\partial u}{\partial y} \tag{1.143}$$

Also, the following nondimensional variables should be introduced:

$$X=\frac{x}{L},\ Y=\frac{y}{L},\ \Omega=\frac{\omega L^2}{\alpha},\ \Psi=\frac{\psi}{\alpha},\ \Theta=\frac{T-T_c}{T_h-T_c},\ \Phi=\frac{\phi-\phi_c}{\phi_h-\phi_c},\ U=\frac{uL}{\alpha},\ V=\frac{vL}{\alpha}$$
$$H=\frac{\overline{H}}{\overline{H}_0},\ H_x=\frac{\overline{H}_x}{\overline{H}_0},\ H_y=\frac{\overline{H}_y}{\overline{H}_0} \tag{1.144}$$

By using these dimensionless parameters, the equations become:

$$\begin{aligned}\left[\frac{\partial\Psi}{\partial Y}\frac{\partial\Omega}{\partial X}-\frac{\partial\Psi}{\partial X}\frac{\partial\Omega}{\partial Y}\right]&=\Pr\left(\frac{\partial^2\Omega}{\partial X^2}+\frac{\partial^2\Omega}{\partial Y^2}\right)+\Pr\mathrm{Ra}\left(\frac{\partial\Theta}{\partial X}-\mathrm{Nr}\frac{\partial\Phi}{\partial X}\right)\\&+\mathrm{Mn_F}\Pr\left\{\frac{\partial H}{\partial X}\frac{\partial\Theta}{\partial Y}-\frac{\partial H}{\partial Y}\frac{\partial\Theta}{\partial X}\right\}H\\&-\mathrm{Ha}^2\Pr\times\left\{\frac{\partial V}{\partial X}H_x^2+V\left(2H_x\frac{\partial H_x}{\partial X}\right)-\frac{\partial U}{\partial X}H_xH_y-U\frac{\partial H_x}{\partial X}H_y-U\frac{\partial H_y}{\partial X}H_x\right.\\&\left.-\frac{\partial U}{\partial Y}H_y^2-U\left(2H_y\frac{\partial H_y}{\partial Y}\right)+\frac{\partial V}{\partial Y}H_xH_y+V\frac{\partial H_x}{\partial Y}H_y+V\frac{\partial H_y}{\partial Y}H_x\right\}\end{aligned} \tag{1.145}$$

$$\begin{aligned}\frac{\partial\Psi}{\partial Y}\frac{\partial\Theta}{\partial X}-\frac{\partial\Psi}{\partial X}\frac{\partial\Theta}{\partial Y}&=\left(\frac{\partial^2\Theta}{\partial X^2}+\frac{\partial^2\Theta}{\partial Y^2}\right)+\mathrm{Nb}\left(\frac{\partial\Phi}{\partial X}\frac{\partial\Theta}{\partial X}+\frac{\partial\Phi}{\partial Y}\frac{\partial\Theta}{\partial Y}\right)+\mathrm{Nt}\left(\left(\frac{\partial\Theta}{\partial X}\right)^2+\left(\frac{\partial\Theta}{\partial Y}\right)^2\right)\\&+\mathrm{Ha}^2\,\mathrm{Ec}\left\{UH_y-VH_x\right\}^2+\mathrm{Mn_F}\mathrm{Ec}\left\{U\frac{\partial H}{\partial X}+V\frac{\partial H}{\partial Y}\right\}H(\varepsilon_1+\Theta)\\&+\mathrm{Ec}\left\{2\left(\frac{\partial U}{\partial X}\right)^2+2\left(\frac{\partial V}{\partial X}\right)^2+\left(\frac{\partial U}{\partial X}+\frac{\partial V}{\partial Y}\right)^2\right\}\end{aligned} \tag{1.146}$$

$$\frac{\partial\Psi}{\partial Y}\frac{\partial\Phi}{\partial X}-\frac{\partial\Psi}{\partial X}\frac{\partial\Phi}{\partial Y}=\frac{1}{\mathrm{Le}}\left(\frac{\partial^2\Phi}{\partial X^2}+\frac{\partial^2\Phi}{\partial Y^2}\right)+\frac{\mathrm{Nt}}{\mathrm{Nb\,Le}}\left(\frac{\partial^2\Theta}{\partial X^2}+\frac{\partial^2\Theta}{\partial Y^2}\right) \tag{1.147}$$

$$\frac{\partial^2\Psi}{\partial X^2}+\frac{\partial^2\Psi}{\partial Y^2}=-\Omega \tag{1.148}$$

where the thermal Rayleigh number, buoyancy ratio number, Prandtl number, Brownian motion parameter, thermophoretic parameter, Lewis number, Hartmann number, Eckert number, and magnetic number arising from FHD of nanofluid are defined as: $\mathrm{Ra}=(1-\phi_c)\rho_{f_0}g\beta L^3(T_h-T_c)/(\mu\alpha)$, $\mathrm{Nr}=(\rho_p-\rho_0)(\phi_h-\phi_c)/[(1-\phi_c)\rho_{f_0}\beta L(T_h-T_c)]$, $\Pr=\mu/\rho_f\alpha$, $\mathrm{Nb}=(\rho c)_p D_B(\phi_h-\phi_c)/((\rho c)_f\alpha)$, $\mathrm{Nt}=(\rho c)_p D_T(T_h-T_c)/[(\rho c)_f\alpha T_c]$, $\mathrm{Le}=\alpha/D_B$, $\mathrm{Ha}=LH_0\mu_0\sqrt{\sigma/\mu}$, $\mathrm{Ec}=(\alpha\mu)/\left[(\rho C_P)\Delta TL^2\right]$ and $\mathrm{Mn_F}=\mu_0H_0^2K'(T_h-T_c)L^2/(\mu\alpha)$, respectively.

1.5.2.2 Mixed convection

To obtain the governing equations for force convection, the nondimensional variables should be introduced as:

$$X=\frac{x}{L},\ Y=\frac{y}{L},\ \Omega=\frac{\omega L}{u_r},\ \Psi=\frac{\psi}{u_rL},\ \Theta=\frac{T-T_c}{T_h-T_c},\ \Phi=\frac{\phi-\phi_c}{\phi_h-\phi_c},\ U=\frac{u}{u_r},\ V=\frac{v}{u_r}$$
$$H=\frac{\overline{H}}{\overline{H}_0},\ H_x=\frac{\overline{H}_x}{\overline{H}_0},\ H_y=\frac{\overline{H}_y}{\overline{H}_0} \tag{1.149}$$

By using these dimensionless parameters, the equations become:

$$\begin{aligned}\left[\frac{\partial\Psi}{\partial Y}\frac{\partial\Omega}{\partial X}-\frac{\partial\Psi}{\partial X}\frac{\partial\Omega}{\partial Y}\right]&=\frac{1}{\mathrm{Re}}\left(\frac{\partial^2\Omega}{\partial X^2}+\frac{\partial^2\Omega}{\partial Y^2}\right)+\frac{\mathrm{Gr}}{\mathrm{Re}^2}\left(\frac{\partial\Theta}{\partial X}-\mathrm{Nr}\frac{\partial\Phi}{\partial X}\right)\\&+\mathrm{Mn_F}\left\{\frac{\partial H}{\partial X}\frac{\partial\Theta}{\partial Y}-\frac{\partial H}{\partial Y}\frac{\partial\Theta}{\partial X}\right\}H\\&-\frac{\mathrm{Ha}^2}{\mathrm{Re}}\times\left\{\frac{\partial V}{\partial X}H_x^2+V\left(2H_x\frac{\partial H_x}{\partial X}\right)-\frac{\partial U}{\partial X}H_xH_y-U\frac{\partial H_x}{\partial X}H_y-U\frac{\partial H_y}{\partial X}H_x\right.\\&\left.-\frac{\partial U}{\partial Y}H_y^2-U\left(2H_y\frac{\partial H_y}{\partial Y}\right)+\frac{\partial V}{\partial Y}H_xH_y+V\frac{\partial H_x}{\partial Y}H_y+V\frac{\partial H_y}{\partial Y}H_x\right\}\end{aligned}\tag{1.150}$$

$$\begin{aligned}\frac{\partial\Psi}{\partial Y}\frac{\partial\Theta}{\partial X}-\frac{\partial\Psi}{\partial X}\frac{\partial\Theta}{\partial Y}&=\frac{1}{\mathrm{Re\,Pr}}\left(\frac{\partial^2\Theta}{\partial X^2}+\frac{\partial^2\Theta}{\partial Y^2}\right)+\frac{\mathrm{Nb}}{\mathrm{Re}}\left(\frac{\partial\Phi}{\partial X}\frac{\partial\Theta}{\partial X}+\frac{\partial\Phi}{\partial Y}\frac{\partial\Theta}{\partial Y}\right)\\&+\frac{\mathrm{Nt}}{\mathrm{Re}}\left(\left(\frac{\partial\Theta}{\partial X}\right)^2+\left(\frac{\partial\Theta}{\partial Y}\right)^2\right)+\frac{\mathrm{Ha}^2\,\mathrm{Ec}}{\mathrm{Re}}\{UH_y-VH_x\}^2\\&+\mathrm{Mn_F}\mathrm{Ec}\left\{U\frac{\partial H}{\partial X}+V\frac{\partial H}{\partial Y}\right\}H(\varepsilon_1+\Theta)\\&+\frac{\mathrm{Ec}}{\mathrm{Re}}\left\{2\left(\frac{\partial U}{\partial X}\right)^2+2\left(\frac{\partial V}{\partial X}\right)^2+\left(\frac{\partial U}{\partial X}+\frac{\partial V}{\partial Y}\right)^2\right\}\end{aligned}\tag{1.151}$$

$$\frac{\partial\Psi}{\partial Y}\frac{\partial\Phi}{\partial X}-\frac{\partial\Psi}{\partial X}\frac{\partial\Phi}{\partial Y}=\frac{1}{\mathrm{Re\,Le}}\left(\frac{\partial^2\Phi}{\partial X^2}+\frac{\partial^2\Phi}{\partial Y^2}\right)+\frac{\mathrm{Nt}}{\mathrm{Re\ Nb\ Le}}\left(\frac{\partial^2\Theta}{\partial X^2}+\frac{\partial^2\Theta}{\partial Y^2}\right)\tag{1.152}$$

$$\frac{\partial^2\Psi}{\partial X^2}+\frac{\partial^2\Psi}{\partial Y^2}=-\Omega\tag{1.153}$$

where the thermal Grashof number, buoyancy ratio number, Prandtl number, Brownian motion parameter, thermophoretic parameter, Lewis number, Reynolds number, Hartmann number, Eckert number, and magnetic number arising from FHD of nanofluid are defined as: $\mathrm{Gr}=(1-\phi_\mathrm{c})\rho_{\mathrm{f_0}}g\beta L^3\rho_\mathrm{f}(T_\mathrm{h}-T_\mathrm{c})/(\mu^2)$, $Nr=(\rho_\mathrm{p}-\rho_0)(\phi_\mathrm{h}-\phi_\mathrm{c})/[(1-\phi_\mathrm{c})\rho_{\mathrm{f_0}}\beta L(T_\mathrm{h}-T_\mathrm{c})]$, $\mathrm{Pr}=\mu/\rho_\mathrm{f}\alpha$, $Nb=(\rho c)_\mathrm{p}D_\mathrm{B}(\phi_\mathrm{h}-\phi_\mathrm{c})/((\rho c)_\mathrm{f}\upsilon)$, $\mathrm{Nt}=(\rho c)_\mathrm{p}D_\mathrm{T}(T_\mathrm{h}-T_\mathrm{c})/[(\rho c)_\mathrm{f}\upsilon T_\mathrm{c}$, $\mathrm{Re}=\dfrac{\rho_\mathrm{f}Lu_\mathrm{r}}{\mu}$, $\mathrm{Le}=\upsilon/D_\mathrm{B}$, $\mathrm{Ha}=LH_0\mu_0\sqrt{\sigma/\mu}$, $\mathrm{Ec}=(u_\mathrm{r}^2)/[(C_\mathrm{P})\Delta T]$ and $\mathrm{Mn_F}=\mu_0H_0^2K'(T_\mathrm{h}-T_\mathrm{c})L^2/(\mu\alpha)$, respectively.

1.6 MAGNETIC FIELD–DEPENDENT VISCOSITY

Realizing the importance of magnetic field–dependent (MFD) viscosity on ferrofluid flows, several studies have been undertaken. The effect of a homogeneous magnetic field on the viscosity of a fluid with solid particles possessing intrinsic magnetic moments has been investigated by Shliomis [69]. The effect of MFD viscosity on the onset of ferroconvection in a rotating ferrofluid layer is discussed by Vaidyanathan et al. [70], with or without dust particles by Sunil and Divya [71], and the nonlinear stability analysis has also been performed by Sunil et al. [72].

1.6.1 MATHEMATICAL MODELING FOR THE SINGLE-PHASE MODEL

1.6.1.1 Natural convection

The flow is considered to be steady, 2D, incompressible, and laminar. Using the Boussinesq approximation, the governing equations of heat transfer and fluid flow for nanofluid can be obtained as follows [73]:

$$\frac{\partial u}{\partial x}+\frac{\partial v}{\partial y}=0 \tag{1.154}$$

$$\rho_{\mathrm{nf}}\left(u\frac{\partial u}{\partial x}+v\frac{\partial u}{\partial y}\right)=-\frac{\partial P}{\partial x}+\eta\left(\frac{\partial^2 u}{\partial x^2}+\frac{\partial^2 u}{\partial y^2}\right)+\mu_0 M\frac{\partial \overline{H}}{\partial x}-\sigma_{\mathrm{nf}}B_y^2 u+\sigma_{\mathrm{nf}}B_x B_y v \tag{1.155}$$

$$\begin{aligned}\rho_{\mathrm{nf}}\left(u\frac{\partial v}{\partial x}+v\frac{\partial v}{\partial y}\right)=&-\frac{\partial P}{\partial y}+\eta\left(\frac{\partial^2 v}{\partial x^2}+\frac{\partial^2 v}{\partial y^2}\right)+\mu_0 M\frac{\partial \overline{H}}{\partial y}-\sigma_{\mathrm{nf}}B_x^2 v+\sigma_{\mathrm{nf}}B_x B_y u\\&+\rho_{\mathrm{nf}}\beta_{\mathrm{nf}}g(T-T_{\mathrm{c}})\end{aligned} \tag{1.156}$$

$$\begin{aligned}(\rho C_{\mathrm{p}})_{\mathrm{nf}}\left(u\frac{\partial T}{\partial x}+v\frac{\partial T}{\partial y}\right)=&k_{\mathrm{nf}}\left(\frac{\partial^2 T}{\partial x^2}+\frac{\partial^2 T}{\partial y^2}\right)+\sigma_{\mathrm{nf}}(uB_y-vB_x)^2\\&-\mu_0 T\frac{\partial M}{\partial T}\left(u\frac{\partial \overline{H}}{\partial x}+v\frac{\partial \overline{H}}{\partial y}\right)+\eta\left\{2\left(\frac{\partial u}{\partial x}\right)^2+2\left(\frac{\partial v}{\partial x}\right)^2+\left(\frac{\partial u}{\partial x}+\frac{\partial v}{\partial y}\right)^2\right\}\end{aligned} \tag{1.157}$$

where $\eta=(1+\vec{\delta}\cdot\vec{B})\mu_{\mathrm{nf}}$, the variation of MFD viscosity (δ) has been taken to be isotropic, $\delta_1=\delta_2=\delta_3=\delta$.

By introducing the following nondimensional variables:

$$\begin{aligned}&X=\frac{x}{L},\ Y=\frac{y}{L},\ \Omega=\frac{\omega L^2}{\alpha_{\mathrm{f}}},\ \Psi=\frac{\psi}{\alpha_{\mathrm{f}}},\ \Theta=\frac{T-T_{\mathrm{c}}}{T_{\mathrm{h}}-T_{\mathrm{c}}},\ U=\frac{uL}{\alpha_{\mathrm{f}}},\ V=\frac{vL}{\alpha_{\mathrm{f}}},\\&H=\frac{\overline{H}}{\overline{\overline{H_0}}},\ H_x=\frac{\overline{H_x}}{\overline{\overline{H_0}}},\ H_y=\frac{\overline{H_y}}{\overline{\overline{H_0}}},\ P=\frac{p}{\rho_{\mathrm{f}}(\alpha_{\mathrm{f}}/\mathrm{L})^2}\end{aligned} \tag{1.158}$$

where in Eq. (1.158) $\overline{H_0}=\overline{H}(\overline{a},0)=\frac{\gamma}{2\pi|b|}$ and $L=r_{\mathrm{out}}-r_{\mathrm{in}}=r_{\mathrm{in}}$. Using the dimensionless parameters, the equations now become:

$$\frac{\partial U}{\partial X}+\frac{\partial V}{\partial Y}=0 \tag{1.159}$$

$$\begin{aligned}U\frac{\partial U}{\partial X}+V\frac{\partial U}{\partial Y}=&-\frac{\partial P}{\partial X}+\Pr\left[\frac{\mu_{\mathrm{nf}}/\mu_{\mathrm{f}}}{\rho_{\mathrm{nf}}/\rho_{\mathrm{f}}}\right](1+\delta^*(H_x+H_y))\left(\frac{\partial^2 U}{\partial X^2}+\frac{\partial^2 U}{\partial Y^2}\right)\\&+\mathrm{Mn_F}\Pr\left(\frac{\rho_{\mathrm{f}}}{\rho_{\mathrm{nf}}}\right)(\varepsilon_2-\varepsilon_1-\Theta)H\frac{\partial H}{\partial X}-\mathrm{Ha}^2\Pr\left[\frac{\sigma_{\mathrm{nf}}/\sigma_{\mathrm{f}}}{\rho_{\mathrm{nf}}/\rho_{\mathrm{f}}}\right](H_y^2 U-H_x H_y V)\end{aligned} \tag{1.160}$$

$$\begin{aligned}U\frac{\partial V}{\partial X}+V\frac{\partial V}{\partial Y}=&-\frac{\partial P}{\partial X}+\Pr\left[\frac{\mu_{\mathrm{nf}}/\mu_{\mathrm{f}}}{\rho_{\mathrm{nf}}/\rho_{\mathrm{f}}}\right](1+\delta^*(H_x+H_y))\left(\frac{\partial^2 V}{\partial X^2}+\frac{\partial^2 V}{\partial Y^2}\right)\\&+\mathrm{Mn_F}\Pr\left(\frac{\rho_{\mathrm{f}}}{\rho_{\mathrm{nf}}}\right)(\varepsilon_2-\varepsilon_1-\Theta)H\frac{\partial H}{\partial Y}-\mathrm{Ha}^2\Pr\left[\frac{\sigma_{\mathrm{nf}}/\sigma_{\mathrm{f}}}{\rho_{\mathrm{nf}}/\rho_{\mathrm{f}}}\right](H_x^2 V-H_x H_y U)\\&+\mathrm{Ra}\Pr\left[\frac{\beta_{\mathrm{nf}}}{\beta_{\mathrm{f}}}\right]\Theta\end{aligned} \tag{1.161}$$

$$U\frac{\partial\Theta}{\partial X}+V\frac{\partial\Theta}{\partial Y}=\left[\frac{\frac{k_{\mathrm{nf}}}{k_{\mathrm{f}}}}{\frac{(\rho C_{\mathrm{P}})_{\mathrm{nf}}}{(\rho C_{\mathrm{P}})_{\mathrm{f}}}}\right]\left(\frac{\partial^2\Theta}{\partial X^2}+\frac{\partial^2\Theta}{\partial Y^2}\right)+Ha^2\,Ec\left[\frac{\frac{\sigma_{\mathrm{nf}}}{\sigma_{\mathrm{f}}}}{\frac{(\rho C_{\mathrm{P}})_{\mathrm{nf}}}{(\rho C_{\mathrm{P}})_{\mathrm{f}}}}\right]\{U\,H_y-V\,H_x\}^2$$
$$+\mathrm{Mn_F}\mathrm{Ec}\frac{(\rho C_{\mathrm{P}})_{\mathrm{f}}}{(\rho C_{\mathrm{P}})_{\mathrm{nf}}}\left\{U\frac{\partial H}{\partial X}+V\frac{\partial H}{\partial Y}\right\}H(\varepsilon_1+\Theta) \tag{1.162}$$
$$+Ec\left[\frac{\frac{\mu_{\mathrm{nf}}}{\mu_{\mathrm{f}}}}{\frac{(\rho C_{\mathrm{P}})_{\mathrm{nf}}}{(\rho C_{\mathrm{P}})_{\mathrm{f}}}}\right]\left(1+\delta^*(H_x+H_y)\right)\left\{2\left(\frac{\partial U}{\partial X}\right)^2+2\left(\frac{\partial V}{\partial Y}\right)^2+\left(\frac{\partial U}{\partial Y}+\frac{\partial V}{\partial X}\right)^2\right\}$$

where $\mathrm{Ra_f}=g\beta_f L^3(T_h-T_c)/(\alpha_f\upsilon_f)$, $\mathrm{Pr_f}=\upsilon_f/\alpha_f$, $\mathrm{Ha}=L\mu_0H_0\sqrt{\sigma_f/\mu_f}$, $\varepsilon_1=T_1/\Delta T$, $\mathrm{Ec}=(\mu_f\alpha_f)/[(\rho C_P)_f\Delta T\,L^2]$, $\delta^*=\delta\mu_0H_0$ and $\mathrm{Mn_F}=\mu_0H_0^2K'(T_h-T_c)L^2/(\mu_f\alpha_f)$ are the Rayleigh number, Prandtl number, Hartmann number arising from MHD, temperature number, Eckert number, viscosity parameter, and magnetic number arising from FHD the for the base fluid, respectively.

1.6.1.2 Mixed convection

The flow is considered to be steady, 2D, incompressible, and laminar. Using the Boussinesq approximation, the governing equations of heat transfer and fluid flow for nanofluid can be obtained as follows:

$$\frac{\partial u}{\partial x}+\frac{\partial v}{\partial y}=0 \tag{1.163}$$

$$\rho_{\mathrm{nf}}\left(u\frac{\partial u}{\partial x}+v\frac{\partial u}{\partial y}\right)=-\frac{\partial P}{\partial x}+\eta\left(\frac{\partial^2 u}{\partial x^2}+\frac{\partial^2 u}{\partial y^2}\right)+\mu_0 M\frac{\partial\overline{H}}{\partial x}-\sigma_{\mathrm{nf}}B_y^2u+\sigma_{\mathrm{nf}}B_xB_yv \tag{1.164}$$

$$\rho_{\mathrm{nf}}\left(u\frac{\partial v}{\partial x}+v\frac{\partial v}{\partial y}\right)=-\frac{\partial P}{\partial y}+\eta\left(\frac{\partial^2 v}{\partial x^2}+\frac{\partial^2 v}{\partial y^2}\right)+\mu_0 M\frac{\partial\overline{H}}{\partial y}-\sigma_{\mathrm{nf}}B_x^2v+\sigma_{\mathrm{nf}}B_xB_yu+\rho_{\mathrm{nf}}\beta_{\mathrm{nf}}g(T-T_c) \tag{1.165}$$

$$(\rho C_p)_{\mathrm{nf}}\left(u\frac{\partial T}{\partial x}+v\frac{\partial T}{\partial y}\right)=k_{\mathrm{nf}}\left(\frac{\partial^2 T}{\partial x^2}+\frac{\partial^2 T}{\partial y^2}\right)+\sigma_{\mathrm{nf}}(uB_y-vB_x)^2-\mu_0T\frac{\partial M}{\partial T}\left(u\frac{\partial\overline{H}}{\partial x}+v\frac{\partial\overline{H}}{\partial y}\right)+\eta\left\{2\left(\frac{\partial u}{\partial x}\right)^2+2\left(\frac{\partial v}{\partial x}\right)^2+\left(\frac{\partial u}{\partial x}+\frac{\partial v}{\partial y}\right)^2\right\} \tag{1.166}$$

where $\eta=(1+\vec{\delta}\cdot\vec{B})\mu_{\mathrm{nf}}$, the variation of MFD viscosity (δ) has been taken to be isotropic, $\delta_1=\delta_2=\delta_3=\delta$.

By introducing the following nondimensional variables:

$$X=\frac{x}{L},\ Y=\frac{y}{L},\ \Theta=\frac{T-T_c}{T_h-T_c},\ U=\frac{u}{U_{\mathrm{Lid}}},\ V=\frac{v}{U_{\mathrm{Lid}}},\ H=\frac{\overline{H}}{\overline{H}_0},\ H_x=\frac{\overline{H}_x}{\overline{H}_0},\ H_y=\frac{\overline{H}_y}{\overline{H}_0},\ P=\frac{p}{\rho_f(U_{\mathrm{Lid}})^2} \tag{1.167}$$

where in Eq. (16) $\overline{H}_0=\overline{H}(\overline{a},0)=\dfrac{\gamma}{2\pi|b|}$ and $L=r_{\text{out}}-r_{\text{in}}=r_{\text{in}}$.

Using the dimensionless parameters, the equations now become:

$$\frac{\partial U}{\partial X}+\frac{\partial V}{\partial Y}=0 \tag{1.168}$$

$$\begin{aligned}U\frac{\partial U}{\partial X}+V\frac{\partial U}{\partial Y}=&-\frac{\partial P}{\partial X}+\frac{1}{\text{Re}}\left[\frac{\mu_{\text{nf}}/\mu_{\text{f}}}{\rho_{\text{nf}}/\rho_{\text{f}}}\right]\left(1+\delta^*\left(H_x+H_y\right)\right)\left(\frac{\partial^2 U}{\partial X^2}+\frac{\partial^2 U}{\partial Y^2}\right)\\&+\text{Mn}_{\text{F}}\left(\frac{\rho_{\text{f}}}{\rho_{\text{nf}}}\right)(\varepsilon_2-\varepsilon_1-\Theta)H\frac{\partial H}{\partial X}-\frac{\text{Ha}^2}{\text{Re}}\left[\frac{\sigma_{\text{nf}}/\sigma_{\text{f}}}{\rho_{\text{nf}}/\rho_{\text{f}}}\right]\left(H_y^2U-H_xH_yV\right)\end{aligned} \tag{1.169}$$

$$\begin{aligned}U\frac{\partial V}{\partial X}+V\frac{\partial V}{\partial Y}=&-\frac{\partial P}{\partial X}+\frac{1}{\text{Re}}\left[\frac{\mu_{\text{nf}}/\mu_{\text{f}}}{\rho_{\text{nf}}/\rho_{\text{f}}}\right]\left(1+\delta^*\left(H_x+H_y\right)\right)\left(\frac{\partial^2 V}{\partial X^2}+\frac{\partial^2 V}{\partial Y^2}\right)\\&+\text{Mn}_{\text{F}}\left(\frac{\rho_{\text{f}}}{\rho_{\text{nf}}}\right)(\varepsilon_2-\varepsilon_1-\Theta)H\frac{\partial H}{\partial Y}-\frac{\text{Ha}^2}{\text{Re}}\left[\frac{\sigma_{\text{nf}}/\sigma_{\text{f}}}{\rho_{\text{nf}}/\rho_{\text{f}}}\right]\left(H_x^2V-H_xH_yU\right)+\frac{\text{Gr}}{\text{Re}^2}\left[\frac{\beta_{\text{nf}}}{\beta_{\text{f}}}\right]\Theta\end{aligned} \tag{1.170}$$

$$\begin{aligned}U\frac{\partial\Theta}{\partial X}+V\frac{\partial\Theta}{\partial Y}=&\frac{1}{\text{Pr}\,\text{Re}}\left[\frac{\frac{k_{\text{nf}}}{k_{\text{f}}}}{\frac{(\rho C_{\text{P}})_{\text{nf}}}{(\rho C_{\text{P}})_{\text{f}}}}\right]\left(\frac{\partial^2\Theta}{\partial X^2}+\frac{\partial^2\Theta}{\partial Y^2}\right)+\text{Ha}^2\frac{Ec}{\text{Re}}\left[\frac{\frac{\sigma_{\text{nf}}}{\sigma_{\text{f}}}}{\frac{(\rho C_{\text{P}})_{\text{nf}}}{(\rho C_{\text{P}})_{\text{f}}}}\right]\left\{UH_y-VH_x\right\}^2\\&+\text{Mn}_{\text{F}}\text{Ec}\frac{(\rho C_{\text{P}})_{\text{f}}}{(\rho C_{\text{P}})_{\text{nf}}}\left\{U\frac{\partial H}{\partial X}+V\frac{\partial H}{\partial Y}\right\}H(\varepsilon_1+\Theta)\\&+\frac{\text{Ec}}{\text{Re}}\left[\frac{\frac{\mu_{\text{nf}}}{\mu_{\text{f}}}}{\frac{(\rho C_{\text{P}})_{\text{nf}}}{(\rho C_{\text{P}})_{\text{f}}}}\right]\left(1+\delta^*\left(H_x+H_y\right)\right)\left\{2\left(\frac{\partial U}{\partial X}\right)^2+2\left(\frac{\partial V}{\partial Y}\right)^2+\left(\frac{\partial U}{\partial Y}+\frac{\partial V}{\partial X}\right)^2\right\}\end{aligned} \tag{1.171}$$

where $\text{Re}=\dfrac{\rho_{\text{f}}LU_{\text{Lid}}}{\mu_{\text{f}}}$, $\text{Gr}=g\beta\Delta TL^3/\upsilon_{\text{f}}^2$, $\text{Ha}=LH_0\mu_0\sqrt{\sigma_{\text{f}}/\mu_{\text{f}}}$, $\varepsilon_1=T_1/\Delta T$, $\text{Ec}=\left(\rho_{\text{f}}U_{\text{Lid}}^2\right)/\left[(\rho C_{\text{P}})_{\text{f}}\Delta T\right]$, $\text{Ri}=\text{Gr}/\text{Re}^2$, $\delta^*=\delta\mu_0H_0$ and $\text{Mn}_{\text{F}}=\mu_0H_0^2K'(T_{\text{h}}-T_{\text{c}})/\left(\rho_{\text{f}}U_{\text{Lid}}^2\right)$ are the Reynolds number, Grashof number, Hartmann number arising from MHD, temperature number, Eckert number, Richardson number, viscosity parameter, and magnetic number arising from FHD the for the base fluid, respectively.

1.6.2 MATHEMATICAL MODELING FOR THE TWO-PHASE MODEL

1.6.2.1 Natural convection

The nanofluid's density ρ is

$$\begin{aligned}\rho&=\phi\rho_{\text{p}}+(1-\phi)\rho_{\text{f}}\\&\cong\phi\rho_{\text{p}}+(1-\phi)\left\{\rho_{\text{f}_0}\left(1-\beta(T-T_{\text{c}})\right)\right\}\end{aligned} \tag{1.172}$$

where ρ_f is the base fluid's density, T_c is a reference temperature, ρ_{f_0} is the base fluid's density at the reference temperature and β is the volumetric coefficient of expansion. Taking the density of base fluid as that of the nanofluid, the density ρ in equation (1.172) thus becomes

$$\rho \cong \phi\rho_p + (1-\phi)\{\rho_0(1-\beta(T-T_c)\} \tag{1.173}$$

where ρ_0 is the nanofluid's density at the reference temperature.

The continuity, momentum under Boussinesq approximation, and energy equations for the laminar and steady-state natural convection in a 2D enclosure can be written in dimensional form as follows:

$$\frac{\partial u}{\partial x} + \frac{\partial v}{\partial y} = 0 \tag{1.174}$$

$$\rho_f\left\{u\frac{\partial u}{\partial x} + v\frac{\partial u}{\partial y}\right\} = -\frac{\partial P}{\partial x} + \eta\left(\frac{\partial^2 u}{\partial x^2} + \frac{\partial^2 u}{\partial y^2}\right) + \mu_0 M\frac{\partial \overline{H}}{\partial x} - \sigma B_y^2 u + \sigma B_x B_y v \tag{1.175}$$

$$\begin{aligned}\rho_f\left\{u\frac{\partial v}{\partial x} + v\frac{\partial v}{\partial y}\right\} = &-\frac{\partial P}{\partial y} + \eta\left(\frac{\partial^2 v}{\partial x^2} + \frac{\partial^2 v}{\partial y^2}\right) - (\phi-\phi_c)(\rho_p - \rho_{f_0})g \\ &+ (1-\phi_c)\rho_{f_0}(T-T_c)g + \mu_0 M\frac{\partial \overline{H}}{\partial y} - \sigma B_x^2 v + \sigma B_x B_y u\end{aligned} \tag{1.176}$$

$$\begin{aligned}u\frac{\partial T}{\partial x} + v\frac{\partial T}{\partial y} = \alpha\left(\frac{\partial^2 T}{\partial x^2} + \frac{\partial^2 T}{\partial y^2}\right) + \frac{(\rho c)_p}{(\rho c)_f}\left[\begin{array}{l} D_B\left\{\frac{\partial \phi}{\partial x}\cdot\frac{\partial T}{\partial x} + \frac{\partial \phi}{\partial y}\cdot\frac{\partial T}{\partial y}\right\} \\ +(D_T/T_c)\left\{\left(\frac{\partial T}{\partial x}\right)^2 + \left(\frac{\partial T}{\partial y}\right)^2\right\}\end{array}\right] + \left(uB_y - vB_x\right)^2 \\ -\mu_0 T\frac{\partial M}{\partial T}\left(u\frac{\partial \overline{H}}{\partial x} + v\frac{\partial \overline{H}}{\partial y}\right) + \eta\left\{2\left(\frac{\partial u}{\partial x}\right)^2 + 2\left(\frac{\partial v}{\partial x}\right)^2 + \left(\frac{\partial u}{\partial x} + \frac{\partial v}{\partial y}\right)^2\right\}\end{aligned} \tag{1.177}$$

$$u\frac{\partial \phi}{\partial x} + v\frac{\partial \phi}{\partial y} = D_B\left\{\frac{\partial^2 \phi}{\partial x^2} + \frac{\partial^2 \phi}{\partial y^2}\right\} + \left(\frac{D_T}{T_c}\right)\left\{\frac{\partial^2 T}{\partial x^2} + \frac{\partial^2 T}{\partial y^2}\right\} \tag{1.178}$$

where $\eta = (1+\vec{\delta}\cdot\vec{B})\mu_{nf}$, the variation of MFD viscosity (δ) has been taken to be isotropic, $\delta_1 = \delta_2 = \delta_3 = \delta$. By introducing the following nondimensional variables:

$$\begin{gathered} X = \frac{x}{L},\ Y = \frac{y}{L},\ \Theta = \frac{T-T_c}{T_h - T_c},\ \Phi = \frac{\phi - \phi_c}{\phi_h - \phi_c},\ U = \frac{uL}{\alpha},\ V = \frac{vL}{\alpha} \\ H = \frac{\overline{H}}{H_0},\ H_x = \frac{\overline{H_x}}{H_0},\ H_y = \frac{\overline{H_y}}{H_0},\ P = \frac{p}{\rho_f(\alpha_f/L)^2}\end{gathered} \tag{1.179}$$

By using these dimensionless parameters, the equations become:

$$\frac{\partial U}{\partial X} + \frac{\partial V}{\partial Y} = 0 \tag{1.180}$$

$$\begin{aligned}\left\{U\frac{\partial U}{\partial X}+V\frac{\partial U}{\partial Y}\right\} &= -\frac{\partial P}{\partial X}+\Pr\left(1+\delta^{*}\left(H_x+H_y\right)\right)\left(\frac{\partial^2 U}{\partial X^2}+\frac{\partial^2 U}{\partial Y^2}\right)\\ &\quad +\mathrm{Mn_F}\Pr\left(\varepsilon_2-\varepsilon_1-\Theta\right)H\frac{\partial H}{\partial X}-\mathrm{Ha}^2\Pr\left(H_y^2U-H_xH_yV\right)\end{aligned} \tag{1.181}$$

$$\begin{aligned}\left\{U\frac{\partial V}{\partial X}+V\frac{\partial V}{\partial Y}\right\} &= -\frac{\partial P}{\partial Y}+\Pr\left(1+\delta^{*}\left(H_x+H_y\right)\right)\left(\frac{\partial^2 V}{\partial X^2}+\frac{\partial^2 V}{\partial Y^2}\right)+\Pr\mathrm{Ra}\left(\Theta-\mathrm{Nr}\,\Phi\right)\\ &\quad +\mathrm{Mn_F}\Pr\left(\varepsilon_2-\varepsilon_1-\Theta\right)H\frac{\partial H}{\partial Y}-\mathrm{Ha}^2\Pr\left(H_x^2V-H_xH_yU\right)\end{aligned} \tag{1.182}$$

$$\begin{aligned}U\frac{\partial \Theta}{\partial X}+V\frac{\partial \Theta}{\partial Y} &= \left(\frac{\partial^2 \Theta}{\partial X^2}+\frac{\partial^2 \Theta}{\partial Y^2}\right)+\mathrm{Nb}\left(\frac{\partial \Phi}{\partial X}\frac{\partial \Theta}{\partial X}+\frac{\partial \Phi}{\partial Y}\frac{\partial \Theta}{\partial Y}\right)+\mathrm{Nt}\left(\left(\frac{\partial \Theta}{\partial X}\right)^2+\left(\frac{\partial \Theta}{\partial Y}\right)^2\right)\\ &\quad +\mathrm{Ha}^2\,\mathrm{Ec}\left\{U H_y-V H_x\right\}^2+\mathrm{Mn_F}\mathrm{Ec}\left\{U\frac{\partial H}{\partial X}+V\frac{\partial H}{\partial Y}\right\}H\left(\varepsilon_1+\Theta\right)\\ &\quad +\mathrm{Ec}\left(1+\delta^{*}\left(H_x+H_y\right)\right)\left\{2\left(\frac{\partial U}{\partial X}\right)^2+2\left(\frac{\partial V}{\partial X}\right)^2+\left(\frac{\partial U}{\partial X}+\frac{\partial V}{\partial Y}\right)^2\right\}\end{aligned} \tag{1.183}$$

$$U\frac{\partial \Phi}{\partial X}+V\frac{\partial \Phi}{\partial Y}=\frac{1}{\mathrm{Le}}\left(\frac{\partial^2 \Phi}{\partial X^2}+\frac{\partial^2 \Phi}{\partial Y^2}\right)+\frac{\mathrm{Nt}}{\mathrm{Nb\,Le}}\left(\frac{\partial^2 \Theta}{\partial X^2}+\frac{\partial^2 \Theta}{\partial Y^2}\right) \tag{1.184}$$

where thermal Rayleigh number, buoyancy ratio number, Prandtl number, Brownian motion parameter, thermophoretic parameter, Lewis number, Hartmann number, Eckert number, viscosity parameter, and magnetic number arising from FHD of nanofluid are defined as: $\mathrm{Ra}=(1-\phi_c)\rho_{f_0}g\beta L^3(T_h-T_c)/(\mu\alpha)$, $\mathrm{Nr}=(\rho_p-\rho_0)(\phi_h-\phi_c)/[(1-\phi_c)\rho_{f_0}\beta L(T_h-T_c)]$, $\Pr=\mu/\rho_f\alpha$, $\mathrm{Nb}=(\rho c)_p D_B(\phi_h-\phi_c)/((\rho c)_f\alpha)$, $\mathrm{Nt}=(\rho c)_p D_T(T_h-T_c)/[(\rho c)_f\alpha T_c]$, $\mathrm{Le}=\alpha/D_B$, $\mathrm{Ha}=LH_0\mu\sqrt{\sigma/\mu}$, $\mathrm{Ec}=(\alpha\mu)/[(\rho C_P)\Delta TL^2]$, $\delta^{*}=\delta\mu_0H_0$ and $\mathrm{Mn_F}=\mu_0H_0^2K'(T_h-T_c)L^2/(\mu\alpha)$, respectively.

1.6.2.2 Mixed convection

The nanofluid's density ρ is

$$\begin{aligned}\rho &= \phi\rho_p+(1-\phi)\rho_f\\ &\cong \phi\rho_p+(1-\phi)\left\{\rho_{f_0}\left(1-\beta(T-T_c)\right)\right\}\end{aligned} \tag{1.185}$$

where ρ_f is the base fluid's density, T_c is a reference temperature, ρ_{f_0} is the base fluid's density at the reference temperature, and β is the volumetric coefficient of expansion. Taking the density of base fluid as that of the nanofluid, the density ρ in equation (1.185) thus becomes

$$\rho\cong\phi\rho_p+(1-\phi)\left\{\rho_0\left(1-\beta(T-T_c)\right)\right\} \tag{1.186}$$

where ρ_0 is the nanofluid's density at the reference temperature.

The continuity, momentum under Boussinesq approximation, and energy equations for the laminar and steady-state natural convection in a 2D enclosure can be written in dimensional form as follows:

$$\frac{\partial u}{\partial x}+\frac{\partial v}{\partial y}=0 \tag{1.187}$$

$$\rho_f\left\{u\frac{\partial u}{\partial x}+v\frac{\partial u}{\partial y}\right\}=-\frac{\partial P}{\partial x}+\eta\left(\frac{\partial^2 u}{\partial x^2}+\frac{\partial^2 u}{\partial y^2}\right)+\mu_0 M\frac{\partial \overline{H}}{\partial x}-\sigma B_y^2 u+\sigma B_x B_y v \tag{1.188}$$

$$\begin{aligned}\rho_f\left\{u\frac{\partial v}{\partial x}+v\frac{\partial v}{\partial y}\right\}=&-\frac{\partial P}{\partial y}+\eta\left(\frac{\partial^2 v}{\partial x^2}+\frac{\partial^2 v}{\partial y^2}\right)-(\phi-\phi_c)(\rho_p-\rho_{f_0})g\\&+(1-\phi_c)\rho_{f_0}(T-T_c)g+\mu_0 M\frac{\partial \overline{H}}{\partial y}-\sigma B_x^2 v+\sigma B_x B_y u\end{aligned} \tag{1.189}$$

$$\begin{aligned}u\frac{\partial T}{\partial x}+v\frac{\partial T}{\partial y}=&\alpha\left(\frac{\partial^2 T}{\partial x^2}+\frac{\partial^2 T}{\partial y^2}\right)+\frac{(\rho c)_p}{(\rho c)_f}\left[\begin{array}{l}D_B\left\{\frac{\partial \phi}{\partial x}\cdot\frac{\partial T}{\partial x}+\frac{\partial \phi}{\partial y}\cdot\frac{\partial T}{\partial y}\right\}\\+(D_T/T_c)\left\{\left(\frac{\partial T}{\partial x}\right)^2+\left(\frac{\partial T}{\partial y}\right)^2\right\}\end{array}\right]+\left(uB_y-vB_x\right)^2\\&-\mu_0 T\frac{\partial M}{\partial T}\left(u\frac{\partial \overline{H}}{\partial x}+v\frac{\partial \overline{H}}{\partial y}\right)+\eta\left\{2\left(\frac{\partial u}{\partial x}\right)^2+2\left(\frac{\partial v}{\partial x}\right)^2+\left(\frac{\partial u}{\partial x}+\frac{\partial v}{\partial y}\right)^2\right\}\end{aligned} \tag{1.190}$$

$$u\frac{\partial \phi}{\partial x}+v\frac{\partial \phi}{\partial y}=D_B\left\{\frac{\partial^2 \phi}{\partial x^2}+\frac{\partial^2 \phi}{\partial y^2}\right\}+\left(\frac{D_T}{T_c}\right)\left\{\frac{\partial^2 T}{\partial x^2}+\frac{\partial^2 T}{\partial y^2}\right\} \tag{1.191}$$

where $\eta=(1+\vec{\delta}\cdot\vec{B})\mu_{nf}$, the variation of MFD viscosity (δ) has been taken to be isotropic, $\delta_1=\delta_2=\delta_3=\delta$. By introducing the following nondimensional variables:

$$\begin{aligned}&X=\frac{x}{L},\ Y=\frac{y}{L},\ \Theta=\frac{T-T_c}{T_h-T_c},\ \Phi=\frac{\phi-\phi_c}{\phi_h-\phi_c},\ U=\frac{u}{u_r},\ V=\frac{v}{u_r}\\&H=\frac{\overline{H}}{\overline{\overline{H_0}}},\ H_x=\frac{\overline{H_x}}{\overline{\overline{H_0}}},\ H_y=\frac{\overline{H_y}}{\overline{\overline{H_0}}},\ P=\frac{p}{\rho_f(u_r)^2}\end{aligned} \tag{1.192}$$

By using these dimensionless parameters, the equations become:

$$\frac{\partial U}{\partial X}+\frac{\partial V}{\partial Y}=0 \tag{1.193}$$

$$\begin{aligned}\left\{U\frac{\partial U}{\partial X}+V\frac{\partial U}{\partial Y}\right\}=&-\frac{\partial P}{\partial X}+\frac{1}{\mathrm{Re}}\left(1+\delta^*\left(H_x+H_y\right)\right)\left(\frac{\partial^2 U}{\partial X^2}+\frac{\partial^2 U}{\partial Y^2}\right)\\&+\mathrm{Mn_F}\left(\varepsilon_2-\varepsilon_1-\Theta\right)H\frac{\partial H}{\partial X}-\frac{\mathrm{Ha}^2}{\mathrm{Re}}\left(H_y^2 U-H_x H_y V\right)\end{aligned} \tag{1.194}$$

$$\begin{aligned}\left\{U\frac{\partial V}{\partial X}+V\frac{\partial V}{\partial Y}\right\}=&-\frac{\partial P}{\partial Y}+\frac{1}{\mathrm{Re}}\left(1+\delta^*\left(H_x+H_y\right)\right)\left(\frac{\partial^2 V}{\partial X^2}+\frac{\partial^2 V}{\partial Y^2}\right)+\frac{\mathrm{Gr}}{\mathrm{Re}^2}(\Theta-\mathrm{Nr}\,\Phi)\\&+\mathrm{Mn_F}\left(\varepsilon_2-\varepsilon_1-\Theta\right)H\frac{\partial H}{\partial Y}-\frac{\mathrm{Ha}^2}{\mathrm{Re}}\left(H_x^2 V-H_x H_y U\right)\end{aligned} \tag{1.195}$$

$$U\frac{\partial \Theta}{\partial X}+V\frac{\partial \Theta}{\partial Y}=\left(\frac{\partial^2 \Theta}{\partial X^2}+\frac{\partial^2 \Theta}{\partial Y^2}\right)+\mathrm{Nb}\left(\frac{\partial \Phi}{\partial X}\frac{\partial \Theta}{\partial X}+\frac{\partial \Phi}{\partial Y}\frac{\partial \Theta}{\partial Y}\right)+\mathrm{Nt}\left(\left(\frac{\partial \Theta}{\partial X}\right)^2+\left(\frac{\partial \Theta}{\partial Y}\right)^2\right)$$
$$+\mathrm{Ha}^2\,\mathrm{Ec}\{U H_y - V H_x\}^2+\mathrm{Mn_F}\mathrm{Ec}\left\{U\frac{\partial H}{\partial X}+V\frac{\partial H}{\partial Y}\right\}H\,(\varepsilon_1+\Theta) \tag{1.196}$$
$$+\mathrm{Ec}\left(1+\delta^*\left(H_x+H_y\right)\right)\left\{2\left(\frac{\partial U}{\partial X}\right)^2+2\left(\frac{\partial V}{\partial X}\right)^2+\left(\frac{\partial U}{\partial X}+\frac{\partial V}{\partial Y}\right)^2\right\}$$

$$U\frac{\partial \Phi}{\partial X}+V\frac{\partial \Phi}{\partial Y}=\frac{1}{\mathrm{Le}}\left(\frac{\partial^2 \Phi}{\partial X^2}+\frac{\partial^2 \Phi}{\partial Y^2}\right)+\frac{\mathrm{Nt}}{\mathrm{Nb\ Le}}\left(\frac{\partial^2 \Theta}{\partial X^2}+\frac{\partial^2 \Theta}{\partial Y^2}\right) \tag{1.197}$$

where the thermal Grashof number, buoyancy ratio number, Prandtl number, Brownian motion parameter, thermophoretic parameter, Lewis number, Reynolds number, Hartmann number, Eckert number, viscosity parameter, and magnetic number arising from FHD of nanofluid are defined as: $\mathrm{Gr}=(1-\phi_c)\rho_{f_0} g\beta L^3 \rho_f (T_h-T_c)/(\mu^2)$, $\mathrm{Nr}=(\rho_p-\rho_0)(\phi_h-\phi_c)/[(1-\phi_c)\rho_{f_0}\beta L(T_h-T_c)]$, $\mathrm{Pr}=\mu/\rho_f\alpha$, $\mathrm{Nb}=(\rho c)_p D_B(\phi_h-\phi_c)/((\rho c)_f \upsilon)$, $\mathrm{Nt}=(\rho c)_p D_T(T_h-T_c)/[(\rho c)_f \upsilon T_c]$, $\mathrm{Re}=\rho_f L u_r/\mu$, $\mathrm{Le}=\upsilon/D_B$, $\mathrm{Ha}=LH_0\mu_0\sqrt{\sigma/\mu}$, $\mathrm{Ec}=(u_r^2)/[(C_P)\Delta T]$, $\delta^*=\delta\mu_0 H_0$ and $\mathrm{Mn_F}=\mu_0 H_0^2 K'(T_h-T_c)/(\rho_{f_0}u_r^2)$, respectively.

REFERENCES

[1] Alfvén H. Existence of electromagnetic-hydrodynamic waves. Nature 1942;150:405.
[2] Rosensweig RE. Ferrohydrodynamic. London: Cambridge University Press; 1985.
[3] Hiegeister R, Andra W, Buske N, Hergt R, Hilger I, Richter U, Kaiser W. Application of magnetite ferrofluids for hyperthermia. J Magn Magn Mater 1999;201:420–2.
[4] Nakatsuka K, Jeyadevan B, Neveu S, Koganezawa H. The magnetic fluid for heat transfer applications. J Magn Magn Mater 2002;252:360–2.
[5] Shuchi S, Sakatani K, Yamaguchi H. An application of a binary mixture of magnetic fluid for heat transport devices. J Magn Magn Mater 2005;289:257–9.
[6] Selimefendigil F, Oztop HF. Effect of a rotating cylinder in forced convection of ferrofluid over a backward facing step. Int J Heat Mass Transfer 2014;71:142–8.
[7] Wrobel W, Fornalik-Wajs E, Szmyd JS. Experimental and numerical analysis of thermo-magnetic convection a vertical annular enclosure. Int J Heat Fluid Flow 2010;31:1019–31.
[8] Tzirtzilakis EE, Xenos M, Loukopoulos VC, Kafoussias NG. Turbulent biomagnetic fluid flow in a rectangular channel under the action of a localized magnetic field. Int J Eng Sci 2006;44:1205–24.
[9] Tzirtzilakis EE, Sakalis VD, Kafoussias NG, Hatzikonstantinou PM. Biomagnetic fluid flow in a 3D rectangular duct. Int J Numer Meth Fluids 2004;44:1279–98.
[10] Aminfar H, Mohammadpourfard M, Narrimani Kahnamouei Y. A 3D numerical simulation of mixed convection of a magnetic nanofluid in the presence of non-uniform magnetic field in a vertical tube using two phase mixture model. J Magn Magn Mater 2011;323:1963–72.

[11] Sheikholeslami M, Gorji-Bandpy M. Free convection of ferrofluid in a cavity heated from below in the presence of an external magnetic field. Powder Technol 2014;256:490–8.
[12] Tzirtzilakis EE. A mathematical model for blood flow in magnetic field. Phys Fluids 2005;17:1–15.
[13] Fertman VE. Magnetic fluids guidebook: properties and applications. New York: Hemisphere; 1990.
[14] Rosensweig RE. Ferrohydrodynamic. Cambridge: Cambridge University Press; 1985.
[15] Andersson HI, Valnes OA. Flow of a heated ferrofluid over a stretching sheet in the presence of a magnetic dipole. Acta Mechanica 1998;128:39.
[16] Khashan SA, Haik Y. Numerical simulation of biomagnetic fluid downstream an eccentric stenotic orifice. Phys Fluids 2006;18:113601.
[17] Higashi T, Yamagishi T, Takeuchi A, Kawaguchi N, Sagawa S, Onishi S, Date M. Orientation of erythrocytes in a strong static magnetic fields. Blood 1993;82:1328.
[18] Shaylgin AN, Norina SB, Kondorsky EI. Behavior of erythrocytes in high gradient magnetic field. J Magn Magn Mater 1983;31(555).
[19] Khanafer K, Vafai K, Lightstone M. Buoyancy-driven heat transfer enhancement in a two-dimensional enclosure utilizing nanofluids. Int J Heat Mass Transfer 2003;46:3639–53.
[20] Abu-Nada E, Masoud Z, Hijazi A. Natural convection heat transfer enhancement in horizontal concentric annuli using nanofluids. Int J Heat Mass Transf 2008;35:657–65.
[21] Hatami M, Sheikholeslami M, Ganji DD. Nanofluid flow and heat transfer in an asymmetric porous channel with expanding or contracting wall. J Mol Liq 2014;195:230–9.
[22] Sheikholeslami M, Ganji DD. Numerical investigation for two phase modeling of nanofluid in a rotating system with permeable sheet. J Mol Liq 2014;194:13–9.
[23] Wasp FJ. Solid–liquid slurry pipeline transportation. Berlin: Trans Tech; 1977.
[24] Abu-Nada E, Oztop HF. Effects of inclination angle on natural convection in enclosures filled with Cu–water nanofluid. Int J Heat Fluid Flow 2009;30:669–78.
[25] Hatami M, Sheikholeslami M, Ganji DD. Laminar flow and heat transfer of nanofluid between contracting and rotating disks by least square method. Powder Technol 2014;253:769–79.
[26] Sheikholeslami M, Hatami M, Ganji DD. Nanofluid flow and heat transfer in a rotating system in the presence of a magnetic field. J Mol Liq 2014;190:112–20.
[27] Ding Y, Wen D. Particle migration in a flow of nanoparticle suspensions. Powder Technol 2005;149(2–3):84–92.
[28] Bianco V, Manca O, Nardini S. Numerical investigation on nanofluids turbulent convection heat transfer inside a circular tube. Int J Therm Sci 2009;29:3632–42.
[29] Mirmasoumi S, Behzadmehr A. Numerical study of laminar mixed convection of a nanofluid in a horizontal tube using two-phase mixture model. Appl Therm Eng 2008;28:717–27.
[30] Akbarinia A, Laur R. Investigation the diameter of solid particles effects on a laminar nanofluid flow in a curved tube using a two phase approach. Int J Heat Fluid Flow 2009;30:706–14.
[31] Buongiorno J. Convective transport in nanofluids. J Heat Transfer 2006;128:240–50.
[32] Behzadmehr A, Saffar-Avval M, Galanis N. Prediction of turbulent forced convection of a nanofluid in a tube with uniform heat flux using a two phase approach. Int J Heat Fluid Flow 2007;28(2):211–9.
[33] Soleimani S, Sheikholeslami M, Ganji DD, Gorji-Bandpay M. Natural convection heat transfer in a nanofluid filled semi-annulus enclosure. Int J Heat Mass Transf 2012;39:565–74.

[34] Pak BC, Cho YI. Hydrodynamic and heat transfer study of dispersed fluids with submicron metallic oxide particles. Exp Heat Transfer 1998;11:151–70.
[35] Eastman JA, Choi SUS, Li S, Soyez G, Thompson LJ, Di Melfi RJ. Novel thermal properties of nanostructured materials. Mater Sci Forum 1999;312–4. 629-34.
[36] Palm SJ, Roy G, Nguyen CT. Heat transfer enhancement with the use of nanofluids in radial flow cooling systems considering temperature dependent properties. Appl Therm Eng 2006;26:2209–18.
[37] Sheikholeslami M, Gorji-Bandpy M, Ganji DD, Soleimani S. Effect of a magnetic field on natural convection in an inclined half-annulus enclosure filled with Cu–water nanofluid using CVFEM. Adv Powder Technol 2013;24:980–91.
[38] Brinkman H. The viscosity of concentrated suspensions and solutions. J Chem Phys 1952;20:571.
[39] Abu-Nada E, Chamkha AJ. Effect of nanofluid variable properties on natural convection in enclosures filled with a CuO–EG–Water nanofluid. Int J Therm Sci 2010;49(12): 2339–52.
[40] Namburu PK, Kulkarni DP, Misra D, Das DK. Viscosity of copper oxide nanoparticles dispersed in ethylene glycol and water mixture. Exp Therm Fluid Sci 2007;32(2): 397–402.
[41] Arefmanesh A, Mahmoodi M. Effects of uncertainties of viscosity models for Al2O3–water nanofluid on mixed convection numerical simulations. Int J Therm Sci 2011;50(9):1706–19.
[42] Maïga SEB, Nguyen CT, Galanis N, Roy G. Heat transfer behaviours of nanofluids in a uniformly heated tube. Superlattice Microst 2004;35(3–6):543–57.
[43] Einstein A. Investigation on the Theory of Brownian Motion. New York: Dover; 1956.
[44] Pak BC, Cho Y. Hydrodynamic and heat transfer study of dispersed fluids with submicron metallic oxide particle. Exp Heat Transfer 1998;11:151–70.
[45] Nguyen CT, Desgranges F, Roy G, Galanis N, Mare T, Boucher S, Angue Minsta H. Temperature and particle-size dependent viscosity data for water-based nanofluids–hysteresis phenomenon. Int J Heat Fluid Flow 2007;28:1492–506.
[46] Abu-Nada E. Effects of variable viscosity and thermal conductivity of Al2O3-water nanofluid on heat transfer enhancement in natural convection. Int J Heat Fluid Flow 2009;30:679–90.
[47] Jang SP, Lee JH, Hwang KS, Choi SUS. Particle concentration and tube size dependence of viscosities of Al_2O_3-water nanofluids flowing through micro and minitubes. Appl Phys Lett 2007;91:243112.
[48] Koo J, Kleinstreuer C. A new thermal conductivity model for nanofluids. J Nanopart Res 2004;6:577–88.
[49] Maiga SEB, Palm SM, Nguyen CT, Roy G, Galanis N. Heat transfer enhancement by using nanofluids in forced convection flows. Int J Heat Fluid Flow 2005;26:530–46.
[50] Orozco D. Hydrodynamic behavior of suspension of polar particles. Encycl Surf Colloid Sci 2005;4:2375–96.
[51] Nguyen CT, Desgranges F, Galanis N, Roy G, Mare T, Boucher S, Mintsa HA. Viscosity data for Al_2O_3-water nanofluid-hysteresis: is heat transfer enhancement using nanofluids reliable? Int J Therm Sci 2008;47:103–11.
[52] Masoumi N, Sohrabi N, Behzadmehr A. A new model for calculating the effective viscosity of nanofluids. J Phys D Appl Phys 2009;42:055501.
[53] Gherasim I, Roy G, Nguyen CT, Vo-Ngoc D. Experimental investigation of nanofluids in confined laminar radial flows. Int J Therm Sci 2009;48:1486–93.

[54] Jang SP, Choi SUS. The role of Brownian motion in the enhanced thermal conductivity of nanofluids. Appl Phys Lett 2004;84:4316–8.
[55] Wang BX, Zhou LR, Peng XF. A fractal model for predicting the effective thermal conductivity of liquid with suspension of nanoparticles. Int J Heat Mass Transfer 2003;46:2665–72.
[56] Chon CH, Kihm KD, Lee SP, Choi SUS. Empirical correlation finding the role of temperature and particle size for nanofluid (Al2O3) thermal conductivity enhancement. Appl Phys Lett 2005;87(15). Article ID 153107.
[57] Charuyakorn P, Sengupta S, Roy SK. Forced convection heat transfer in micro encapsulated phase change material slurries. Int J Heat Mass Transfer 1991;34:819–33.
[58] Eastman JA, Phillpot SR, Choi US, Keblinski P. Thermal transport in nanofluids. Annu Rev Mater Res 2000;34:219–46.
[59] Yu W, Choi SUS. The role of interfacial layers in the enhanced thermal conductivity of nanofluids: a renovated Maxwell model. J Nanopart Res 2003;5:167–71.
[60] Patel HE, Sundarrajan T, Pradeep T, Dasgupta A, Dasgupta N, Das SK. A microconvection model for thermal conductivity of nanofluid. Pramana J Phys 2005;65:863–9.
[61] Angue Mintsa H, Roy G, Nguyen CT, Doucet D. New temperature dependent thermal conductivity data for water-based nanofluids. Int J Therm Sci 2009;48:363–71.
[62] Sheikholeslami M, Gorji-Bandpy M, Ganji DD, Soleimani S. Natural convection heat transfer in a nanofluid filled inclined L-shaped enclosure, IJST. T Mech Eng 2014;38: 1–5.
[63] Sheikholeslami M, Gorji-Bandpy M, Domairry G. Free convection of nanofluid filled enclosure using lattice Boltzmann method (LBM). Appl Math Mech 2013;34(7):1–15.
[64] Sheikholeslami M, Gorji-Bandpy M, Ganji DD, Soleimani S. MHD natural convection in a nanofluid filled inclined enclosure with sinusoidal wall using CVFEM. Neural Comput Appl 2014;24:873–82.
[65] Sheikholeslami M, Gorji-Bandpy M, Ganji DD, Soleimani S. Natural convection heat transfer in a cavity with sinusoidal wall filled with CuO-water nanofluid in presence of magnetic field. J Taiwan Inst Chem Eng 2014;45:40–9.
[66] Sheikholeslami M, Gorji-Bandpy M, Soleimani S. Two phase simulation of nanofluid flow and heat transfer using heatline analysis. Int J Heat Mass Transf 2013;47:73–81.
[67] Sheikholeslami M, Gorji-Bandpy M, Ganji DD, Soleimani S. Heat flux boundary condition for nanofluid filled enclosure in presence of magnetic field. J Mol Liq 2014;193: 174–84.
[68] Sheikholeslami M, Ganji DD. Ferrohydrodynamic and magnetohydrodynamic effects on ferrofluid flow and convective heat transfer. Energy 2014;75:400–10.
[69] Shliomis MI. Effect of viscosity of magnetic suspensions. Soviet Phys Uspekhi 1972;34:1291–4.
[70] Vaidyanathan G, Sekar R, Ramanathan A. Effect of magnetic field dependent viscosity on ferroconvection in rotating medium. Indian J Pure Appl Phys 2002;40:159–65.
[71] Sharma S, Divya RC. The effect of magnetic-field-dependent viscosity on ferroconvection in a porous medium in the presence of dust particles. J Geophy Eng 2004;1:277–86.
[72] Sunil P, Sharma A, Mahajan. A nonlinear stability analysis for thermoconvective magnetized ferrofluid with magnetic field dependent viscosity. Int J Heat Mass Transf 2008;35:1281–7.
[73] Sheikholeslami M, Rashidi MM. Ferrofluid heat transfer treatment in the presence of variable magnetic field. Eur Phys J Plus 2015;130:115.

CHAPTER

The control volume finite element method: application for magnetohydrodynamic nanofluid hydrothermal behavior

2

CHAPTER OUTLINE

2.1 INTRODUCTION

The control volume finite element method (CVFEM) is a scheme that uses the advantages of both the finite volume and finite element methods for simulation of multiphysics problems in complex geometries [1,2]. Soleimani et al. [3] studied natural convection heat transfer in a semi-annulus enclosure filled with nanofluid using CVFEM. They found that the angle of turn has an important effect on the streamlines, isotherms, and maximum or minimum values of the local Nusselt number. Sheikholeslami et al. [4] studied the problem of natural convection between a circular enclosure and a sinusoidal cylinder. They concluded that streamlines; isotherms; and the number, size, and formation of the cells inside the enclosure strongly depend on the Rayleigh number, values of amplitude, and the number of undulations of the enclosure. The magnetic field effect on natural convection heat transfer in a curved-shape enclosure was studied by Sheikholeslami et al. [5] using CVFEM. Their obtained results showed the significant effects of the Hartmann number on the fluid flow and temperature distribution inside the enclosure. Also, it was found that the Nusselt number decreases with an increase in the Hartmann number. Sheikholeslami et al. [6] used heatline analysis to simulate two-phase simulation of nanofluid flow and heat transfer. Their results indicated that the average Nusselt number decreases as the buoyancy ratio number increases until it reaches a minimum value and then starts increasing. Sheikholeslami et al. [7] applied CVFEM to simulate free convection heat transfer in an enclosure filled with nanofluid. Their results indicated that Nusselt number is an increasing function of buoyancy ratio number, and it is a decreasing function of Lewis number and angle of turn.

CVFEM has been applied to investigate flow and heat transfer of CuO–water nanofluid in the presence of a magnetic field by Sheikholeslami et al. [8]. Their results showed that the Nusselt number is an increasing function of nanoparticles' volume fraction, dimensionless amplitude of the sinusoidal wall, and Rayleigh number, and it is a decreasing function of Hartmann number. Sheikholeslami et al. [9] studied magnetohydrodynamic (MHD) flow in a nanofluid-filled inclined enclosure with a sinusoidal wall. They reported that for all values of Hartmann number, at Ra = 10^4 and 10^5, maximum values of E are obtained at $\gamma = 60°$ and $\gamma = 0°$, respectively. Sheikholeslami et al. [10] used CVFEM to simulate MHD effect on natural convection heat transfer of Cu–water nanofluid. The MHD effect on natural convection heat transfer in an inclined L-shape enclosure filled with nanofluid was studied by Sheikholeslami et al. [11]. They found that enhancement in heat transfer has a reverse relationship with the Hartmann number and Rayleigh number. Sheikholeslami et al. [12] studied the natural convection heat transfer of nanofluid in an inclined L-shaped enclosure. They indicated that adding nanoparticles to the base fluid can enhance the rate of heat transfer. Sheikholeslami et al. [13] studied the problem of the heat flux boundary condition for nanofluid-filled enclosure in presence of a magnetic field. They found that enhancement in heat transfer increases with increase of Hartmann number and aspect ratio, but it decreases with augment of Rayleigh number. The effect of MHD on natural convection heat transfer in an enclosure filled with nanofluid was studied by Sheikholeslami et al. [14]. Their results indicated that the Nusselt number is an increasing function of the buoyancy ratio number, but it is a decreasing function of the Lewis number and Hartmann number. Sheikholeslami et al. [15] used CVFEM to simulate the effect of a magnetic field on natural convection in an inclined half-annulus enclosure filled with Cu–water nanofluid. Their results indicated that Hartmann number and the inclination angle of the enclosure can be considered control parameters at different Rayleigh numbers. Sheikholeslami et al. [16] investigated ferrofluid hydrothermal behavior in the presence of thermal radiation. They showed that the Nusselt number increases with an increase in Lorentz forces.

2.2 BASIC IDEA OF THE CONTROL VOLUME FINITE ELEMENT METHOD

Triangular elements are considered the building blocks of the discretization using CVEM. The values of variables are approximated with linear interpolation within the elements. A control volume is created by joining the center of each element in the support to the midpoints of the element sides that pass through the central node i, which creates a close polygonal control volume (see Fig. 2.1). To illustrate the

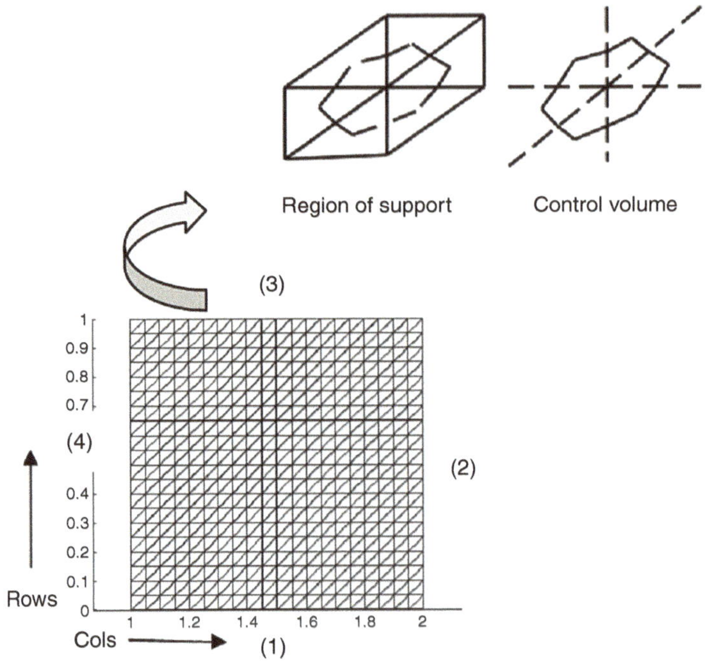

FIGURE 2.1 The Mesh of Enclosure Considered in This Work

solution procedure using CVFEM, one can consider the general form of advection–diffusion equation for node i in integral form [4]:

$$-\int_V Q\, dV - \int_A k\nabla\phi\cdot n\, dA + \int_A (v\cdot n)\phi\, dA = 0 \tag{2.1}$$

or point form

$$-\nabla\cdot(k\nabla\phi)+\nabla\cdot(v\phi)-Q=0, \tag{2.2}$$

which can be represented by the system of CVFEM discrete equations as:

$$\left[a_i + Q_{C_i} + B_{C_i}\right]\phi_i = \sum_{j=1}^{n_i} a_{i,j}\phi_{S_{i,j}} + Q_{B_i} + B_{B_i} \tag{2.3}$$

In the aforementioned, the as are the coefficients, the index (i, j) indicates the jth node in the support of node i, the index $S_{i,j}$ provides the node number of the jth node in the support, the Bs account for boundary conditions, and the Qs account for source terms. For the selected triangular element shown in Fig. 2.2, this approximation without considering the source term leads to

$$-\left(a_1^{\mathrm{k}} + a_1^{\mathrm{u}}\right)\phi_i + \left(a_2^{\mathrm{k}} + a_2^{\mathrm{u}}\right)\phi_{S_{i,3}} + \left(a_2^{\mathrm{k}} + a_2^{\mathrm{u}}\right)\phi_{S_{i,4}} = 0 \tag{2.4}$$

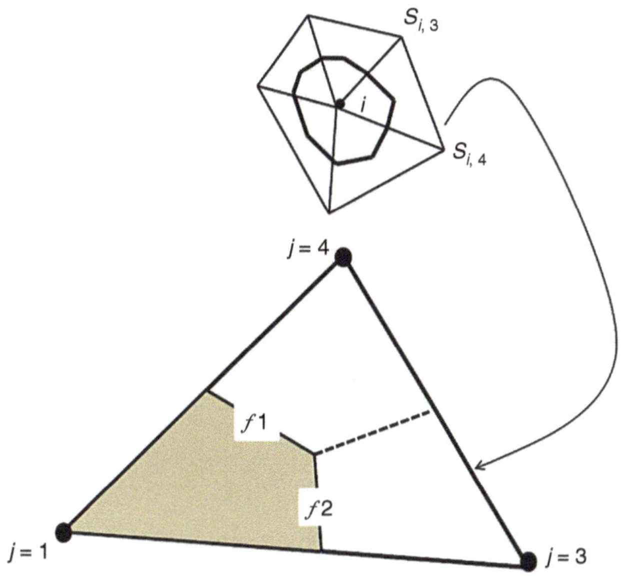

FIGURE 2.2 A Sample Triangular Element and Its Corresponding Control Volume

Using upwinding the advective coefficients, identified with the superscripts $(\ )^{\mathrm{u}}$, are given by

$$\begin{aligned} a_1^{\mathrm{u}} &= \max[q_{\mathrm{f1}}, 0] + \max[q_{\mathrm{f2}}, 0] \\ a_2^{\mathrm{u}} &= \max[-q_{\mathrm{f1}}, 0] \\ a_3^{\mathrm{u}} &= \max[-q_{\mathrm{f2}}, 0] \end{aligned} \tag{2.5}$$

And the diffusion coefficients, identified with the superscripts $(\ )^{\mathrm{k}}$, are given by

$$\begin{aligned} a_1^{\mathrm{k}} &= -k_{\mathrm{f1}} N_{1x} \Delta\vec{y}_{\mathrm{f1}} + k_{\mathrm{f1}} N_{1y} \Delta\vec{x}_{f1} - k_{\mathrm{f2}} N_{1x} \Delta\vec{y}_{\mathrm{f2}} + k_{\mathrm{f2}} N_{1y} \Delta\vec{x}_{\mathrm{f2}} \\ a_2^{\mathrm{k}} &= -k_{\mathrm{f1}} N_{2x} \Delta\vec{y}_{\mathrm{f1}} + k_{\mathrm{f1}} N_{2y} \Delta\vec{x}_{\mathrm{f1}} - k_{\mathrm{f2}} N_{2x} \Delta\vec{y}_{\mathrm{f2}} + k_{\mathrm{f2}} N_{2y} \Delta\vec{x}_{\mathrm{f2}} \\ a_2^{\mathrm{k}} &= -k_{\mathrm{f1}} N_{3x} \Delta\vec{y}_{\mathrm{f1}} + k_{\mathrm{f1}} N_{3y} \Delta\vec{x}_{\mathrm{f1}} - k_{\mathrm{f2}} N_{3x} \Delta\vec{y}_{\mathrm{f2}} + k_{\mathrm{f2}} N_{3y} \Delta\vec{x}_{\mathrm{f2}} \end{aligned} \tag{2.6}$$

In Eq. (2.5), the volume flow across face 1 and 2 in the direction of the outward normal is

$$\begin{aligned} q_{\mathrm{f1}} &= v \cdot n\, A|_{\mathrm{f1}} = v_x^{f1} \Delta\vec{y}_{\mathrm{f1}} - v_y^{f1} \Delta\vec{y}_{\mathrm{f1}} \\ q_{\mathrm{f2}} &= v \cdot n\, A|_{\mathrm{f2}} = v_x^{f2} \Delta\vec{y}_{\mathrm{f2}} - v_y^{f2} \Delta\vec{y}_{\mathrm{f2}} \end{aligned} \tag{2.7}$$

The value of the diffusivity at the midpoint of face 1 can be obtained as

$$k_{\mathrm{f1}} = [N_1 k_1 + N_2 k_2 + N_3 k_3]_{\mathrm{f1}} = \frac{5}{12} k_1 + \frac{5}{12} k_2 + \frac{2}{12} k_3 \tag{2.8}$$

and at the midpoint of face 2

$$k_{f2} = \left[N_1 k_1 + N_2 k_2 + N_3 k_3\right]_{f2} = \frac{5}{12}k_1 + \frac{2}{12}k_2 + \frac{5}{12}k_3 \tag{2.9}$$

The velocity components at the midpoint of face 1 are:

$$\begin{aligned} v_x^{f1} &= \frac{5}{12}v_{x_1} + \frac{5}{12}v_{x_2} + \frac{2}{12}v_{x_3} \\ v_y^{f1} &= \frac{5}{12}v_{y_1} + \frac{5}{12}v_{y_2} + \frac{2}{12}v_{y_3} \end{aligned} \tag{2.10}$$

And on face 2:

$$\begin{aligned} v_x^{f2} &= \frac{5}{12}v_{x_1} + \frac{2}{12}v_{x_2} + \frac{5}{12}v_{x_3} \\ v_y^{f2} &= \frac{5}{12}v_{y_1} + \frac{2}{12}v_{y_2} + \frac{5}{12}v_{y_3} \end{aligned} \tag{2.11}$$

These values can be used to *update* the *i*th support coefficients through the following equation:

$$\begin{aligned} a_i &= a_i + a_1^{k} \\ a_{i,3} &= a_{i,3} + a_2^{k} \\ a_{i,4} &= a_{i,4} + a_3^{k} \end{aligned} \tag{2.12}$$

In Eq. (2.6), moving counterclockwise around node i, the signed distances are

$$\begin{aligned} \Delta\vec{x}_{f1} &= \frac{x_3}{3} - \frac{x_2}{6} - \frac{x_1}{6}, \quad \Delta\vec{x}_{f2} = -\frac{x_2}{3} + \frac{x_3}{6} + \frac{x_1}{6} \\ \Delta\vec{y}_{f1} &= \frac{y_3}{3} - \frac{y_2}{6} - \frac{y_1}{6}, \quad \Delta\vec{y}_{f2} = -\frac{y_2}{3} + \frac{y_3}{6} + \frac{y_1}{6} \end{aligned} \tag{2.13}$$

the derivatives of the shape functions are

$$\begin{aligned} N_{1x} &= \frac{\partial N_1}{\partial x} = \frac{(y_2 - y_3)}{2V^{ele}}, \quad N_{1y} = \frac{\partial N_1}{\partial y} = \frac{(x_3 - x_2)}{2V^{ele}} \\ N_{2x} &= \frac{\partial N_2}{\partial x} = \frac{(y_3 - y_1)}{2V^{ele}}, \quad N_{2y} = \frac{\partial N_1}{\partial y} = \frac{(x_1 - x_3)}{2V^{ele}} \\ N_{3x} &= \frac{\partial N_2}{\partial x} = \frac{(y_1 - y_2)}{2V^{ele}}, \quad N_{3y} = \frac{\partial N_3}{\partial y} = \frac{(x_2 - x_1)}{2V^{ele}} \end{aligned} \tag{2.14}$$

and the volume of the element is

$$V^{ele} = \frac{(x_2 y_3 - x_3 y_2) + x_1(y_2 - y_3) + y_1(x_3 - x_2)}{2} \tag{2.15}$$

The obtained algebreic equations from the discretization procedure using CVFEM are solved by the Gauss–Seidel method.

2.3 IMPLEMENTATION OF SOURCE TERMS AND BOUNDARY CONDITIONS

The boundary conditions for the present problem can be enforced using B_{B_i} and B_{C_i} as follows:

$$\text{Insulated boundary: } B_{B_i} = 0 \text{ and } B_{C_i} = 0 \tag{2.16}$$

$$\text{Fixed value boundary: } B_{B_i} = \phi_{\text{value}} \times 10^{16} \text{ and } B_{C_i} = 10^{16} \tag{2.17}$$

$$\text{Fixed flux boundary: } B_{B_i} = A_k \times q'' \text{ and } B_{C_i} = 0 \tag{2.18}$$

where ϕ_{value} is the prescribed value on the boundary and A_k is the length of the control volume surface on the boundary segment. The volume source terms can be applied to Eq. (2.2) as:

$$\sum_{j=1}^{\text{elements}} \int_{V_j} Q\,dV \approx Q_i V_i \tag{2.19}$$

or after linearizing the as source term

$$Q_i V_i = -Q_{C_i}\phi_i + Q_{B_i} \tag{2.20}$$

2.4 CVFEM FOR STEADY TWO-DIMENSIONAL PURE DIFFUSION AND ADVECTION-DIFFUSION

2.4.1 STEADY TWO-DIMENSIONAL PURE DIFFUSION

The steady-state two-dimensional (2D) pure diffusion equation can be defined as follows:

$$\frac{d^2T}{dx^2} + \frac{d^2T}{dy^2} = 0 \tag{2.21}$$

In this example, we supposed that the left wall is hot ($T = 1$), the right wall is cold ($T = 0$), and the other walls are adiabatic. Fig. 2.3 shows the isotherms for this problem. In Appendix A1, a FORTRAN code for the CVFEM solution of a steady-state 2D pure diffusion equation is provided.

2.4.2 STEADY TWO-DIMENSIONAL ADVECTION–DIFFUSION

The steady-state 2D advection–diffusion equation can be defined as follows:

$$\alpha\left[\frac{d^2T}{dx^2} + \frac{d^2T}{dy^2}\right] = u\frac{\partial T}{\partial x} + v\frac{\partial T}{\partial y} \tag{2.22}$$

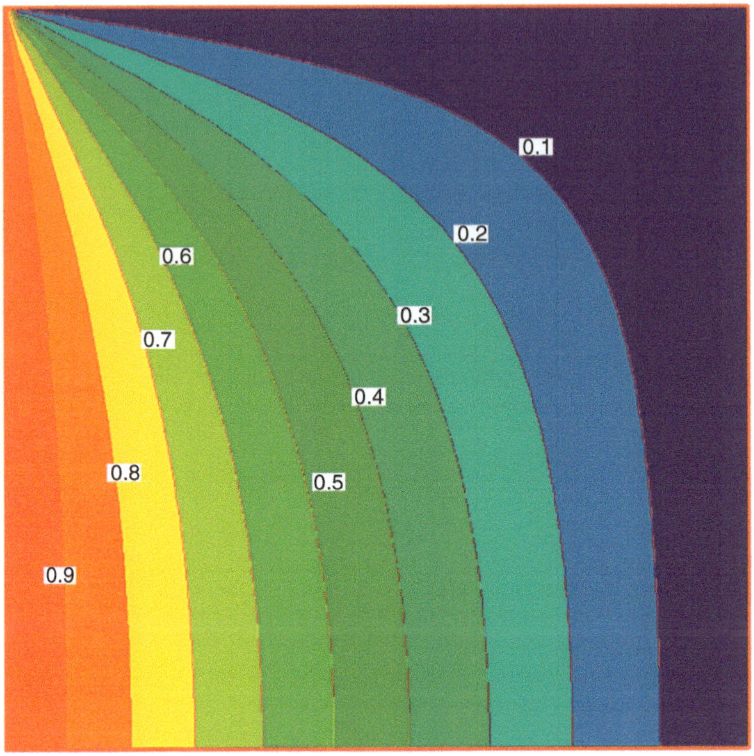

FIGURE 2.3 Isotherms for Two-Dimensional Pure Diffusion

In this example, we supposed that the left wall is hot ($T = 1$) and the right wall is cold wall ($T = 0$) and other walls are adiabatic. Also we have $\alpha = 0.01$, $u = 0.1$, $v = 0.2$. Fig. 2.4 shows the isotherms for this problem. In Appendix A2, a FORTRAN code for the CVFEM solution of a steady-state 2D advection–diffusion equation is provided.

2.5 APPLICATION OF CVFEM FOR NANOFLUID HYDROTHERMAL BEHAVIOR IN THE PRESENCE OF A MAGNETIC FIELD

2.5.1 VALIDATION OF THIS CODE

The convergence criterion for the termination of all computations is:

$$\max_{\text{grid}} \left| \frac{\Gamma^{n+1} - \Gamma^{n}}{\Gamma^{n+1}} \right| \leq 10^{-7} \tag{2.23}$$

where n is the iteration number and Γ stands for the independent variables (U,V,T). The results have been validated for the natural convection flow in an enclosed cavity filled by a pure fluid, as reported by Khanafer et al. [17] and De Vahl Davis [18] to observe a good agreement; see Table 2.1. Another test for validation of the current

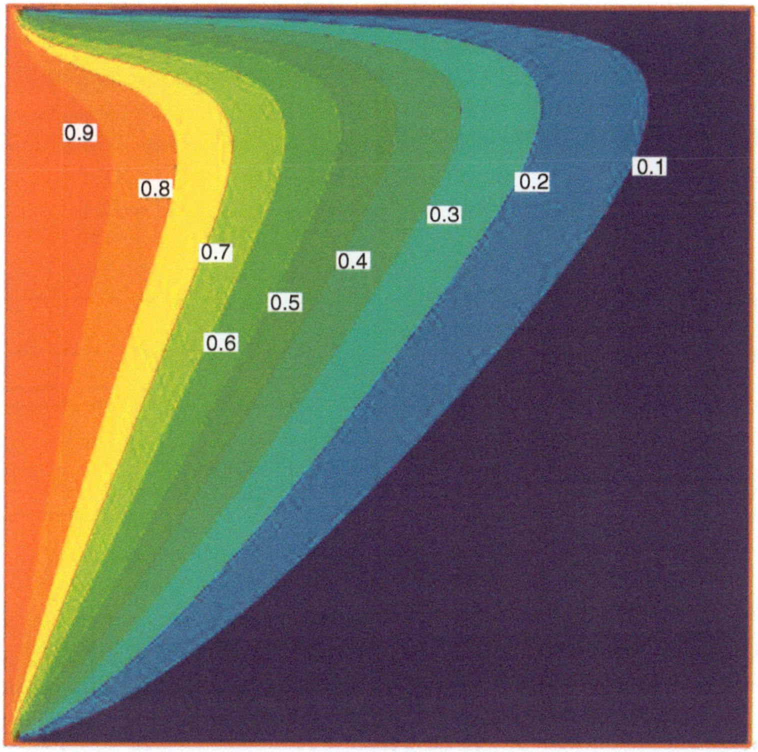

FIGURE 2.4 Isotherms for Two-Dimensional Pure Diffusion

Table 2.1 Comparison of the Present Results with Previous Work for Different Rayleigh Numbers When Pr = 0.7

Ra	Present Results	Khanafer et al. [17]	De Vahl Davis [18]
10^3	1.1432	1.118	1.118
10^4	2.2749	2.245	2.243
10^5	4.5199	4.522	4.519

Ra, Rayleigh number.

code was performed for the case of natural convection in a square enclosure in the presence of a magnetic field. In this test case, the average Nusselt number using different Gr and Ha numbers has been compared with those obtained by Rudraiah et al. [19] as shown in Table 2.2.

Fig. 2.5a illustrates a comparison between the calculated average Nusselt number on the bottom wall with constant heat flux and the numerical results of Sharif et al. [20] for cavity inclination angle equals zero, dimensionless length of the heat source 0.6, and different Grashof numbers. In Fig. 2.5b, the present computation also is validated against the results of Khanafer et al. [17] carried for natural convection in an enclosure

Table 2.2 Average Nusselt Number Versus Different Grashof Number Under Various Strengths of the Magnetic Field at Pr = 0.733

	$Gr = 2 \times 10^4$		$Gr = 2 \times 10^5$	
Ha	**Present Results**	**Rudraiah et al. [19]**	**Present Results**	**Rudraiah et al. [19]**
0	2.5665	2.5188	5.093205	4.9198
10	2.26626	2.2234	4.9047	4.8053
50	1.09954	1.0856	2.67911	2.8442
100	1.02218	1.011	1.46048	1.4317

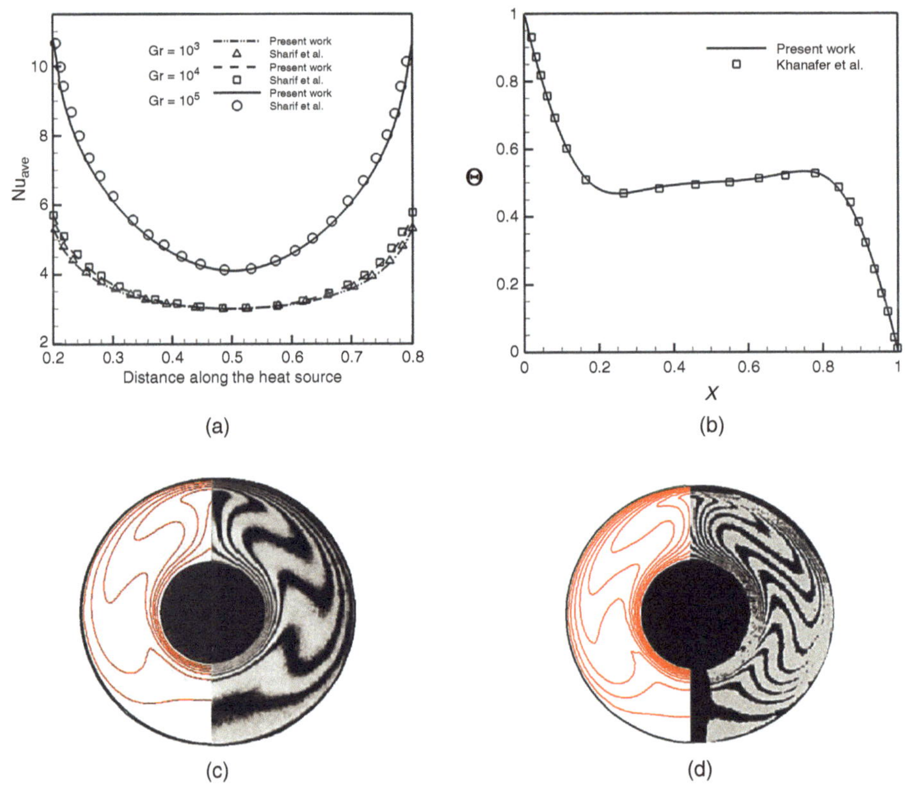

FIGURE 2.5

(a) Comparison of the temperature on axial midline between the present results and the numerical results by Sharif et al. [20]. (b) Comparison of the average Nusselt number between the present results and the numerical results by Khanafer et al. [17] $Gr = 10^4$, $\phi = 0.1$, and Pr = 6.2(Cu – Water) Comparison of isotherms between the present work and experimental study of (c) Kuehn and Goldstein [21] and (d) Laboni and Guj [22] for viscous flow ($\phi = 0$, Pr = 0.71) when (c) $R_o/R_i = 2.6$, $Ra = 4.7 \times 10^4$; (d) $R_o/R_i = 2.36$, $Ra = 0.9 \times 10^5$.

filled with Cu–water nanofluid for different Grashof numbers. Furthermore, comparisons of isotherms between the present work and an experimental study by Kuehn and Goldstein [21] and Laboniai and Guj [22] at the different Rayleigh numbers are shown in Fig. 2.5c. All of the previous comparisons indicate the accuracy of the present code.

2.5.2 NATURAL CONVECTION OF NANOFLUIDS IN AN ENCLOSURE BETWEEN A CIRCULAR AND A SINUSOIDAL CYLINDER IN THE PRESENCE OF A MAGNETIC FIELD

2.5.2.1 Problem definition

The physical model and the corresponding triangular elements used in the present CVFEM program are shown in Fig. 2.6 [23]. The inner and outer walls are maintained at constant temperatures T_h and T_c, respectively. The shape of inner cylinder profile is assumed to mimic the following pattern:

$$r = r_{in} + A\cos\left(N\left(\zeta - \zeta_0\right)\right) \tag{2.24}$$

where r_{in} is the base circle radius; r_{out} is the radius of outer cylinder; A and N are amplitude and number of undulations, respectively; and ζ is the rotation angle. It is also assumed that the uniform magnetic field $\vec{B} = B_x\vec{e}_x + B_y\vec{e}_y$ of constant magnitude $B = \sqrt{B_x^2 + B_y^2}$ is applied horizontally. The electric current J and the electromagnetic force F are defined by $J = \sigma\left(\vec{V} \times \vec{B}\right)$ and $F = \sigma\left(\vec{V} \times \vec{B}\right) \times \vec{B}$, respectively.

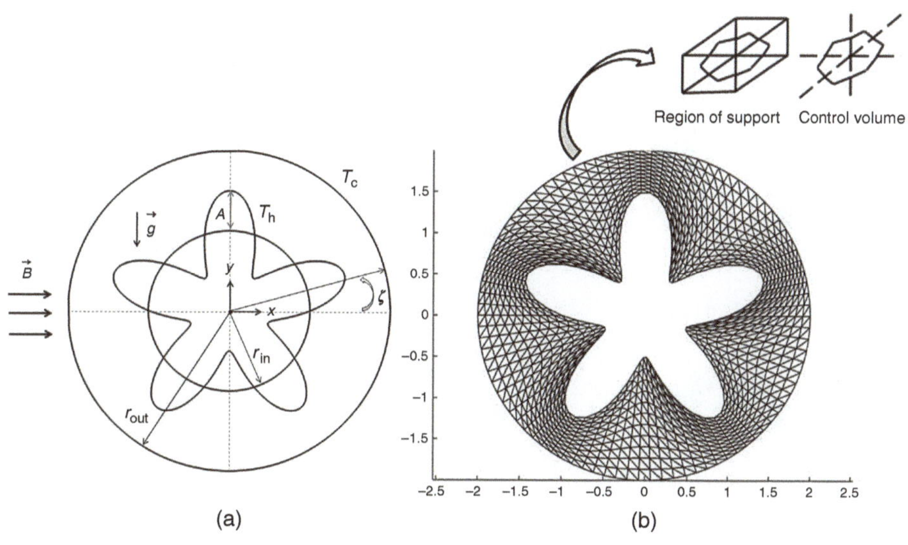

FIGURE 2.6

(a) Geometry and the boundary conditions with (b) the mesh of geometry considered in this work.

2.5.2.2 Effects of active parameters

Calculations are made for various values of Hartmann number (Ha = 0, 20, 60, and 100), Rayleigh number (Ra = 10^3, 10^4, and 10^5), volume fraction of nanoparticles (ϕ = 0, 2, 4, and 6%), and number of undulations (N = 3 and 6) at constant amplitude (A = 0.5) and Prandtl number (Pr = 6.2).

Comparison of the isotherm (left) and streamline (right) contours for different values of Hartmann number and Raleigh number for $N = 3$ and $N = 6$ is shown in Figs 2.7

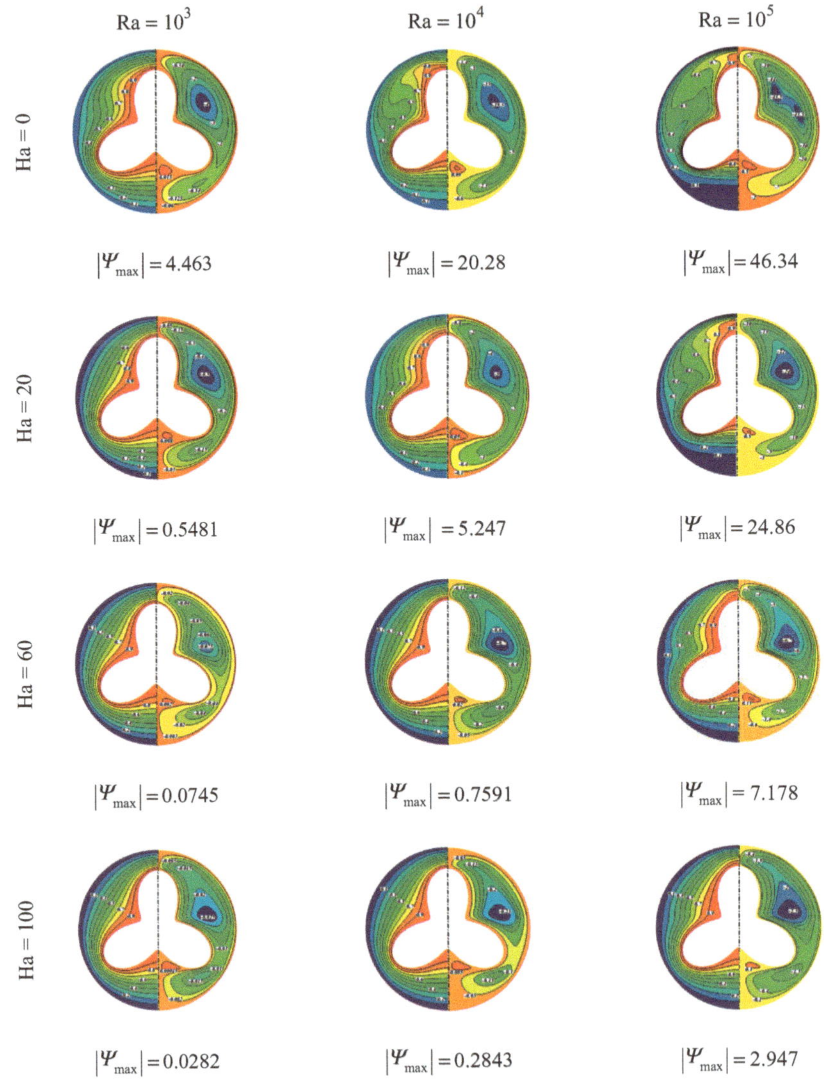

FIGURE 2.7 Comparison of the Isotherm (Left) and Streamline (Right) Contours for Different Values of Hartmann Number and Raleigh Number at $A = 0.5$, $N = 3$

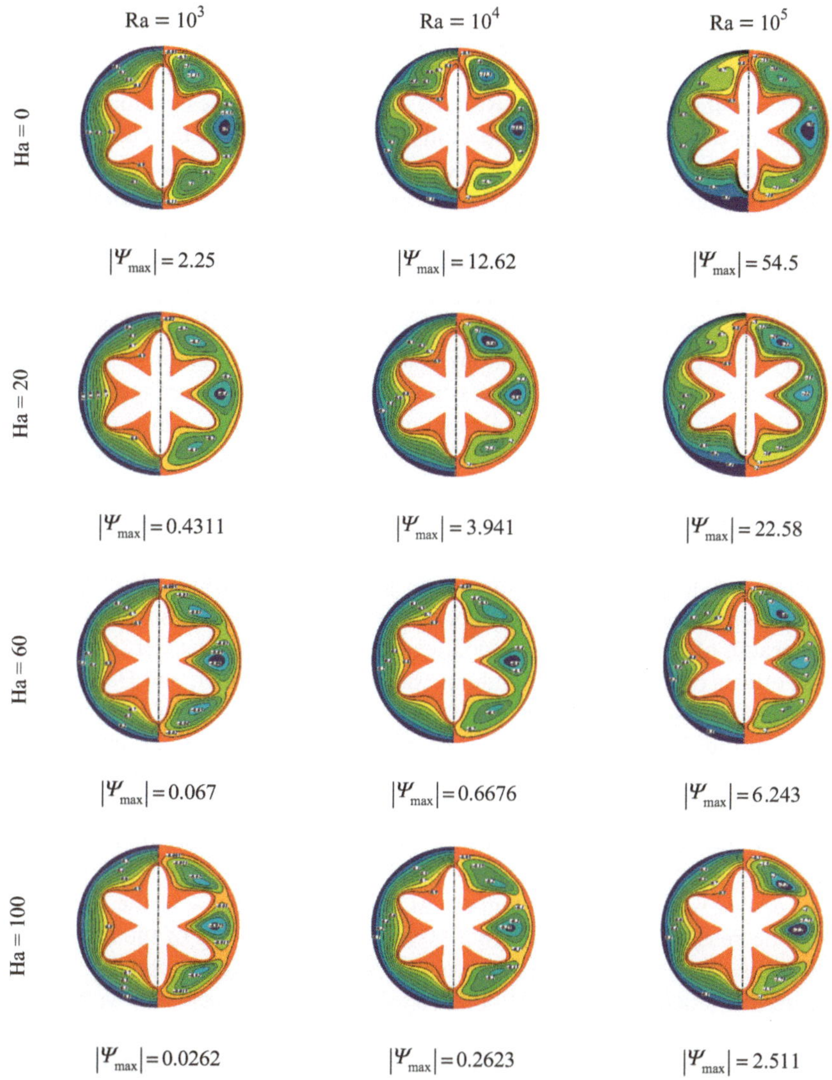

FIGURE 2.8 Comparison of the Isotherm (Left) and Streamline (Right) Contours for Different Values of Hartmann Number and Raleigh Number at $A = 0.5$, $N = 6$

and 2.8, respectively. As seen, the absolute value of stream function increases as Rayleigh number enhances, and it decreases as Hartmann number increases. For both cases at $Ra = 10^3$, the isotherms are uniforms and follow the shape of inner and outer walls, and the values of stream function are relatively small, which indicates the domination of the conduction heat transfer mechanism. For $N = 3$, under the hot surface a very small secondary eddy appears near the vertical centerline; it is due to the resistance of the crest against the flow circulation between the inner and outer cylinders. But for $N = 6$, no small secondary eddy at the bottom half of the enclosure is observed because the lower crest of the inner surface is parallel to the flow. As seen, the isotherms starts to disturbed at $Ra = 10^4$, and thermal plumes gradually appear on the hot surface of inner cylinder, especially at the upper half of the enclosure, which indicates that the convective heat transfer becomes comparable with conduction in this Rayleigh number. In addition, the thermal boundary layer between the crests of inner cylinder becomes thinner in comparison with those of $Ra = 10^3$, which indicates that the unventilated region between the crests diminishes; hence, the thermal boundary layer is suppressed toward the inner hot cylinder at these areas. At $Ra = 10^5$, the isotherms are crowded near the crests and depict the diminution of thermal boundary layer thickness at the bottom of the enclosure. At the top of the inner cylinder ($\zeta = 90°$), a strong plume drives that the flow strongly impinges on the outer wall of the enclosure. It is worthwhile mentioning that the effect of the magnetic field is to decrease the value of the velocity magnitude throughout the enclosure because the presence of the magnetic field introduces a force called the Lorentz force, which acts against the flow if the magnetic field is applied in the normal direction. This type of resisting force slows down the fluid velocity. Increasing the Hartmann number makes the core of vortices move toward the horizontal centerline. Also, the magnetic field causes the thermal plume to disappear and makes the isotherms parallel to each other because of domination of the conduction mode of heat transfer.

The effects of the number of undulations, Rayleigh number, and Hartmann number on the local Nusselt number over the cold wall and between $\zeta = 90$ and $270°$ is depicted in Fig. 2.9. The number of extremum in the local Nusselt number profile corresponds to the number of undulations of the inner hot surface. For $N = 3$, there are one local maximum and one local minimum; for $N = 6$, there are two maximum and three minimum. In the presence of a magnetic field, the extremum points of the local Nusselt number profiles occur at smaller ζ because of the change in the position of eddies with Hartmann number. At $Ra = 10^5$ and $Ha = 0$, the local Nusselt number decreases as ζ increases from 90 and 270°; at the top of the inner cylinder ($\zeta = 90°$), the flow accompanying with cold fluid impinging the upper cold surface, which results in a thicker thermal boundary layer over the enclosure wall; but when we approaches $\zeta = 270°$, the hot surface located at the top of the cold wall and the conduction mechanism become more pronounced; hence, the local Nusselt number decrease noticeably.

Fig. 2.10 shows the effects of the number of undulations, nanoparticle volume fraction, Rayleigh number, and Hartmann number on the average Nusselt number. As a whole, the average Nusselt number increases with an increase in the Rayleigh number and nanoparticle volume fraction, but it decreases as the Hartmann number increases. Adding high thermal conductivity nanoparticles leads to an increase in the conduction. So because of the domination of conduction at a low Rayleigh number, the effect of nanoparticle volume fraction on the average Nusselt number is more pronounced at a low Rayleigh number than at a high Rayleigh number. Also, this figure depicts that increasing the number of undulations

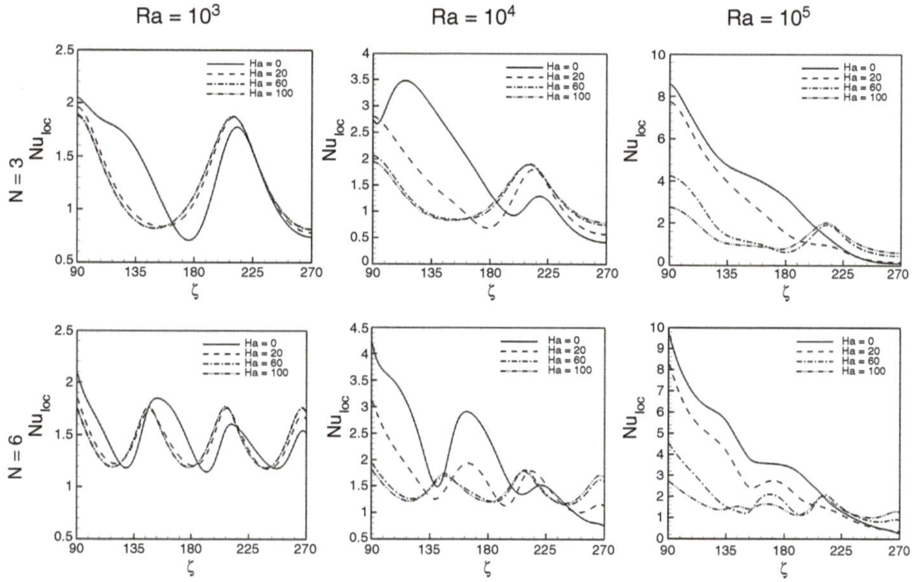

FIGURE 2.9 Effects of the Number of Undulations, Rayleigh Number, and Hartmann Number on the Local Nusselt Number at $\phi = 0.06$, $A = 0.5$

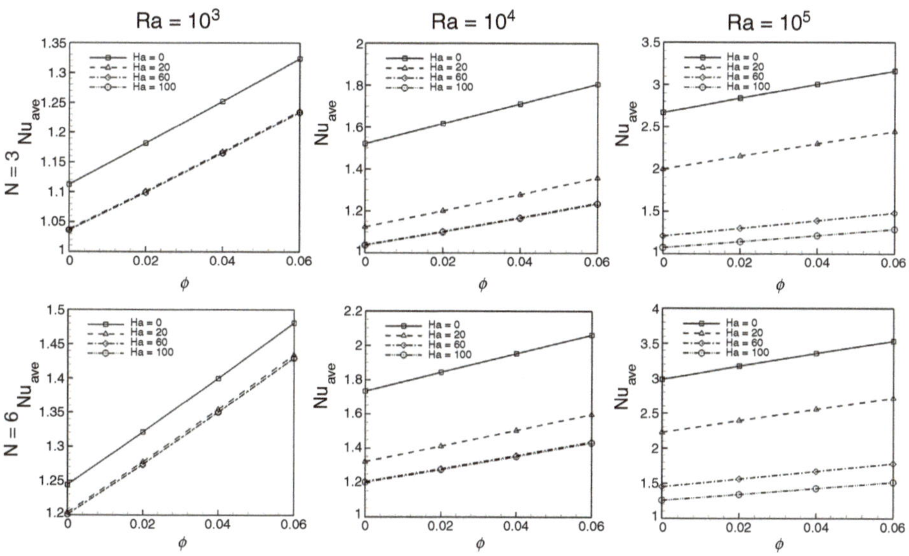

FIGURE 2.10 Effects of the Number of Undulations, Nanoparticle Volume Fraction, Rayleigh Number, and Hartmann Number on the Average Nusselt Number at $A = 0.5$

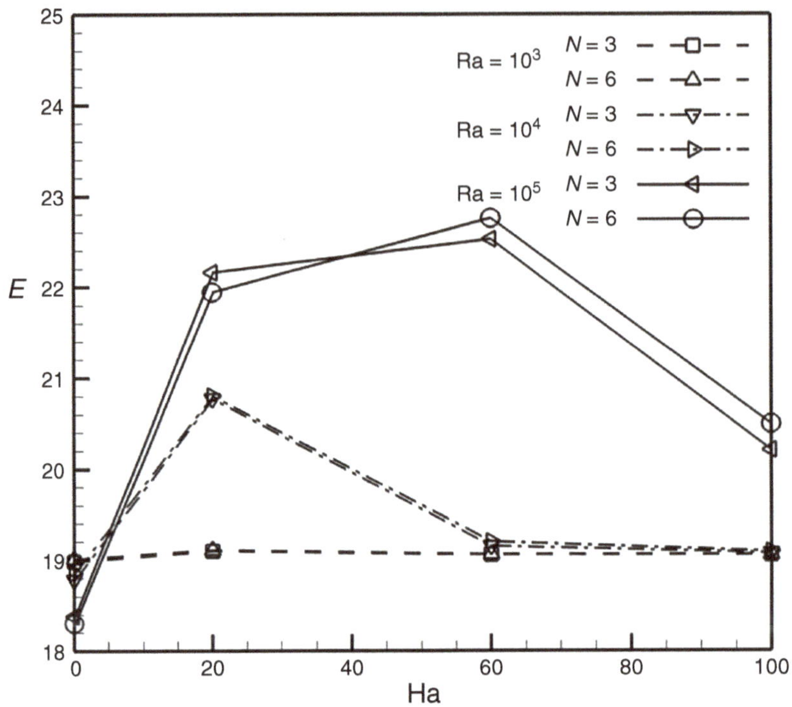

FIGURE 2.11 Effects of the Number of Undulations, Hartmann Number, and Rayleigh Number on the Ratio of Heat Transfer Enhancement Caused by Addition of Nanoparticles

causes the average Nusselt number to increase. To estimate the enhancement of heat transfer between the case of $\phi = 0.06$ and the pure fluid (base fluid) case, the enhancement is defined as:

$$E = \frac{\mathrm{Nu}(\phi = 0.06) - \mathrm{Nu}(\text{base fluid})}{\mathrm{Nu}(\text{base fluid})} \times 100 \tag{2.25}$$

Fig. 2.11 depicts the effect of the number of undulations, Hartmann number, and Rayleigh number on the heat transfer enhancement ratio. There is an insignificant change in enhancement at Ra = 10^3. In the absence of a magnetic field, the enhancement ratio decreases as the Rayleigh number increases; the opposite behavior is observed in the presence of a magnetic field. The maximum values of enhancement for Ra = 10^4 and 10^5 are observed at Ha = 20 and 60, respectively. It is an interesting observation that at Ra = 10^5 the enhancement in heat transfer for case of $N = 6$ is greater than those of obtained for case of $N = 3$ when Ha > 40, while for other values of Rayleigh number the number of undulations has no effect on enhancement.

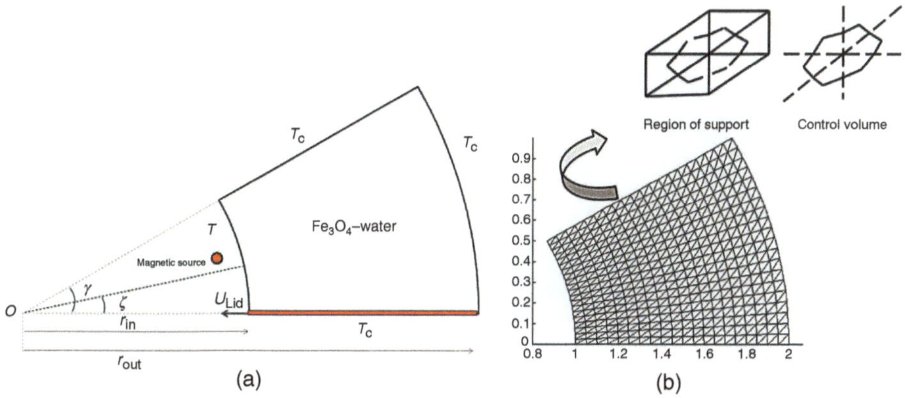

FIGURE 1.12

(a) Geometry and the boundary conditions; (b) the mesh of enclosure considered in this work.

2.5.3 EFFECT OF A NON-UNIFORM MAGNETIC FIELD ON THE FORCED CONVECTION HEAT TRANSFER OF Fe_3O_4–WATER NANOFLUID

2.5.3.1 Problem definition

The schematic diagram and the mesh of the semi-annulus enclosure used in the present CVFEM program are shown in Fig. 2.12 [24]. The inner wall is maintained at constant temperature of T_h, and the other walls are maintained at constant temperature of T_c ($T_h > T_c$). For the expression of the magnetic field strength, it can be considered that the magnetic source represents a magnetic wire placed vertically to the x–y plane at the point $(\overline{a}, \overline{b})$. The contours of the magnetic field strength are shown in Fig. 2.13. In this study, the magnetic source is located at (–0.01 cols, 0.5 rows). The lower wall is Lid driven with a velocity of U_{Lid}. In this example, γ is equal to 45°.

2.5.3.2 Effects of active parameters

The working fluid is Fe_3O_4–water nanofluid. Calculations are made for various values of volume fraction of nanoparticles (ϕ = 0% and 4%), Reynolds number (Re = 10,100 and 1000), and Hartmann number (Ha = 0, 5,10 and 20). In all calculations, the Prandtl number (Pr), temperature number (ε_1), and Eckert number (Ec) are set to 6.8, 0.0, and 10^{-5}, respectively.

Isotherms and streamlines contour for different values of Reynolds number and Hartmann are depicted in Figs 2.14–2.16. At a low Reynolds number, two clockwise eddies exist in streamline, which are symmetric respect to $\zeta = 22.5°$. By applying the magnetic field, Lorentz forces retard the flow, and another eddy generates, which is rotated counterclockwise. As the Reynolds number increases, two main eddies turn into three eddies in which the middle one is rotates in the reverse direction. So a thermal plume appears near $\zeta = 18°$, and in turn, the minimum rate of heat transfer occurs at this point. Increasing the Reynolds number causes the thermal boundary

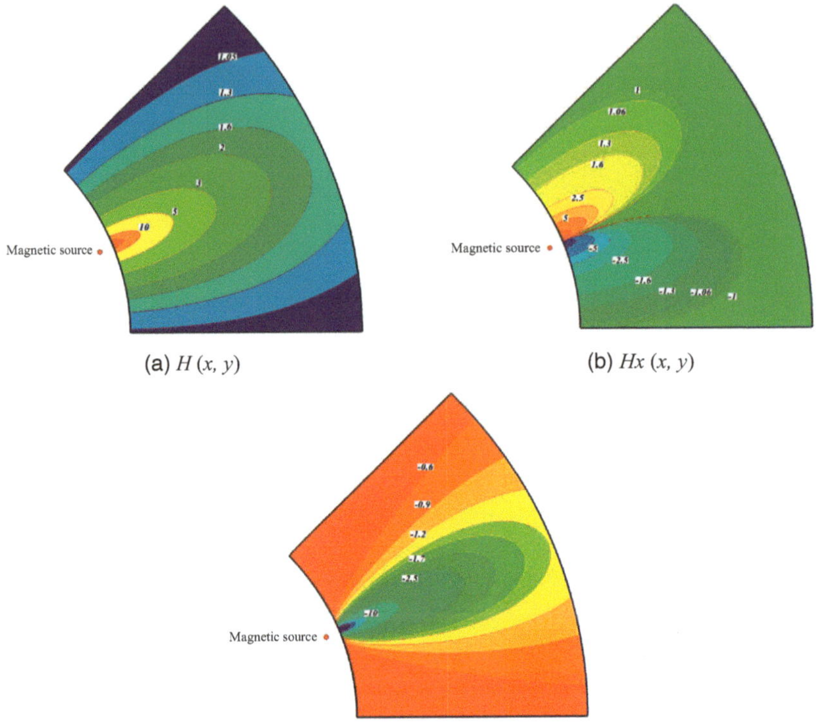

(a) $H(x, y)$ (b) $Hx(x, y)$

(c) $Hy(x, y)$

FIGURE 2.13 Contours of (a) the Magnetic Field Strength *H*; (b) the Magnetic Field Intensity Component in *x* Direction *Hx*; (c) the Magnetic Field Intensity Component in *y* Direction *Hy*

layer thickness to decrease. As the Hartmann number increases, the strength of the middle eddy decreases, and in turn, the thermal plume becomes weaker. Therefore, the Lorentz forces decrease the rate of heat transfer.

The effects of the Hartmann number and Reynolds number on local and average Nusselt number are shown in Figs 2.17 and 2.18. The thermal boundary layer thickness increases with an increase in the Hartmann number, and it decreases with an augment of the Reynolds number. Therefore, the Nusselt number increases with an increase in the Reynolds number, and it decreases with an increase in the Hartmann number. The existence of thermal plumes leads to appear extreme for the local Nusselt number profile.

The effects of the Reynolds number and Hartmann number on heat transfer enhancement are shown in Fig. 2.19. The effect of nanoparticles is more pronounced at a high Hartmann number than at a low Hartmann number because of the greater enhancement rate. This observation can be explained by noting that at a high Hartmann number, the heat transfer is dominant by conduction. Therefore, the addition of high thermal conductivity nanoparticles will increase the conduction and make

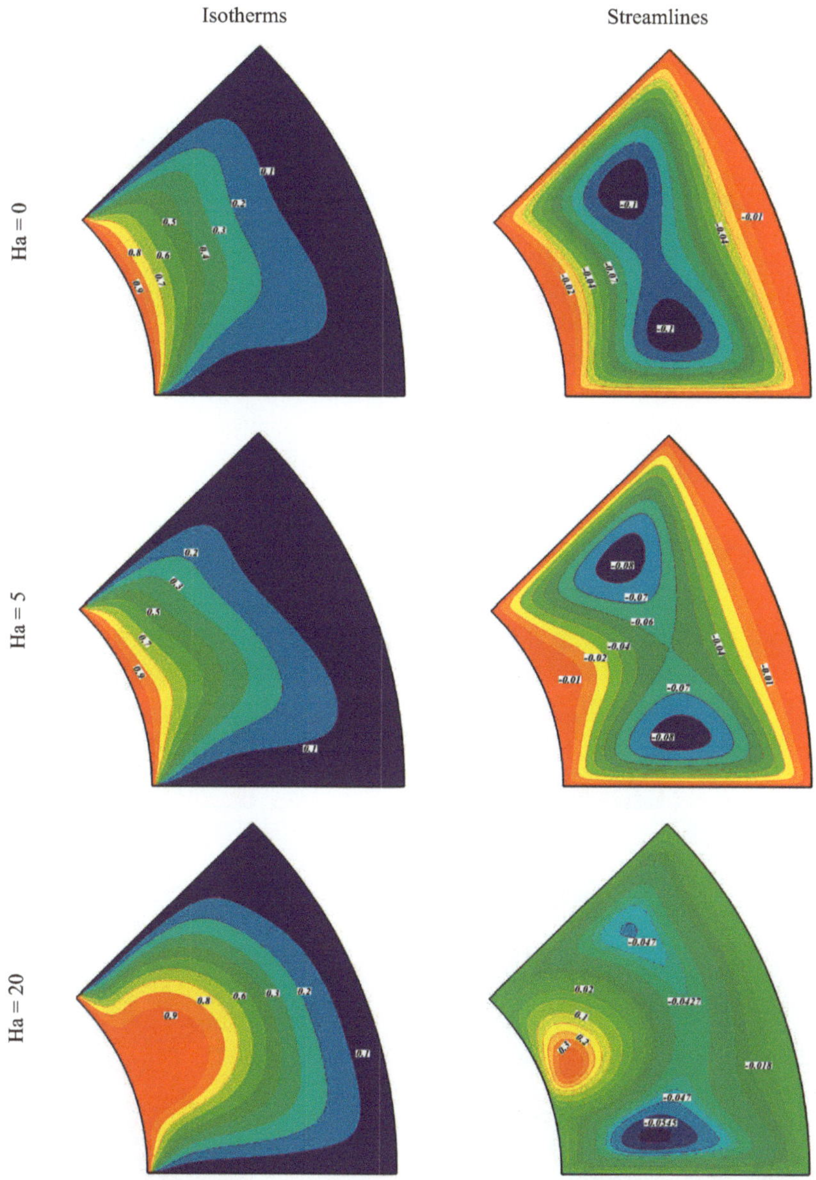

FIGURE 2.14 Isotherm (Left) and Streamline (right) Contours for Different Values of Hartmann Number when Re = 10

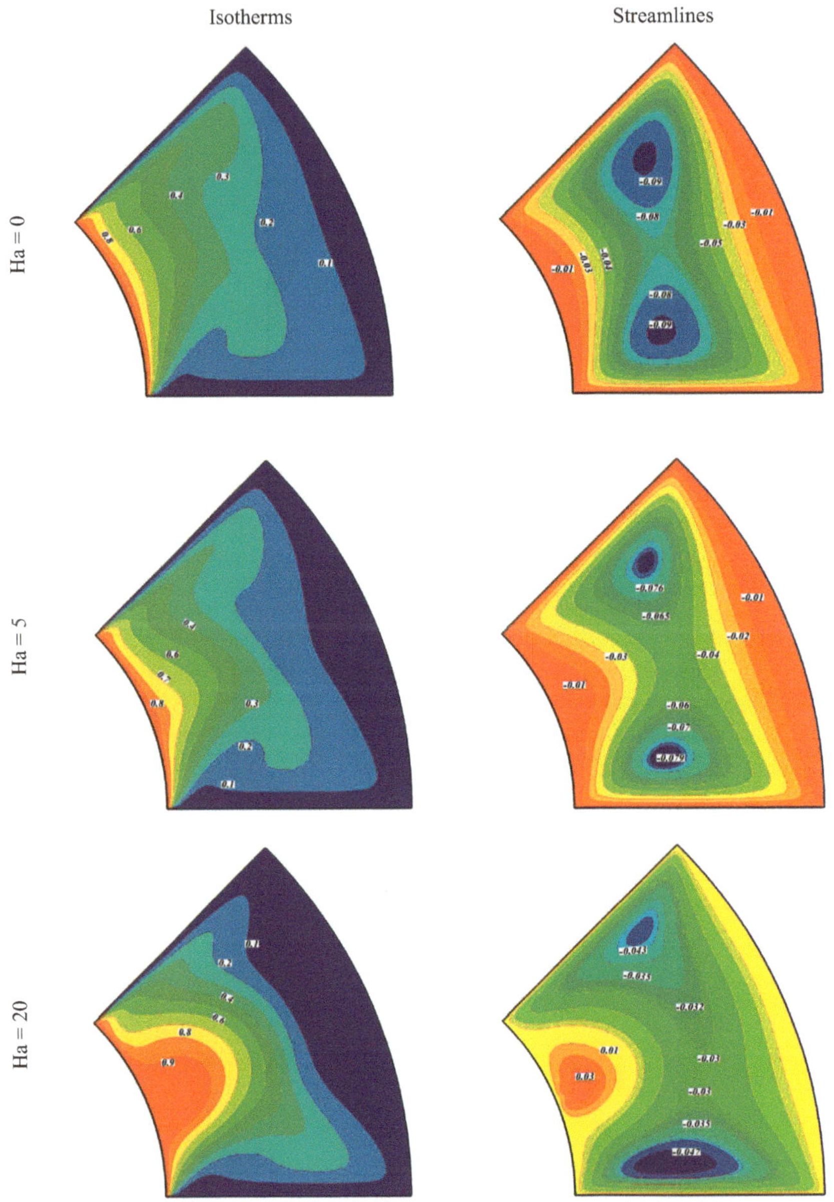

FIGURE 2.15 Isotherm (Left) and Streamline (Right) Contours for Different Values of Hartmann Number when Re = 100

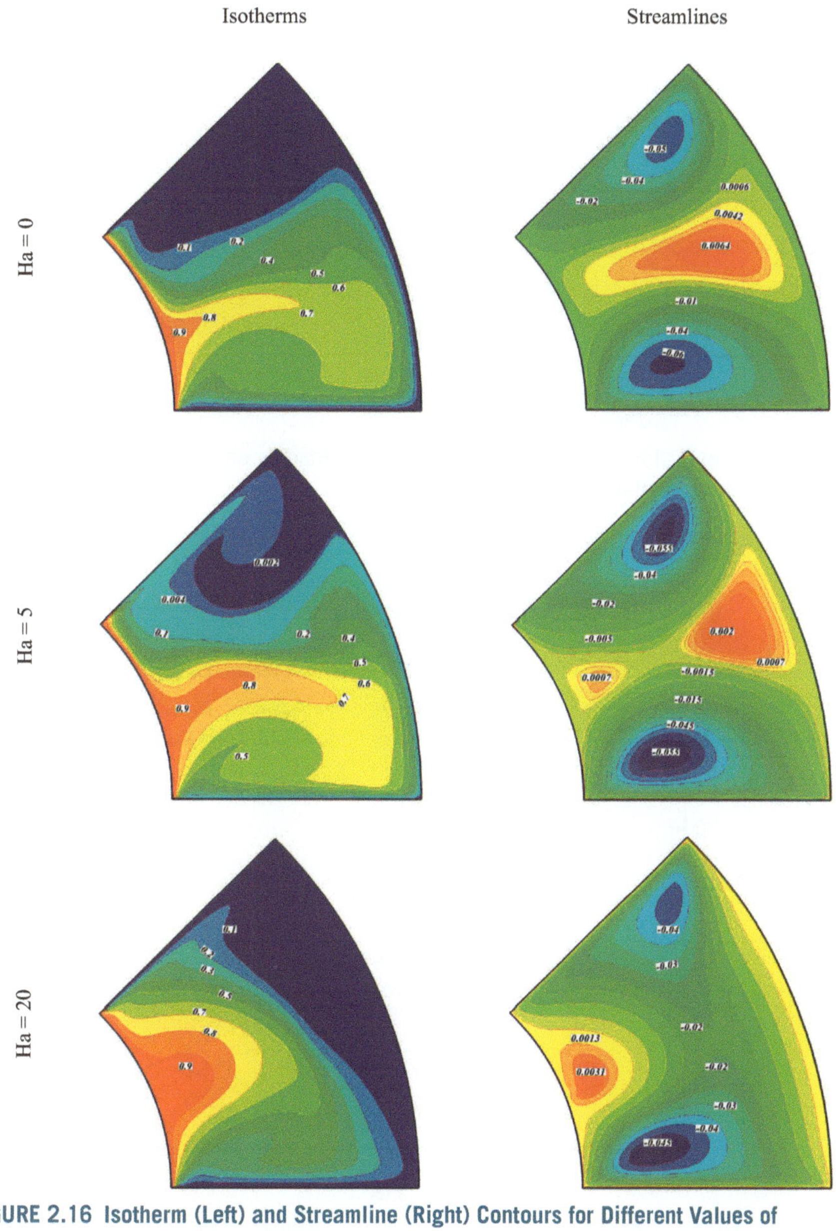

FIGURE 2.16 Isotherm (Left) and Streamline (Right) Contours for Different Values of Hartmann Number when Re = 1000

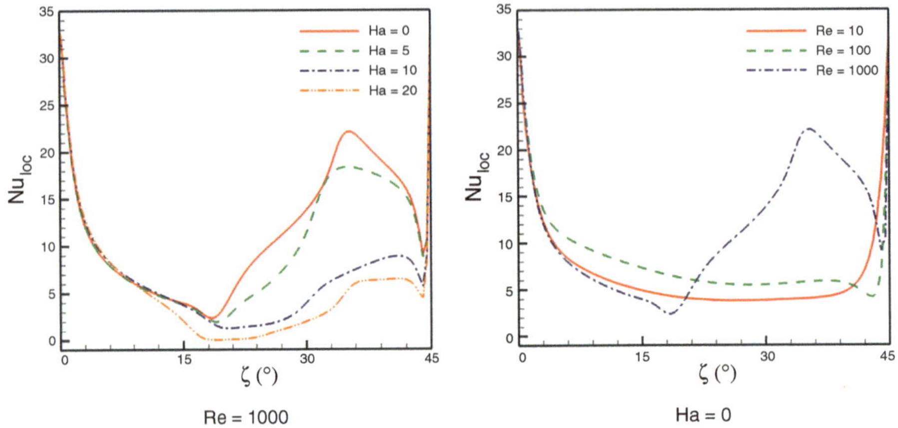

FIGURE 2.17 Effects of Hartmann Number and Reynolds Number on the Local Nusselt Number Nu_{loc} Along the Hot Wall When $\phi = 0.04$

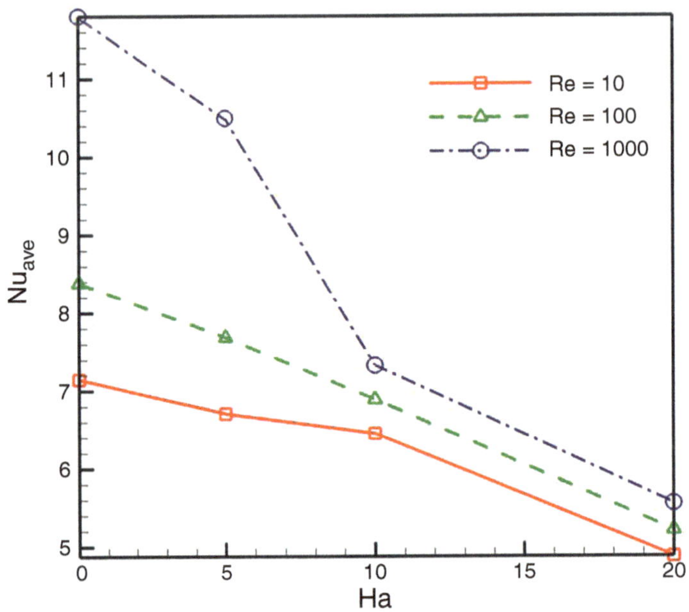

FIGURE 2.18 Effects of Hartmann Number and Reynolds Number on the Local Nusselt Number Nu_{ave} Along the Hot Wall When $\phi = 0.04$

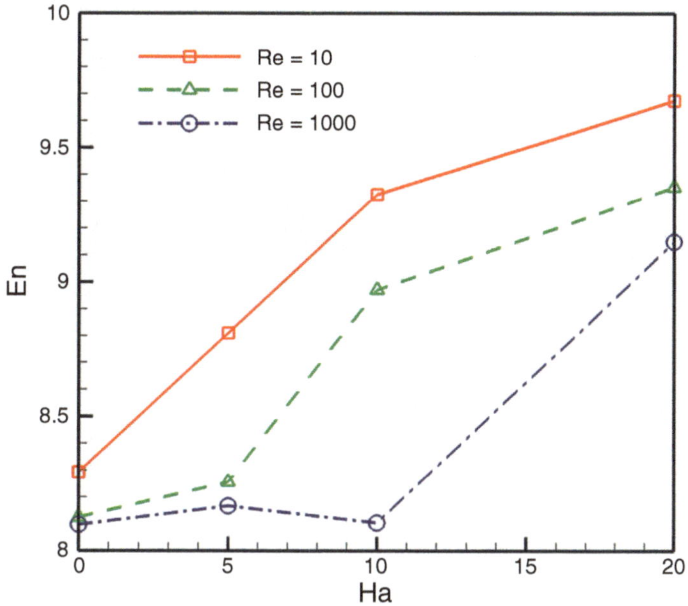

FIGURE 2.19 Effects of the Reynolds Number and Hartmann Number on the Heat Transfer Enhancement

the enhancement more effective. The heat transfer enhancement decreases with increases of the Reynolds number.

2.5.4 EFFECT OF A SPACE-DEPENDENT MAGNETIC FIELD ON FREE CONVECTION OF Fe_3O_4–WATER NANOFLUID

2.5.4.1 Problem definition

The physical model along with the important geometrical parameters and the mesh of the enclosure used in the present CVFEM program are shown in Fig. 2.20 [25]. The inner circular wall is maintained at a constant hot temperature T_h, the two horizontal walls are thermally isolated, and the other walls are maintained at a constant cold temperature T_c ($T_h > T_c$). To reach the shape of the inner circular and outer rectangular boundary, which consists of the right and top walls, a superelliptic function can be used as follows:

$$\left(\frac{X}{a}\right)^{2n} + \left(\frac{Y}{b}\right)^{2n} = 1 \tag{2.26}$$

When $a = b$ and $n = 1$, the geometry becomes a circle. As n increases from 1, the geometry would approach a rectangle for $a \neq b$ and square for $a = b$. In this study, we assumed that $r_{in}/r_{out} = 0.6$.

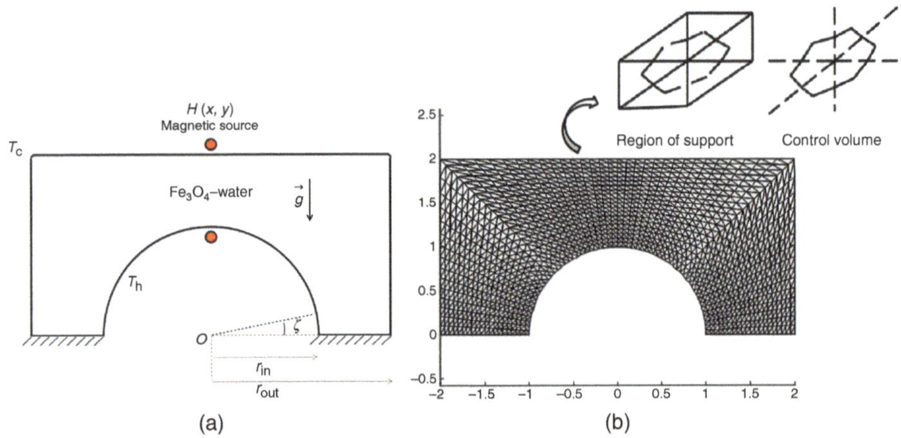

FIGURE 2.20

(a) Geometry and the boundary conditions; (b) the mesh of enclosure considered in this work; (c) a sample triangular element and its corresponding control volume.

For the expression of the magnetic field strength, it can be considered that two magnetic sources at points $\left(\overline{a}_1,\overline{b}_1\right)$ and $\left(\overline{a}_2,\overline{b}_2\right)$. The components of the magnetic field intensity $\left(\overline{H}_x,\overline{H}_y\right)$ and the magnetic field strength $(\overline{H})$ can be considered as [27]:

$$\overline{H}_x=\frac{\gamma}{2\pi}\frac{1}{\left(x-\overline{a}_1\right)^2+\left(y-\overline{b}_1\right)^2}\left(y-\overline{b}_1\right)-\frac{\gamma}{2\pi}\frac{1}{\left(x-\overline{a}_2\right)^2+\left(y-\overline{b}_2\right)^2}\left(y-\overline{b}_2\right) \tag{2.27}$$

$$\overline{H}_y=-\frac{\gamma}{2\pi}\frac{1}{\left(x-\overline{a}_1\right)^2+\left(y-\overline{b}_1\right)^2}\left(x-\overline{a}_1\right)-\frac{\gamma}{2\pi}\frac{1}{\left(x-\overline{a}_2\right)^2+\left(y-\overline{b}_2\right)^2}\left(x-\overline{a}_2\right) \tag{2.28}$$

$$\overline{H}=\sqrt{\overline{H}_x^2+\overline{H}_y^2} \tag{2.29}$$

where γ is the magnetic field strength at the source (of the wire).

The contours of the magnetic field strength are shown in Fig. 2.21. In this study, the magnetic source is located at $\left(\overline{a}_1=1.05\text{ cols},\ \overline{b}_1=0.5\text{ rows}\right)$ and $\left(\overline{a}_2=-0.05\text{ cols},\ \overline{b}_2=0.5\text{ rows}\right)$.

The local Nusselt number of the nanofluid along the hot wall can be expressed as:

$$\mathrm{Nu}_{\mathrm{loc}}=\left(\frac{k_{\mathrm{nf}}}{k_{\mathrm{f}}}\right)\frac{\partial\Theta}{\partial r} \tag{2.30}$$

where r is the radial direction. The average Nusselt number on the hot circular wall is evaluated as:

$$\mathrm{Nu}_{\mathrm{ave}}=\frac{1}{\gamma}\int_0^{\gamma}\mathrm{Nu}_{\mathrm{loc}}(\zeta)\,d\zeta \tag{2.31}$$

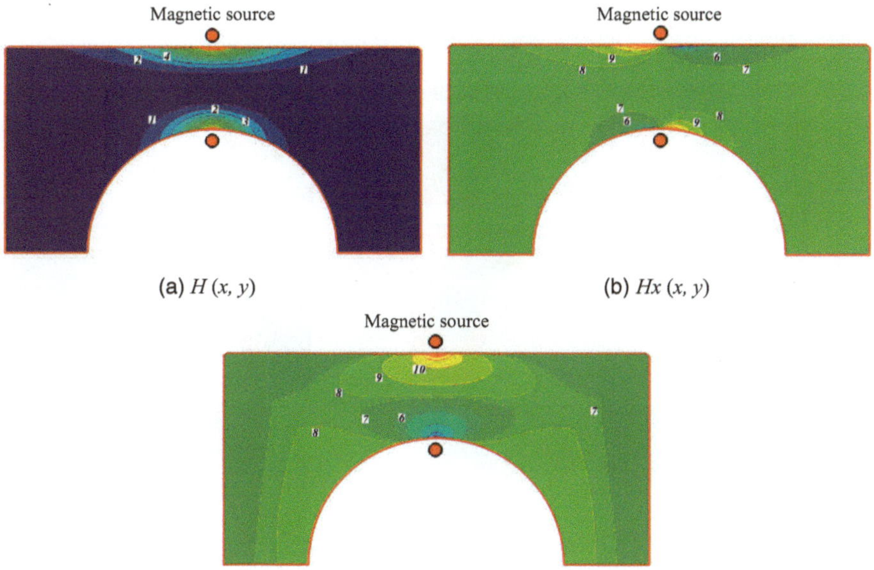

(a) $H\,(x, y)$ (b) $Hx\,(x, y)$

(c) $Hy\,(x, y)$

FIGURE 2.21 Contours of (a) the Magnetic Field Strength *H*; (b) the Magnetic Field Intensity Component in *x* Direction *Hx*; (c) the Magnetic Field Intensity Component in *y* Direction *Hy*

To estimate the enhancement of heat transfer between the case of $\phi = 0.04$ and the pure fluid (base fluid) case, the heat transfer enhancement is defined as:

$$\mathrm{En} = \frac{\mathrm{Nu}\,(\phi = 0.04) - \mathrm{Nu}\,(\text{base fluid})}{\mathrm{Nu}\,(\text{base fluid})} \times 100 \tag{2.32}$$

2.5.4.2 Effects of active parameters

Calculations are made for various values of volume fraction of nanoparticles (ϕ = 0.0 and 0.04), Rayleigh numbers (Ra = 10^3,10^4, and 10^5), magnetic number arising from ferrohydrodynamic (FHD) (Mn_F = 0, 20, 60, and 100), and Hartmann number arising from MHD (Ha = 0, 5, and 10). In all calculations, the Pr, ε_1, and Ec are set to 6.8, 0.0, and 1.0e-6, respectively.

Fig. 2.22 depicts the effects of the Rayleigh number, Hartmann number, and magnetic number on isotherms and streamlines. $|\Psi_{max}|$ increases with increase of Rayleigh number, and it decreases with augment of the Hartmann number and magnetic number. When Ra = 10^3, the heat transfer in the enclosure is mainly dominated by the conduction mode. With an absence of the magnetic field, the streamlines show one rotating eddy. When the magnetic parameter increases, another vortex appears near the centerline, which rotates in the reverse direction compared with the primary vortex. The Lorentz force has no significant effect on the isotherm and streamline

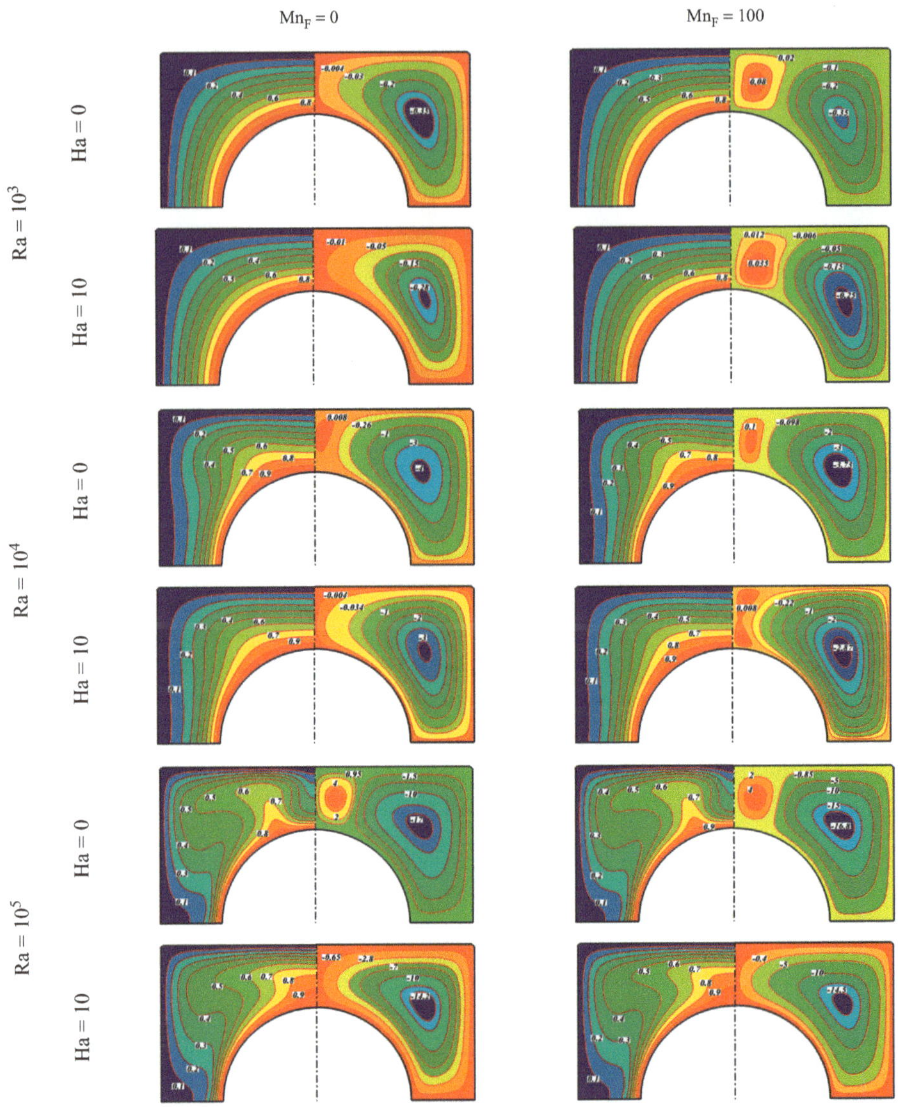

FIGURE 2.22 Isotherm (Left) and Streamline (Right) Contours for Different Values of Rayleigh Number, Hartmann Number, and Magnetic Number When Pr = 6.8, $\phi = 0.04$

in a low Rayleigh number. As the Rayleigh number increases up to 10^4, the role of convection in heat transfer becomes more important, and consequently, the thermal boundary layer on the surface of the inner wall becomes thinner. In addition, a plume starts to appear on the top of the inner circular wall. It is interesting to see that at $Mn_F = 100$, Ha = 10, and the smaller primary vortex turns to two smaller eddies near

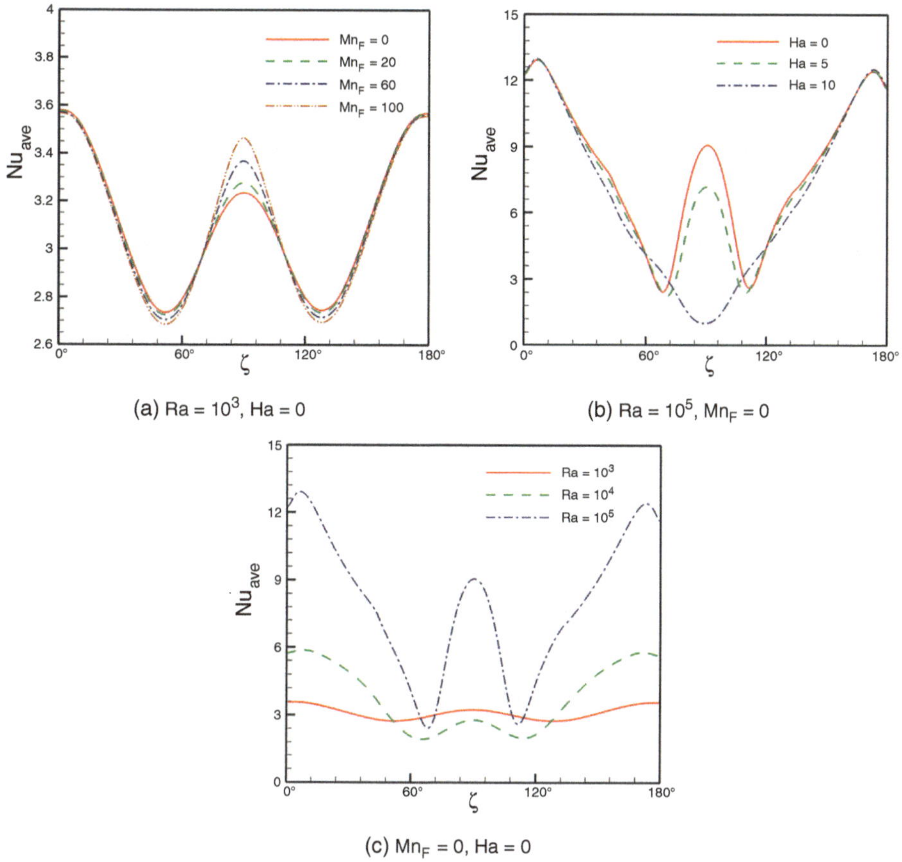

FIGURE 2.23 Effects of the Magnetic number, Hartmann number, and Rayleigh number on the local Nusselt number Nu_{loc} along the hot wall when Pr = 6.8, $\phi = 0.04$

the magnetic sources. At Ra = 10^5, two reverse plumes are observed in the absence of FHD and MHD effects. As the Hartmann number increases, one of the plumes disappears because of Lorentz forces.

Fig. 2.23 shows the effects of the magnetic number, Hartmann number, and Rayleigh number on the local Nusselt number along the hot wall. As the Hartmann number increases, the Nusselt number decreases, but the opposite trend is observed for the magnetic number and Rayleigh number. The existence of maximum and minimum points in the local Nusselt number profile corresponds to the presence of thermal plumes. Fig. 2.24a–c shows the effects of Hartmann number and Rayleigh number on the average Nusselt number along the hot wall. The effects of the magnetic number, Rayleigh number, and Hartmann number on heat transfer enhancement are shown in Fig. 2.24d–f. The corresponding polynomial

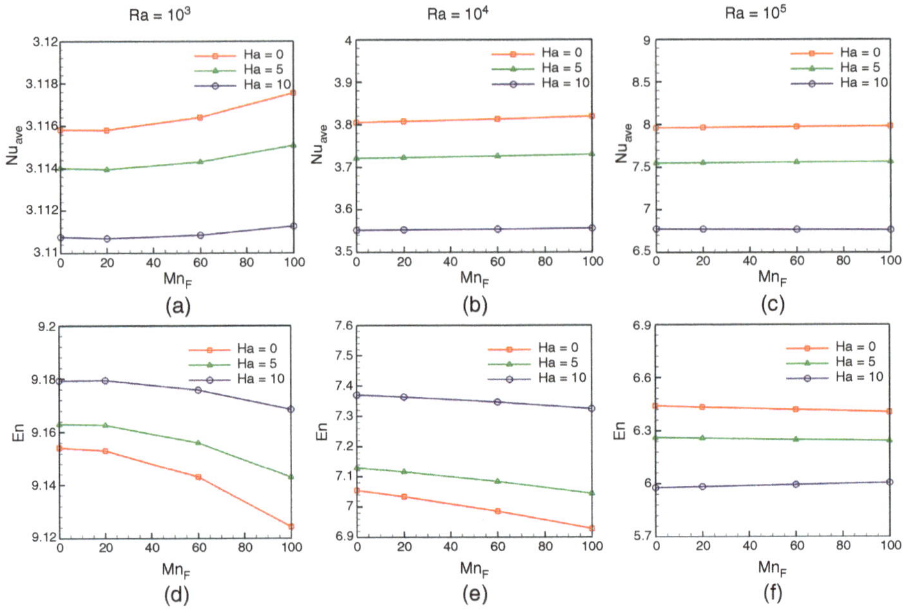

FIGURE 2.24

(a–c) Effects of the Magnetic parameter, Hartmann number, and Rayleigh number on the average Nusselt number Nu_{ave} along the hot wall when Pr = 6.8, $\phi = 0.04$ (d–f) Effects of Magnetic number, Rayleigh number and Hartmann number on the heat transfer enhancement when Pr = 6.8

representations for the Nusselt number and enhancement in heat transfer are as follows:

$$\begin{aligned}
Nu_{ave} &= a_{15} + a_{25}Y_3 + a_{35}Y_4 + a_{45}Y_3^2 + a_{55}Y_4^2 + a_{65}Y_4Y_3 \\
Y_1 &= a_{11} + a_{21}\,Ra^* + a_{31}\,Mn_F^* + a_{41}\,Ra^{*2} + a_{51}\,Mn_F^{*2} + a_{61}\,Ra^*\,Mn_F^* \\
Y_2 &= a_{12} + a_{22}\phi + a_{32}\,Ra^* + a_{42}\phi^2 + a_{52}\,Ra^{*2} + a_{62}\phi\,Ra^* \\
Y_3 &= a_{13} + a_{23}Y_1 + a_{33}\,Ha^* + a_{43}Y_1^2 + a_{53}\,Ha^{*2} + a_{63}Y_1\,Ha^* \\
Y_4 &= a_{14} + a_{24}\,Ha^* + a_{34}Y_2 + a_{44}\,Ha^{*2} + a_{54}Y_2^2 + a_{64}\,Ha^*\,Y_2
\end{aligned} \tag{2.33}$$

$$\begin{aligned}
En &= b_{13} + b_{23}Y_1 + b_{33}Y_2 + b_{43}Y_1^2 + b_{53}Y_2^2 + b_{63}Y_1Y_2 \\
Y_1 &= b_{11} + b_{21}\,Ra^* + b_{31}\,Ha^* + b_{41}\,Ra^{*2} + b_{51}\,Ha^{*2} + b_{61}\,Ra^*\,Ha^* \\
Y_2 &= b_{12} + b_{22}\,Ra^* + b_{32}\,Mn_F^* + b_{42}\,Ra^{*2} + b_{52}\,Mn_F^{*2} + b_{62}\,Re\,Mn_F^*
\end{aligned} \tag{2.34}$$

where $Ra^* = \log(Ra), Ha^* = Ha/10, Mn_F^* = Mn_F/100$. Also, a_{ij}, b_{ij} can be found in Table 2.3 and 2.4. We used the group method of data handling (GMDH) method to obtain these correlations [26].

As the buoyancy force (caused by the temperature gradient) and Kelvin force (caused by the external non-uniform magnetic field) increase, the rate of heat

Table 2.3 Constant Coefficient for Using Eq. (2.33)

a_{ij}	$i=1$	$i=2$	$i=3$	$i=4$	$i=5$	$i=6$
$j=1$	19.51647	−10.0868	−0.00804	1.525112	0.001086	0.003349
$j=2$	19.53085	−0.87952	−10.1291	−0.03518	1.525112	2.186806
$j=3$	−0.29704	1.107821	0.654435	−0.00017	0.041509	−0.24967
$j=4$	−0.15202	0.647229	1.043281	0.044365	0.005999	−0.24883
$j=5$	−0.07156	0.161803	0.861233	1.818237	1.846604	−3.66821

Table 2.4 Constant Coefficient for Using Eq. (2.34)

b_{ij}	$i=1$	$i=2$	$i=3$	$i=4$	$i=5$	$i=6$
$j=1$	21.33634	−5.71464	0.798586	0.542492	0.064253	−0.21888
$j=2$	21.72597	−5.81042	−0.0488	0.542492	−0.01392	0.006107
$j=3$	0.061649	4.86387	−3.87977	−0.38679	0.13164	0.25619

transfer increases; the opposite trend is observed for Lorentz forces. This means that increasing Lorentz forces lead to a decrease of the Nusselt number. Also, it can be concluded that the effect of the magnetic parameter on Nu_{ave} is more pronounced at a low Rayleigh number. At a low Rayleigh number, heat transfer enhancement increases with an increase of the Hartmann number; the opposite trend is observed for the magnetic parameter. At a high Rayleigh number, the magnetic parameter has no significant effect on enhancement, but an increasing Hartmann number causes the heat transfer enhancement to decrease. The effect of nanoparticles is more pronounced at a low Rayleigh number than at high Rayleigh number because of domination of the conduction heat transfer mechanism.

2.5.5 FERROHYDRODYNAMIC AND MAGNETOHYDRODYNAMIC EFFECTS ON FERROFLUID FLOW AND CONVECTIVE HEAT TRANSFER

2.5.5.1 Problem definition

The physical model along with the important geometrical parameters and the mesh of the enclosure used in the present CVFEM program are shown in Fig. 2.25 [27]. The inner and outer walls are maintained at constant temperatures T_h and T_c, respectively. The shape of the inner cylinder profile is assumed to mimic the following pattern:

$$r = r_{in} + A\cos(N(\zeta)) \tag{2.35}$$

in which r_{in} is the base circle radius; r_{out} is the radius of outer cylinder; A and N are amplitude and number of undulations, respectively; and ζ is the rotation angle. In this study, A and N are equal 0.2 and 4, respectively. For the expression of the magnetic field strength, it can be considered that the magnetic source represents a magnetic wire placed vertically to the x–y plane at the point $(\bar{a}, \bar{b})$. The contours of

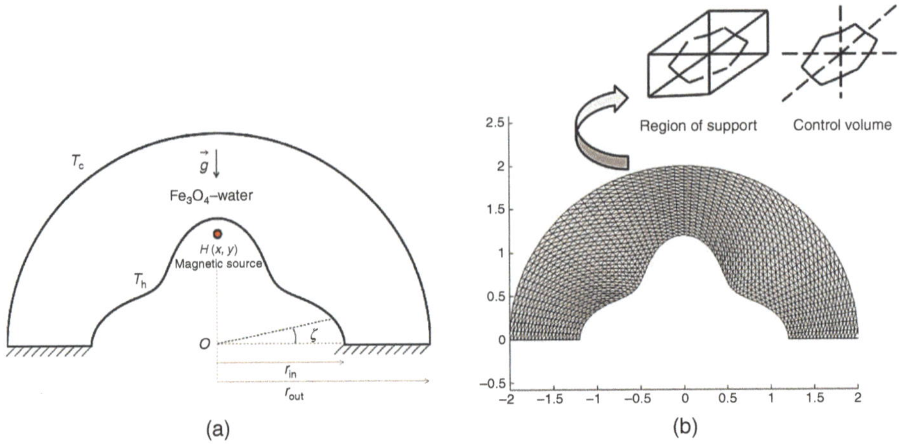

FIGURE 2.25

(a) Geometry and the boundary conditions; (b) the mesh of enclosure considered in this work.

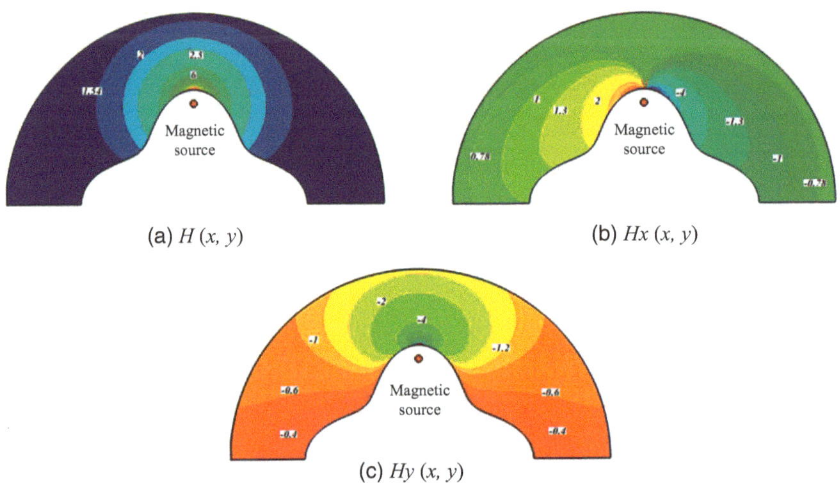

FIGURE 2.26 Contours of (a) the Magnetic Field Strength *H*; (b) the Magnetic Field Intensity Component in *x* Direction *Hx*; (c) the Magnetic Field Intensity Component in *y* Direction *Hy*

the magnetic field strength are shown in Fig. 2.26. In this study, the magnetic source is located at $(-0.05\,\text{cols}, 0.5\,\text{rows})$.

2.5.5.2 Effects of active parameters

The effect of an external magnetic field on ferrofluid flow and heat transfer in a semi-annulus enclosure with a sinusoidal hot wall is studied. The mathematical model used for the formulation of the problem is consistent with the principles of FHD and

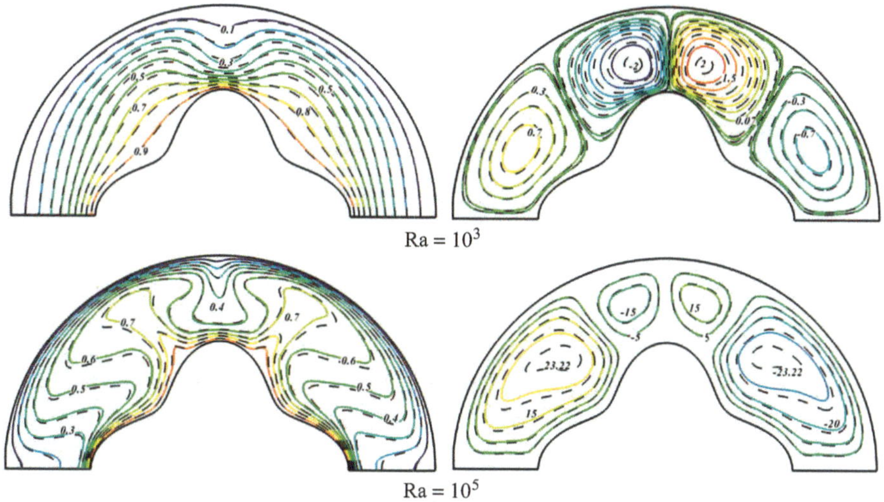

FIGURE 2.27 Comparison of the Isotherm and Streamlines Between Nanofluid ($\phi = 0.04$) (- - -) and Pure Fluid ($\phi = 0$) (—) When $Mn_F = 500$, Ha = 5, and Pr = 6.8

MHD. Calculations are made for various values of volume fraction of nanoparticles ($\phi = 0$ and 4%), Rayleigh numbers (Ra = 10^3, 10^4, and 10^5), magnetic number arising from FHD ($Mn_F = 0$, 50, and 500), and Hartmann number arising from MHD (Ha = 0, 1, 3, and 5). In all calculations, the Pr, ε_1, and Ec are set to 6.8, 0.0, and 1e-5, respectively.

Comparisons of the isotherm and streamlines between nanofluid and pure fluid are shown in Fig. 2.27. The velocity components of nanofluid are increased because of an augment in the energy transport in the fluid. Thus, the absolute values of stream functions increase with augment of the volume fraction of nanofluid. Also, it can be seen that there is no sensible change on isotherm.

The effects of the Rayleigh number, Hartmann number, and magnetic number on isotherms and streamlines are shown in Figs 2.28–2.30. At a low Rayleigh number, the conduction heat transfer mechanism is dominant in the absence of a magnetic field. As the Hartmann number increases, the absolute value of the maximum stream function decreases. As the magnetic number increases, three thermal plume appear over the hot wall. Also, the primary vortex in each side turns into two smaller vortexes. As the Rayleigh number increases, the role of convection in heat transfer becomes more significant, and consequently, the thermal boundary layer on the surface of the inner wall becomes thinner. In addition, a plume starts to appear on the top of the inner circular wall at $\zeta = 90°$. As the Lorentz forces increases, the thermal plume converts to three plumes. Also, it can be seen that at a high Hartmann number, two secondary eddies exist near the vertical center line. At Ra = 10^5, the effects of Kelvin forces is not sensible, but increasing the Lorentz forces leads to appear two equal eddies near the vertical center line.

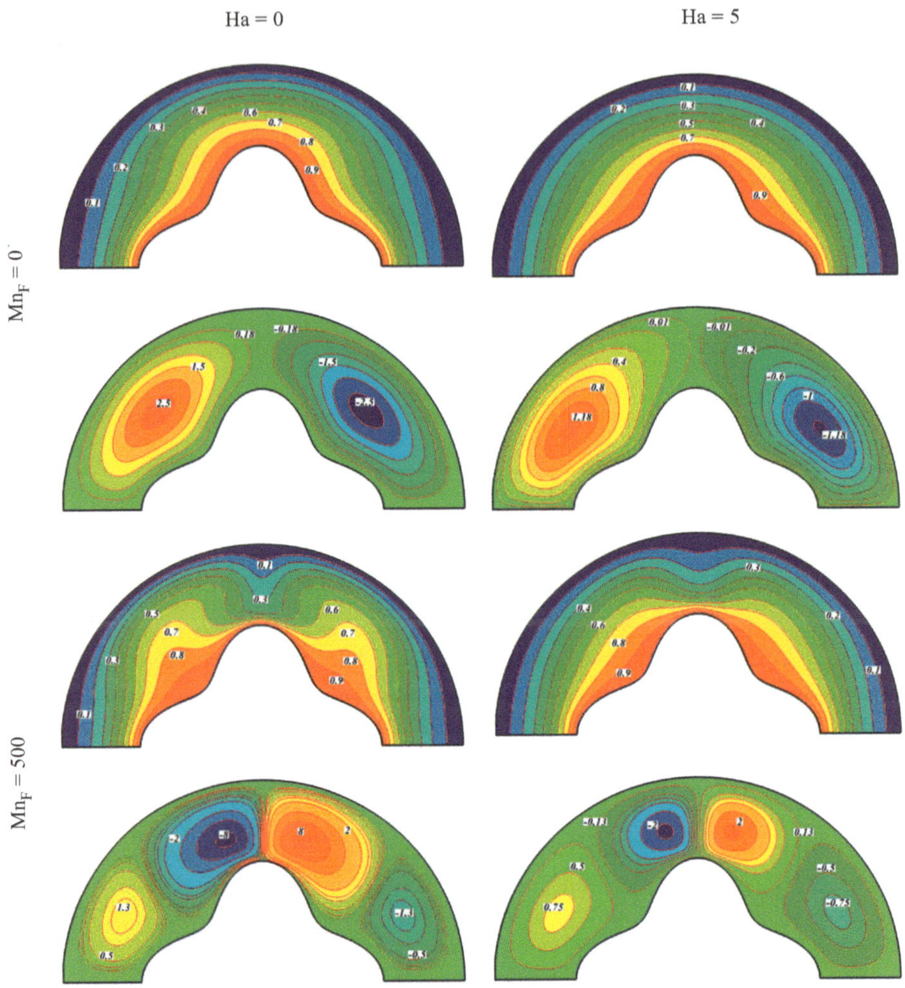

FIGURE 2.28 Isotherm (Up) and Streamline (Down) Contours for Different Values of Hartmann Number and Magnetic Number When Ra = 10^3, $\phi = 0.04$, and Pr = 6.8

The effects of the magnetic number, Hartmann number, and Rayleigh number on the local and average Nusselt number are depicted in Fig. 2.31 and Table 2.5. The Nusselt number increases with an increase of the Rayleigh number, and it decreases with an increase of the Hartmann number. An increasing magnetic number causes the Nusselt number to increase except for Ra = 10^5. Table 2.6 illustrates the effects of the magnetic number, Hartmann number, and Rayleigh number on heat transfer enhancement. For a low Rayleigh number, enhancement increases with an increase of the Hartmann number, but it decrease with an increase of the magnetic number. The opposite trend is observed for the highest Rayleigh number.

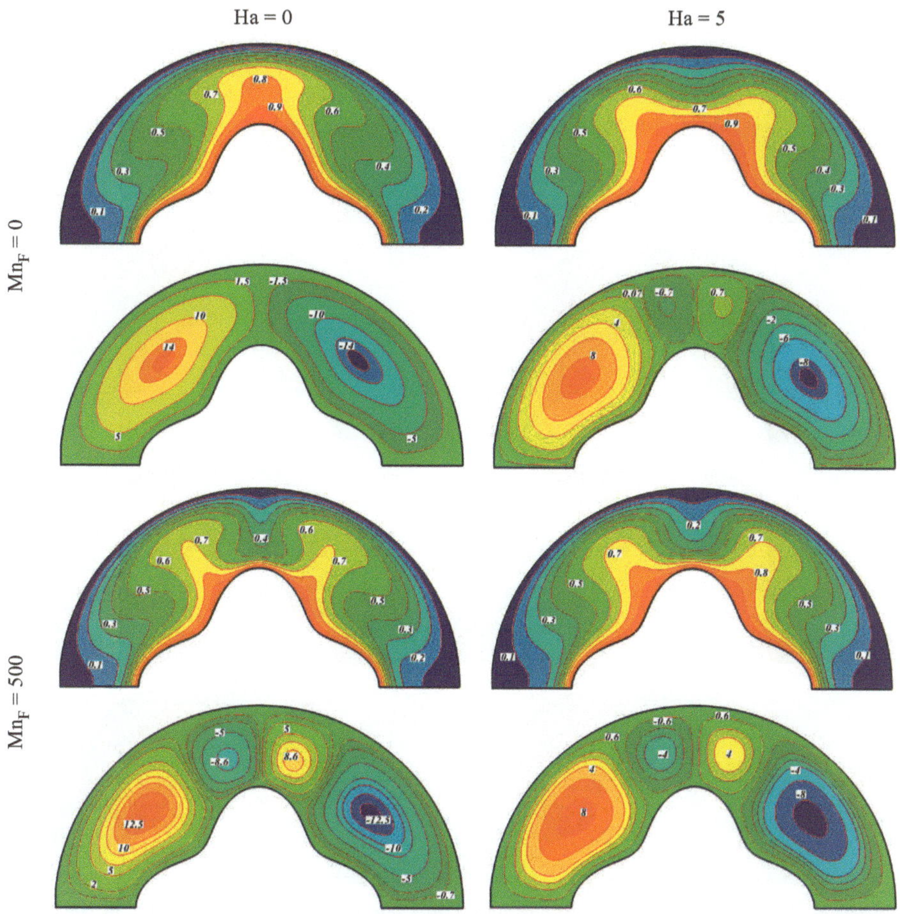

FIGURE 2.29 Isotherm (Up) and Streamline (Down) Contours for Different Values of Hartmann Number and Magnetic Number When Ra = 10^4, $\phi = 0.04$, and Pr = 6.8

2.5.6 EFFECT OF A SPATIALLY VARIABLE MAGNETIC FIELD ON FERROFLUID FLOW AND HEAT TRANSFER CONSIDERING THE CONSTANT HEAT FLUX BOUNDARY CONDITION

2.5.6.1 Problem definition

The physical model along with the important geometrical parameters and the mesh of the enclosure used in the present CVFEM program are shown in Fig. 2.32 [28]. The two horizontal walls with the length of are thermally isolated; whereas the outer cylinder is maintained at a constant cold temperature T_c, the inner circular wall is under constant heat flux. The contours of the magnetic field strength are shown in Fig. 2.33.

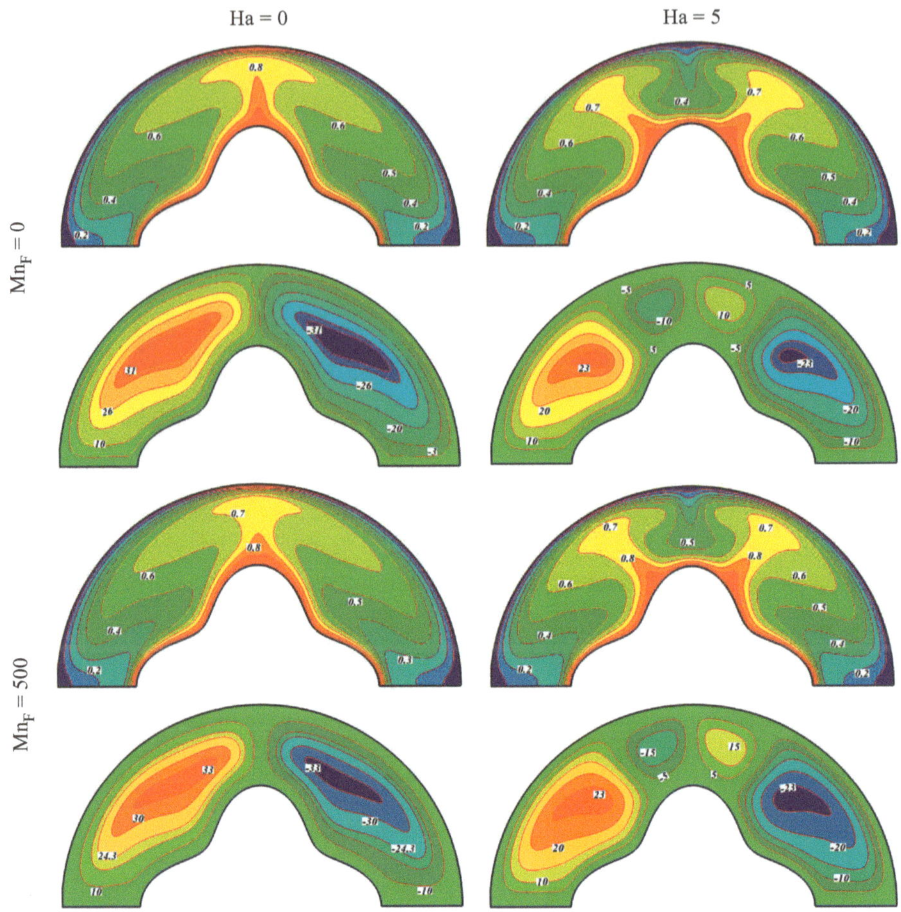

FIGURE 2.30 Isotherm (Up) and Streamline (Down) Contours for Different Values of Hartmann Number and Magnetic Number When Ra = 10^5, $\phi = 0.04$, and Pr = 6.8

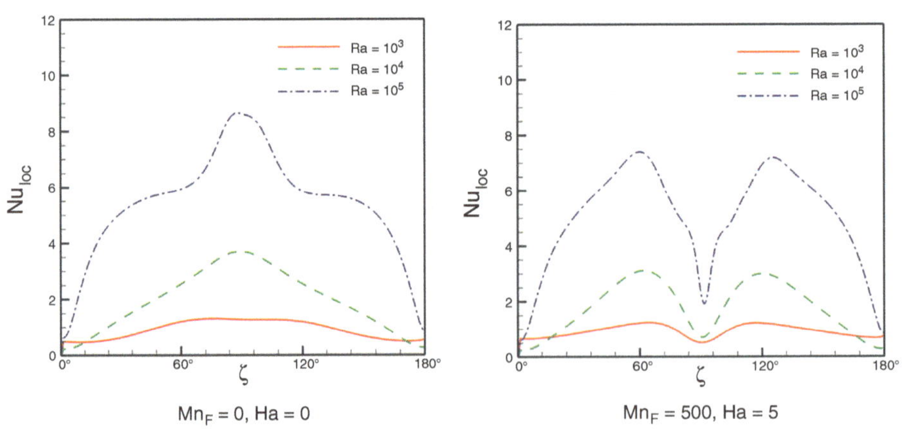

FIGURE 2.31 Effects of Magnetic Number, Hartmann Number, and Rayleigh Number on Local Nusselt Number Nu_{loc} Along the Hot Wall When $\phi = 0.04$

Table 2.5 Effects of Rayleigh Number, Hartmann Number, and Magnetic Number on Average Nusselt Number When $\phi = 0.04$

Ra	Mn_F	Ha	Nu_{ave}
10^3	0	0	0.984445
10^3	0	5	0.899518
10^3	500	0	1.478063
10^3	500	5	0.999565
10^4	0	0	2.135067
10^4	0	5	1.702675
10^4	500	0	2.365715
10^4	500	5	1.926001
10^5	0	0	5.926064
10^5	0	5	5.490067
10^5	500	0	5.843355
10^5	500	5	5.495555

Table 2.6 Effects of Rayleigh Number, Hartmann Number, and Magnetic Number on Heat Transfer Enhancement

Ra	Mn_F	Ha	En
10^3	0	0	7.461432
10^3	0	5	8.552658
10^3	500	0	4.123763
10^3	500	5	2.677329
10^4	0	0	7.135186
10^4	0	5	5.86332
10^4	500	0	6.471199
10^4	500	5	4.675113
10^5	0	0	11.38515
10^5	0	5	11.32785
10^5	500	0	11.60219
10^5	500	5	10.95681

The local Nusselt number of the nanofluid along the inner wall can be expressed as:

$$\mathrm{Nu}_{\mathrm{local}} = \left(\frac{k_{\mathrm{nf}}}{k_{\mathrm{f}}}\right)\frac{1}{\theta}\bigg|_{\text{inner wall}} \tag{2.36}$$

where r is the radial direction. The average Nusselt number on the hot circular wall is evaluated as:

$$\mathrm{Nu}_{\mathrm{ave}} = \frac{1}{\gamma}\int_0^{\gamma} \mathrm{Nu}_{\mathrm{loc}}(\zeta)\, d\zeta \tag{2.37}$$

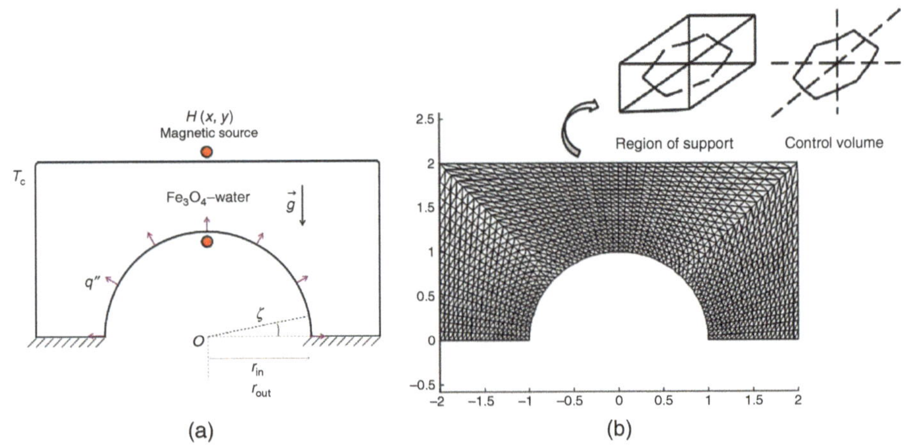

FIGURE 2.32

(a) Geometry and the boundary conditions; (b) the mesh of enclosure considered in this work; (c) a sample triangular element and its corresponding control volume.

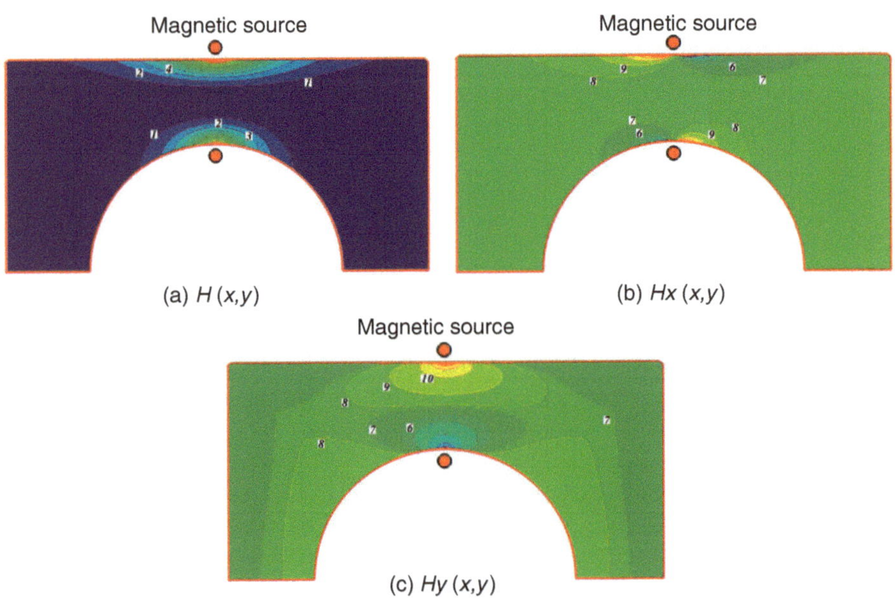

FIGURE 2.33 Contours of (a) the Magnetic Field Strength *H*; (b) the Magnetic Field Intensity Component in *x* Direction *Hx*; (c) the Magnetic Field Intensity Component in *y* Direction *Hy*

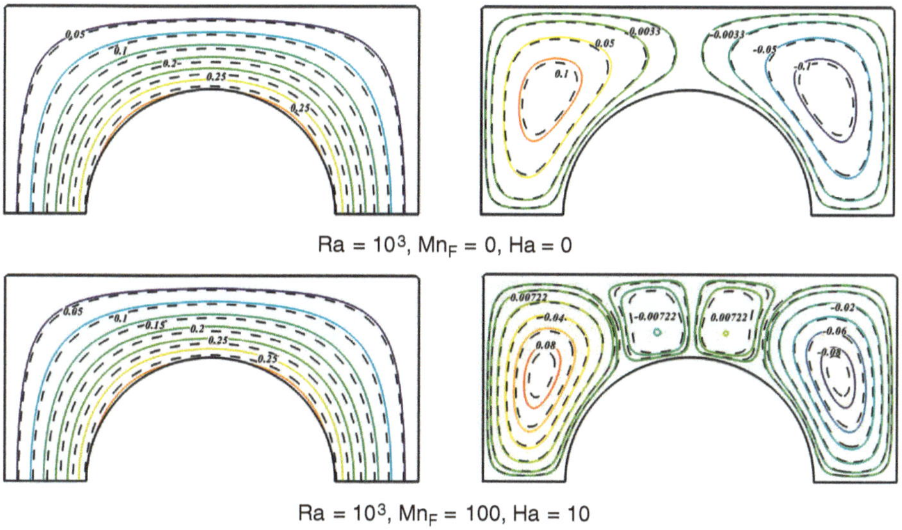

FIGURE 2.34 Comparison of the Isotherms and Streamlines Between Nanofluid ($\phi = 0.04$) (– – –) and Pure Fluid ($\phi = 0$) (—) When Pr = 6.8

2.5.6.2 Effects of active parameters

Calculations are made for various values of volume fraction of nanoparticles (ϕ = 0 and 4%), Rayleigh numbers (Ra = 10^3, 10^4, and 10^5), magnetic number arising from FHD (Mn_F = 0, 20, 60, and 100), and Hartmann number arising from MHD (Ha = 0, 5, and 10). In all calculations, the Pr, ε_1, and Ec are set to 6.8, 0.0, and 10^{-6}, respectively.

Fig. 2.34 shows a comparison of the isotherms and streamlines between nanofluid and pure fluid. The velocity components of nanofluid are increased because of an increase in the energy transport in the fluid with the increasing of volume fraction. Thus, the absolute values of stream functions indicate that the strength of flow increases with increasing the volume fraction of nanofluid. Also, it can be seen that thermal boundary layer thickness around the hot wall decreases with an increase of the nanoparticle volume fraction.

Fig. 2.35 depicts the effects of the Rayleigh number, Hartmann number, and magnetic number on isotherms and streamlines. Absolute value of stream function increases with an increase of Rayleigh number, and it decreases with an increase of the Hartmann number and magnetic number. When Ra = 10^3, the heat transfer in the enclosure is mainly dominated by the conduction mode. At Mn_F = 0, Ha = 0, the streamlines show one rotating eddy. When the magnetic parameter increases, another vortex appears near the centerline, which rotates in the reverse direction compared with the primary vortex. The Lorentz force has no significant effect on the isotherm and streamline in this Rayleigh number. As the Rayleigh number increases up to 10^5, the role of convection in heat transfer becomes more significant, and consequently, the thermal boundary layer on the surface of the inner wall becomes thinner. In addition, a plume starts to appear on the top of the inner circular wall. The effect of the magnetic field on hydrothermal behavior is more pronounced for a low Rayleigh number.

Fig. 2.36 shows the effects of the magnetic parameter, Hartmann number, and Rayleigh number on the average Nusselt number along the inner wall. The Nusselt

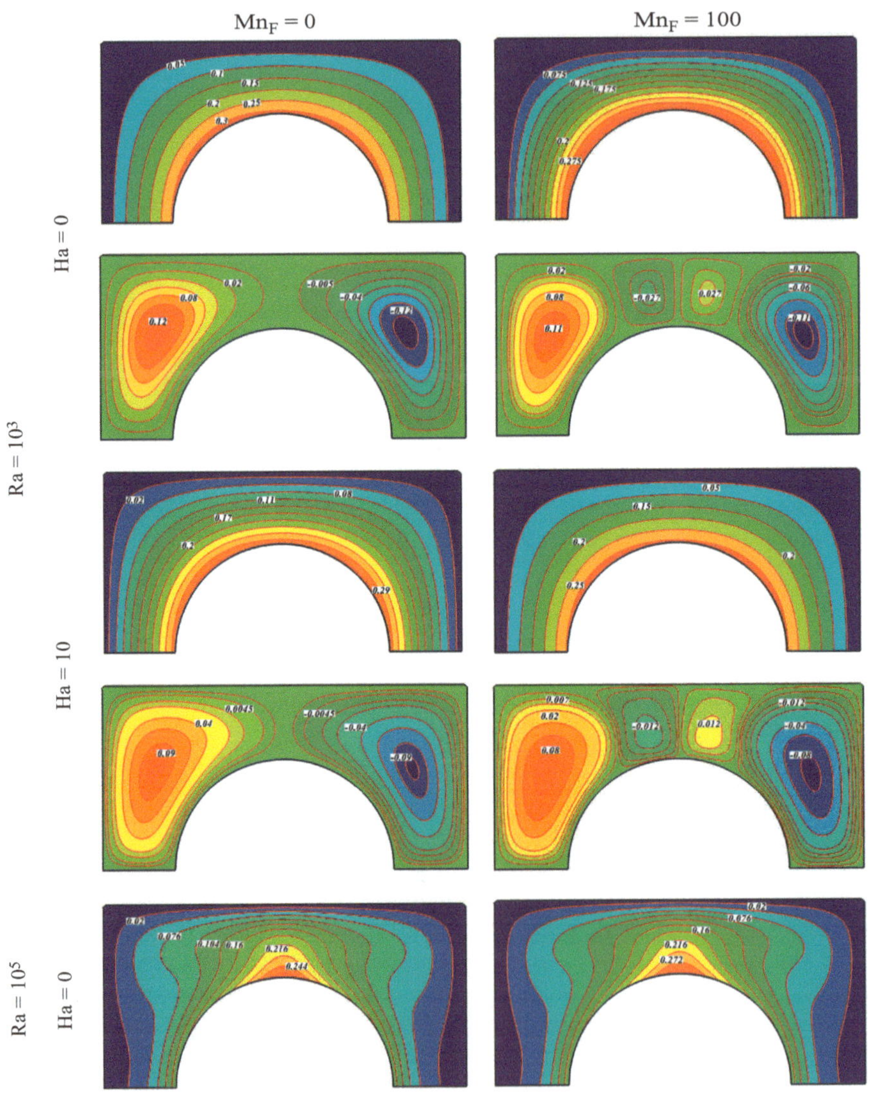

FIGURE 2.35 Isotherms (Left) and Streamlines (Right) Contours for Different Values of Rayleigh Number, Hartmann Number, and Magnetic Number When Pr = 6.8, $\phi = 0.04$

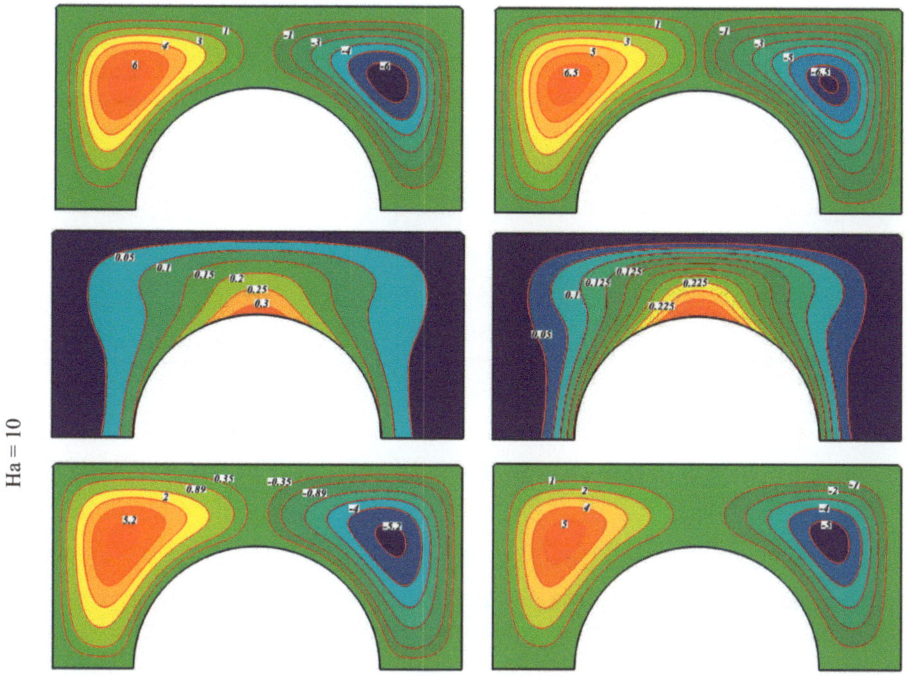

FIGURE 2.35 *(Continued)*

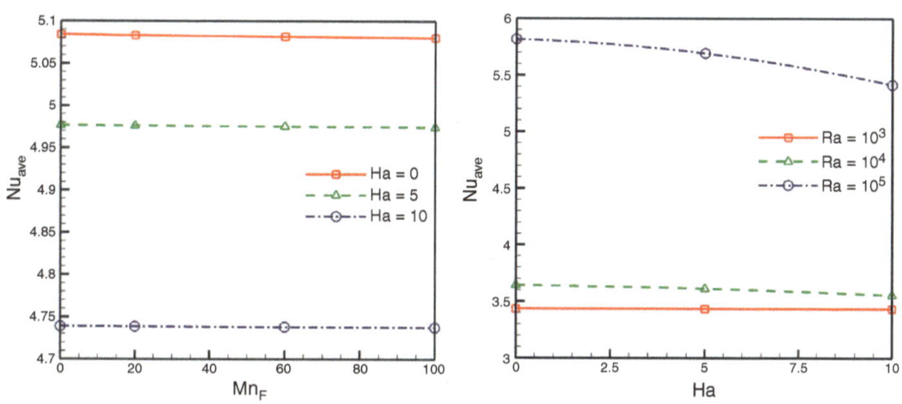

(a) Ra = 10^5

(b) Mn_F = 100

FIGURE 2.36 Effects of Magnetic Parameter, Hartmann Number, and Rayleigh Number on Average Nusselt Number Nu_{ave} along the Hot Wall When Pr = 6.8, ϕ = 0.04

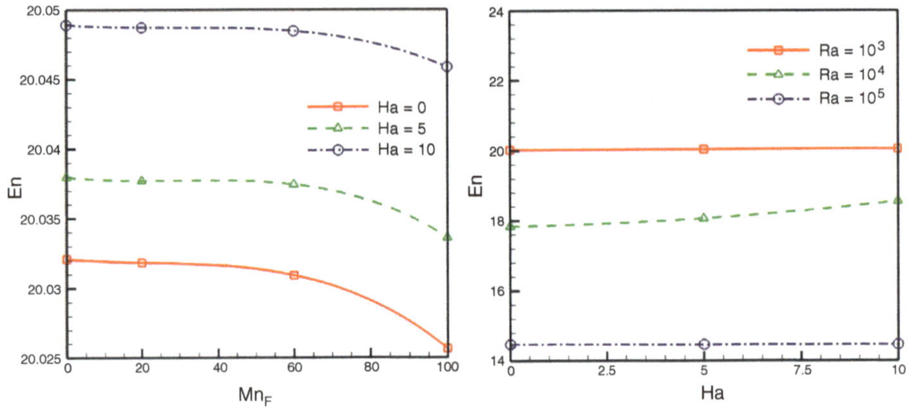

(a) Ra = 10^3 (b) $Mn_F = 0$

FIGURE 2.37 Effects of Magnetic Number, Rayleigh Number and Hartmann Number on the Heat Transfer Enhancement When $Pr = 6.8$

number increases with increases of the Rayleigh number and magnetic parameter, and it decreases with an increase of Hartmann number. The effects of the magnetic number, Rayleigh number, and Hartmann number on the heat transfer enhancement are shown in Fig. 2.37. Heat transfer enhancement is an increasing function of the Hartmann number, and it is a decreasing function of the magnetic number and Rayleigh number. The effect of nanoparticles is more pronounced at a low Rayleigh number than at a high Rayleigh number because of the greater enhancement rate. This observation can be explained by noting that at a low Rayleigh number, the heat transfer is dominant by conduction.

2.5.7 FERROFLUID HEAT TRANSFER TREATMENT IN THE PRESENCE OF VARIABLE MAGNETIC FIELD CONSIDERING MAGNETIC FIELD–DEPENDENT VISCOSITY

2.5.7.1 Problem definition

The schematic diagram and the mesh of the semi-annulus enclosure used in the present CVFEM program are shown in Fig. 2.38 [29]. The inner wall is maintained at a constant temperature T_h, and the other walls are maintained at constant temperature T_c ($T_h > T_c$). The contours of the magnetic field strength are shown in Fig. 2.39. In this study, the magnetic source is located at $\left(-0.01\ \text{cols}, 0.5\ \text{rows}\right)$. The upper wall is Lid driven with a velocity of U_{Lid}. In this study, γ is equal to 45°.

2.5.7.2 Effects of active parameters

Calculations are made for various values of volume fraction of nanoparticles ($\phi = 0$ and 4%), Richardson numbers (Ri = 0.001, 1, and 10), magnetic number (Mn_F = 0, 2, 4, 6, and 10), Hartmann number (Ha = 0, 5, and 10), and viscosity parameter

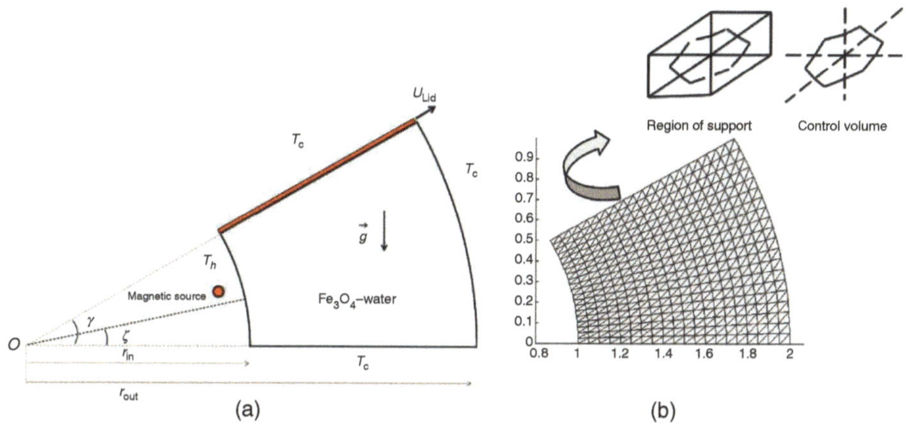

FIGURE 2.38

(a) Geometry and the boundary conditions; (b) the mesh of enclosure considered in this work; (c) a sample triangular element and its corresponding control volume.

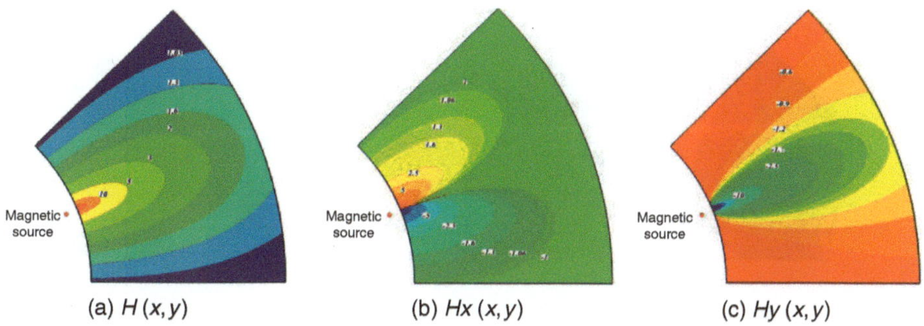

FIGURE 2.39 Contours of (a) the Magnetic Field Strength *H*; (b) the Magnetic Field Intensity Component in *x* Direction *Hx*; (c) the Magnetic Field Intensity Component in *y* Direction *Hy*

($\delta^* = 0$, 0.2, 0.4 and 0.6). In all calculations, the Pr Pr, ε_1, Ec, and Re are set to 6.8, 0.0, 10^{-5}, and 100, respectively.

A comparison of the streamlines between nanofluid and pure fluid is shown in Fig. 2.40. The velocity components of nanofluid are increased because of an increase in the energy transport in the fluid with the increasing of the volume fraction. Thermal boundary layer thickness decreases with an increase of the nanofluid volume fraction.

Isotherm and streamline contours for different values of viscosity parameter and Richardson, Hartmann, and magnetic numbers are shown in Figs 2.41–2.43. When Ri = 0.01, the heat transfer in the enclosure is mainly dominated by the conduction mode. At $Mn_F = 0$, $Ha = 0$, the streamlines show one rotating eddy. When the Hartmann number increases a small vortex generates near the magnetic source, so thermal plume

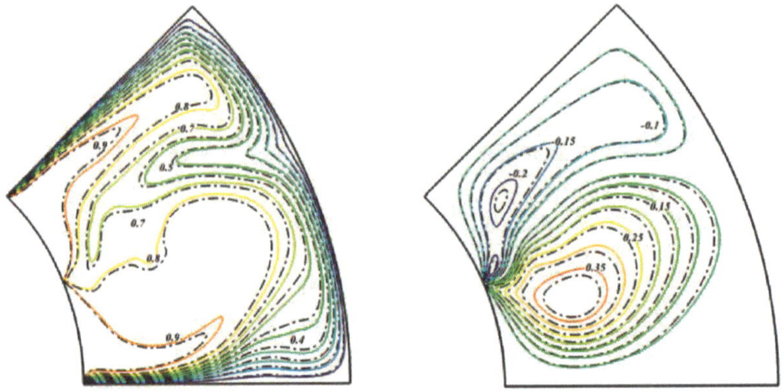

FIGURE 2.40 Comparison of the Streamlines Between Nanofluid ($\phi = 0.04$) (– – –) and Pure Fluid ($\phi = 0$) (—) When Ri = 10, Re = 100, $Mn_F = 10$, Ha = 10, Ec = 10^{-5}, $\varepsilon_1 = 0$, $\delta^* = 0$, and Pr = 6.8

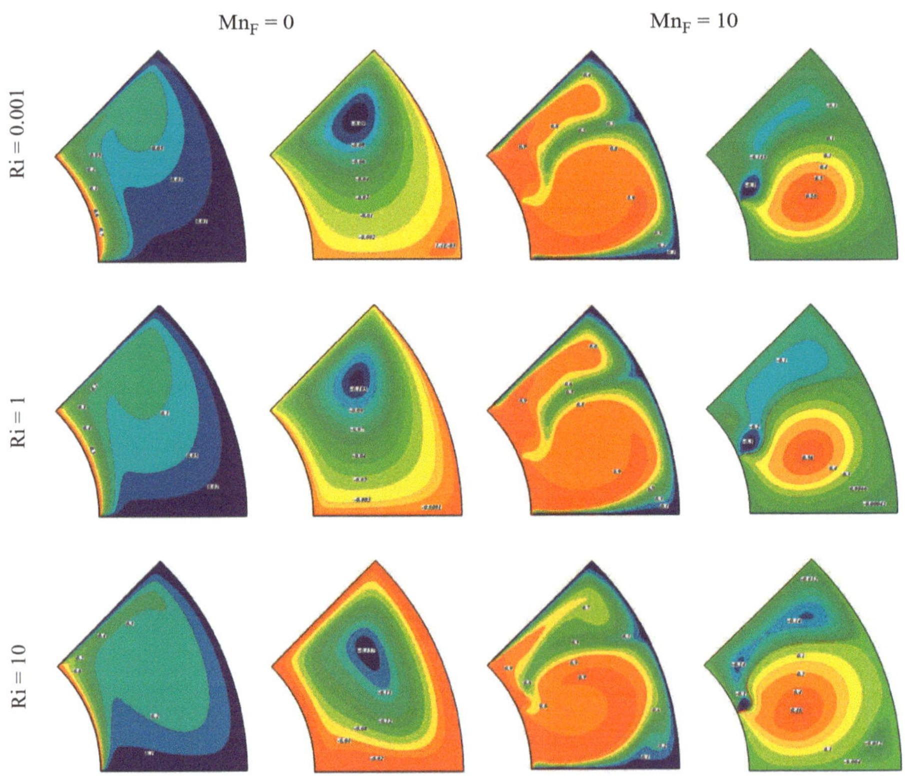

FIGURE 2.41 Isotherm (Left) and Streamlines (Right) contours for different values of Richardson number and magnetic number when Ha = 0, $\delta^* = 0$

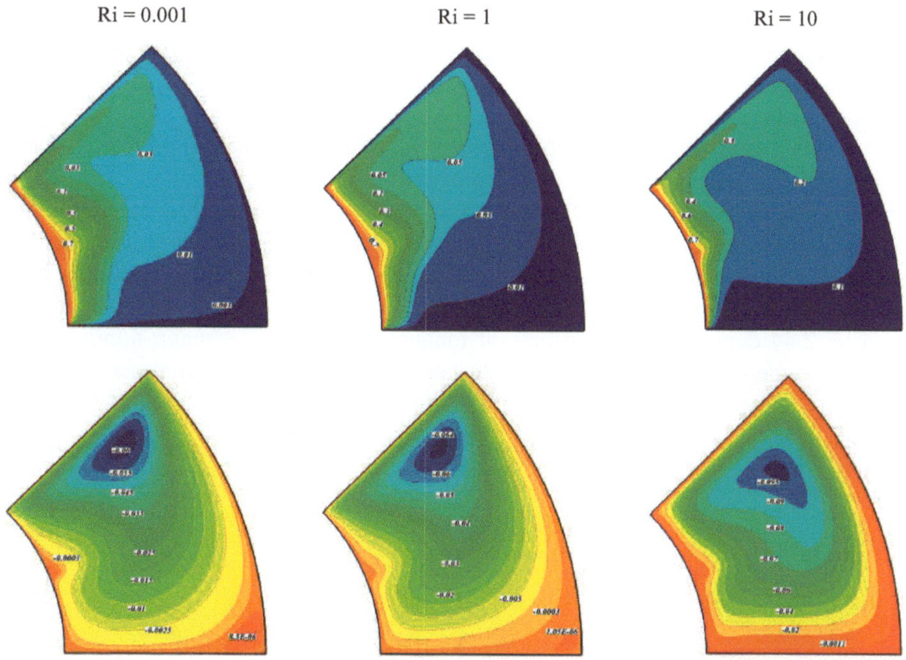

FIGURE 2.42 Isotherms (Up) and Streamlines (Down) Contours for Different Values of Richardson Number When $Mn_F = 0$, $Ha = 10$, and $\delta^* = 0$

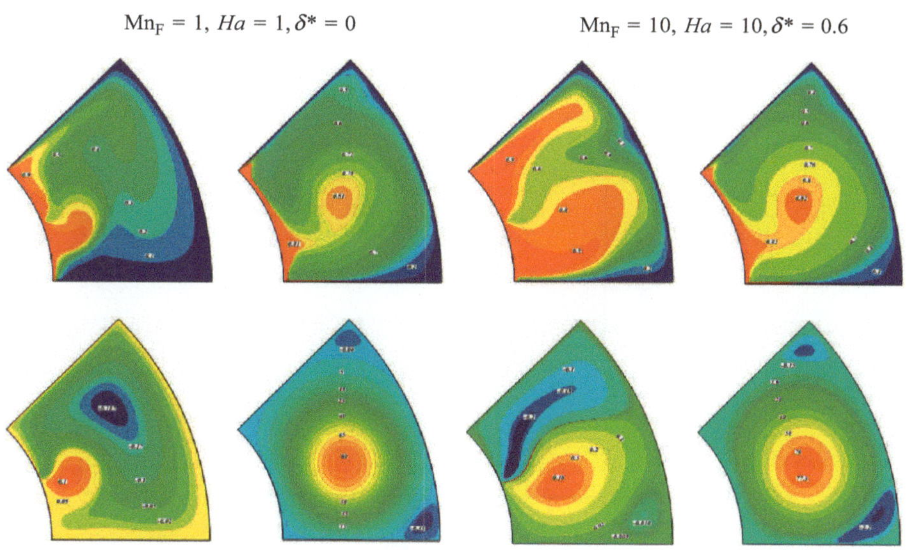

FIGURE 2.43 Isotherms (Left) and Streamlines (Right) Contours for Different Values of δ^*, Ha, and Mn_F When $Ri = 10$

appears in this region. As the magnetic number increases, the main vortex turns into two smaller vortexes, and in turn, a reverse thermal plume appears near the location of the magnetic source. As the Richardson number increases, the role of convection in heat transfer becomes more significant. At Ri = 10, the center of the main vortex moves downward, and the thermal boundary layer thickness near the inner wall becomes thinner. When the Hartmann and magnetic numbers increase up to 10, the main vortex turns into three eddies. Three thermal plumes generate over the inner wall because of the existence of eddies that are rotated in different directions. Also, it can be seen that as the viscosity parameter increases, the thermal boundary layer thickness decreases, and one counterclockwise eddy generates at the center of the enclosure.

Fig. 2.44 depicts that the effects of magnetic number, Hartmann number, viscosity parameter, and Richardson number on the average Nusselt number. Thermal boundary layer thickness increases with increase of magnetic number and Hartmann

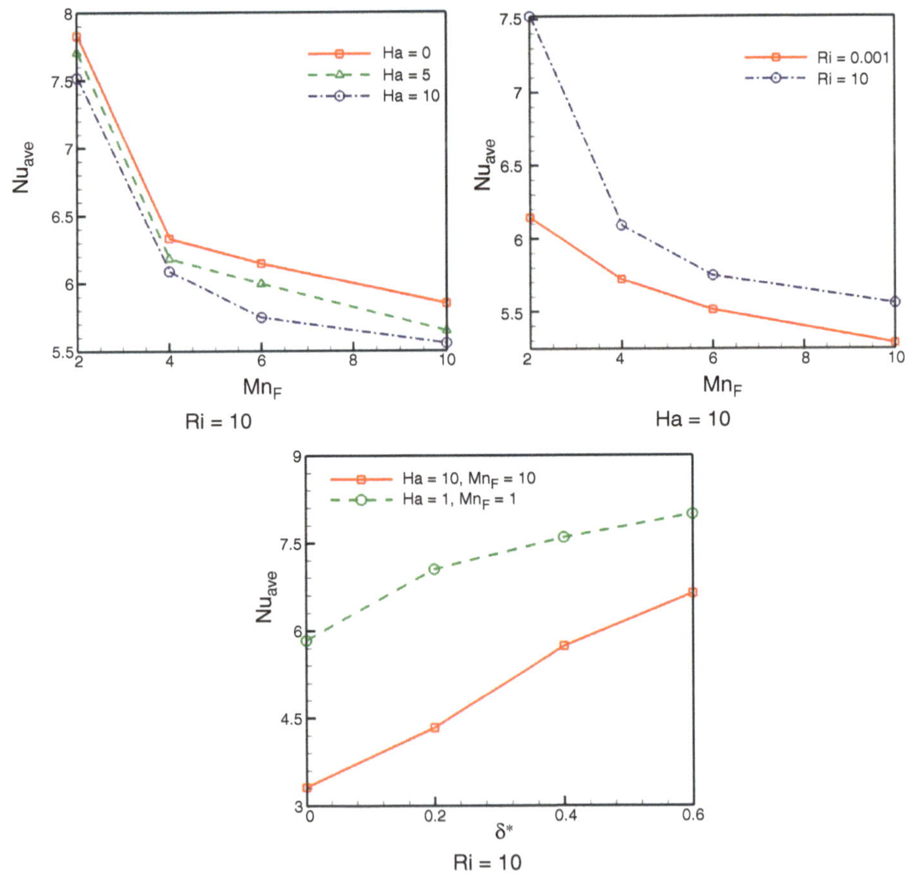

FIGURE 2.44 Effects of Ha, Mn_F, Ri, and δ^* on the Average Nusselt Number Nu_{ave} Along the Hot Wall

number, and it decreases with an augment of the Richardson number. So, the average Nusselt number increases with an increase of the Richardson number, and it decreases with an increase of the Hartmann number and magnetic number. Also, it can be concluded that by considering the effect of the magnetic field on the viscosity of the fluid, the rate of heat transfer increases.

2.5.8 MHD EFFECT ON NATURAL CONVECTION HEAT TRANSFER IN AN INCLINED L-SHAPED ENCLOSURE

2.5.8.1 Problem definition

The physical model along with the important geometrical parameters and the mesh of the enclosure used in the present CVFEM program are shown in Fig. 2.45 [11]. The width and height of the enclosure is H. The right and top walls of the enclosure are maintained at constant cold temperature T_c , the inner circular hot wall is maintained at constant hot temperature T_h, and the two bottom and left walls with the length of $H/2$ are thermally isolated. Under all cases $T_h > T_c$ condition is maintained. To assess the shape of inner circular and outer rectangular boundary, which consists of the right and top walls, a supper elliptic function can be used as follows:

$$\left(\frac{X}{a}\right)^{2n} + \left(\frac{Y}{b}\right)^{2n} = 1 \tag{2.38}$$

When $a = b$ and $n = 1$, the geometry becomes a circle. As n increases from 1, the geometry would approach a rectangle for $a \neq b$ and square for $a = b$. In this study, $r_{in}/r_{out} = 0.75$. It is also assumed that the uniform magnetic field ($\vec{B} = B_x \vec{e}_x + B_y \vec{e}_y$) of constant magnitude $B = \sqrt{B_x^2 + B_y^2}$ is applied, where $\vec{e}_x$ and $\vec{e}_y$ are unit vectors in the Cartesian coordinate system. The orientation of the magnetic field forms an angle λ with a horizontal axis such that $\lambda = B_x/B_y$. In this study, λ equals to zero. The

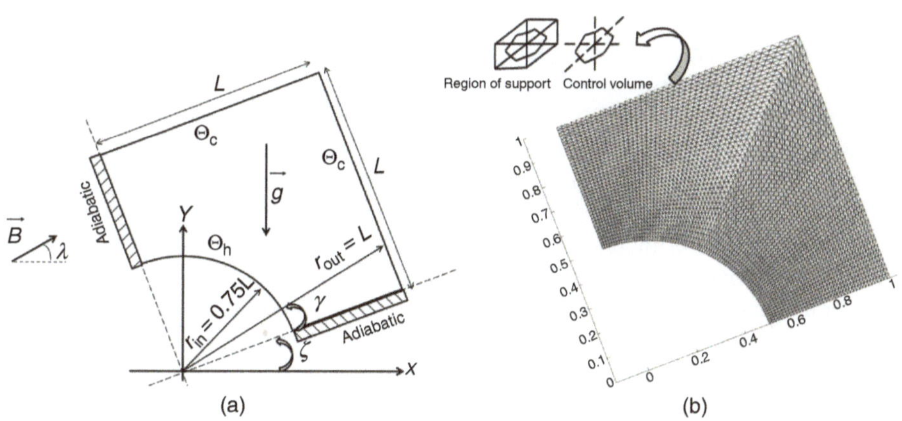

FIGURE 2.45

(a) Geometry and the boundary conditions with (b) the mesh of enclosure considered in this work.

electric current J and the electromagnetic force F are defined by $J = \sigma\left(\vec{V} \times \vec{B}\right)$ and $F = \sigma\left(\vec{V} \times \vec{B}\right) \times \vec{B}$, respectively.

The heatlines are adequate tools for visualization and analysis of 2D convection heat transfer through an extension of the heat flux line concept to include the advection terms. The heat function (H) is defined in terms of the energy equation as

$$\frac{\partial H}{\partial Y} = U\Theta - \frac{\partial \Theta}{\partial X}, \tag{2.39}$$

$$-\frac{\partial H}{\partial X} = V\Theta - \frac{\partial \Theta}{\partial Y} \tag{2.40}$$

2.5.8.2 Effects of active parameters

In this study, the MHD effect on natural convection heat transfer in an L-shaped inclined enclosure filled with nanofluid is investigated numerically using the CVFEM. The fluid in the enclosure is Al_2O_3–water nanofluid. Calculations are made for various values of Hartmann number (Ha = 0, 20, 60 and 100), volume fraction of nanoparticle ($\phi = 0$ and 4%), Rayleigh number (Ra = 10^3, 10^4, and 10^5), inclination angle ($\gamma = -90, -60, -30$, and 0°), and constant Prandtl number (Pr = 6.2).

Comparisons of the isotherms, streamlines, and heatlines for different values of Hartmann number, inclined angles, and Rayleigh number are shown in Figs 2.46 and 2.47. At Ra = 10^3, the isotherms are parallel to each other and take the shape of an enclosure, which is the main characteristic of the conduction heat transfer mechanism. As the Rayleigh number increases, the isotherms become more distorted, and the stream function values enhance, which is caused by the domination of the convective heat transfer mechanism at higher Rayleigh numbers. At Ra = 10^5, a thermal plume appears on the hot circular wall, and three vortices exist in streamline. By increasing $|\zeta|$, these vortices merge to one eddy. As $|\zeta|$ increases, the hot wall locates on the cold wall. Therefore, convective heat transfer becomes weak, and in turn, the Nusselt number decreases with increase of $|\zeta|$.

When the magnetic field is imposed on the enclosure, the velocity field is suppressed owing to the retarding effect of the Lorenz force. So, intensity of convection weakens significantly. The braking effect of the magnetic field is observed from the maximum stream function value. The core vortex is shifted upward vertically as the Hartmann number increases. Also, imposing a magnetic field leads to omit the thermal plume over the inner wall. At a high Hartmann number, the conduction heat transfer mechanism is more pronounced. For this reason, the isotherms are parallel to each other. The increase of Ra causes the clustering of heatlines from hot to the cold wall and generates passive heat transfer area in which heat is rotated without having a significant effect on heat transfer between walls.

Fig. 2.48 shows the distribution of local Nusselt numbers along the surface of the inner circular wall for different inclination angles, Rayleigh numbers, and Hartmann

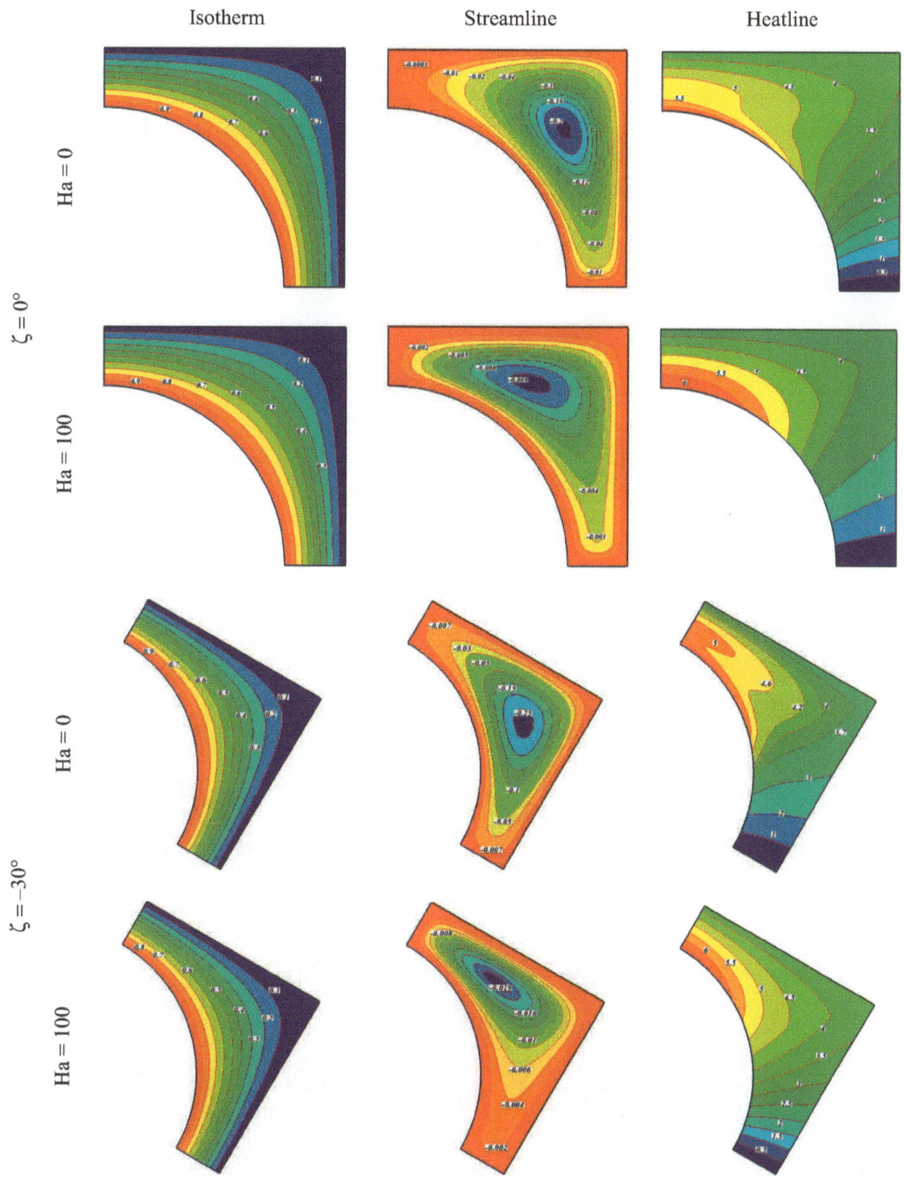

FIGURE 2.46 Comparison of the Isotherms, Streamlines, and Heatlines for Different Values of Hartmann Number and Inclination Angle at Ra = 10^3, $\phi = 0.04$, and Pr = 6.2

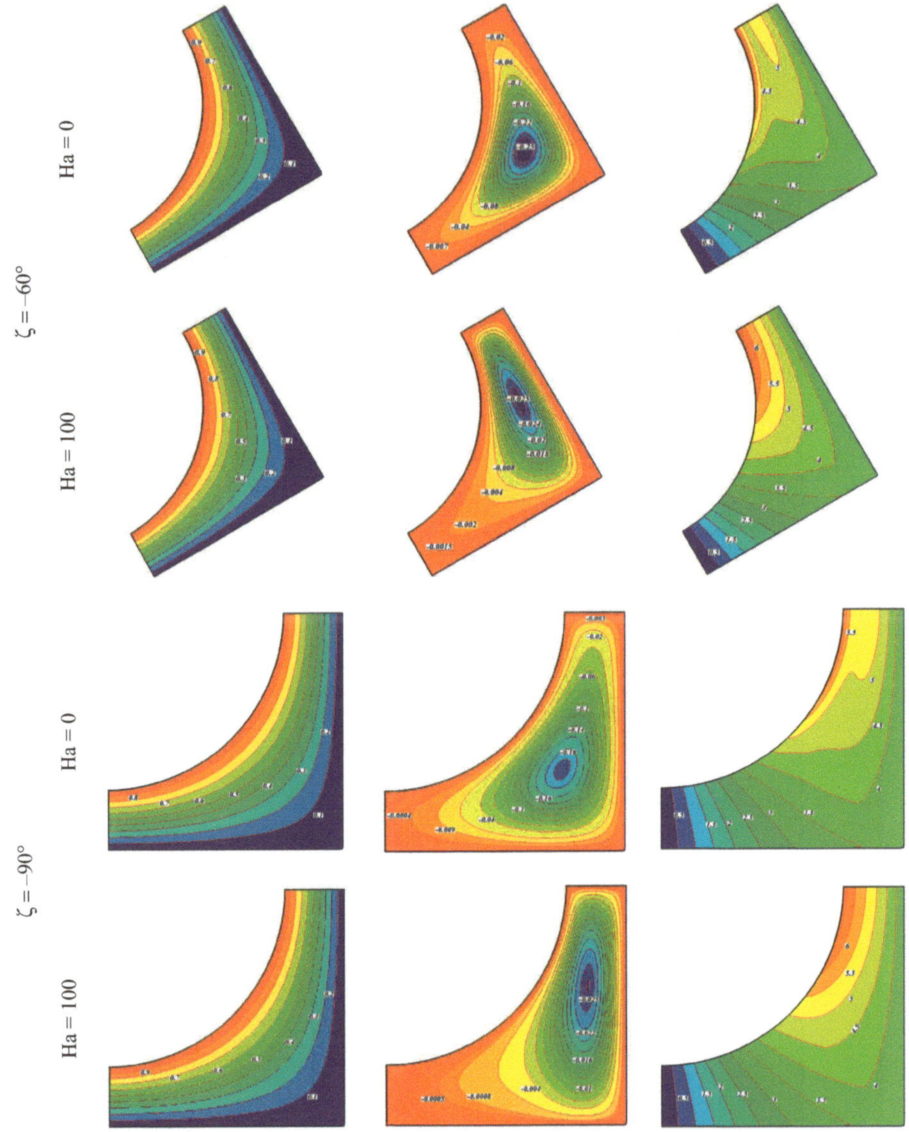

FIGURE 2.46 ***(Continued)***

numbers. Increasing the Rayleigh number leads to an increase in the local Nusselt number, but increasing the Hartmann number and inclination angle cause the local Nusselt number to decrease. At $Ra = 10^3$, because of the domination conduction heat transfer mechanism, the distribution of the local Nusselt numbers along the surface of inner circular shows a symmetric shape. It is interesting to notice that at a high Rayleigh number, the local Nusselt number profiles are more complex because of

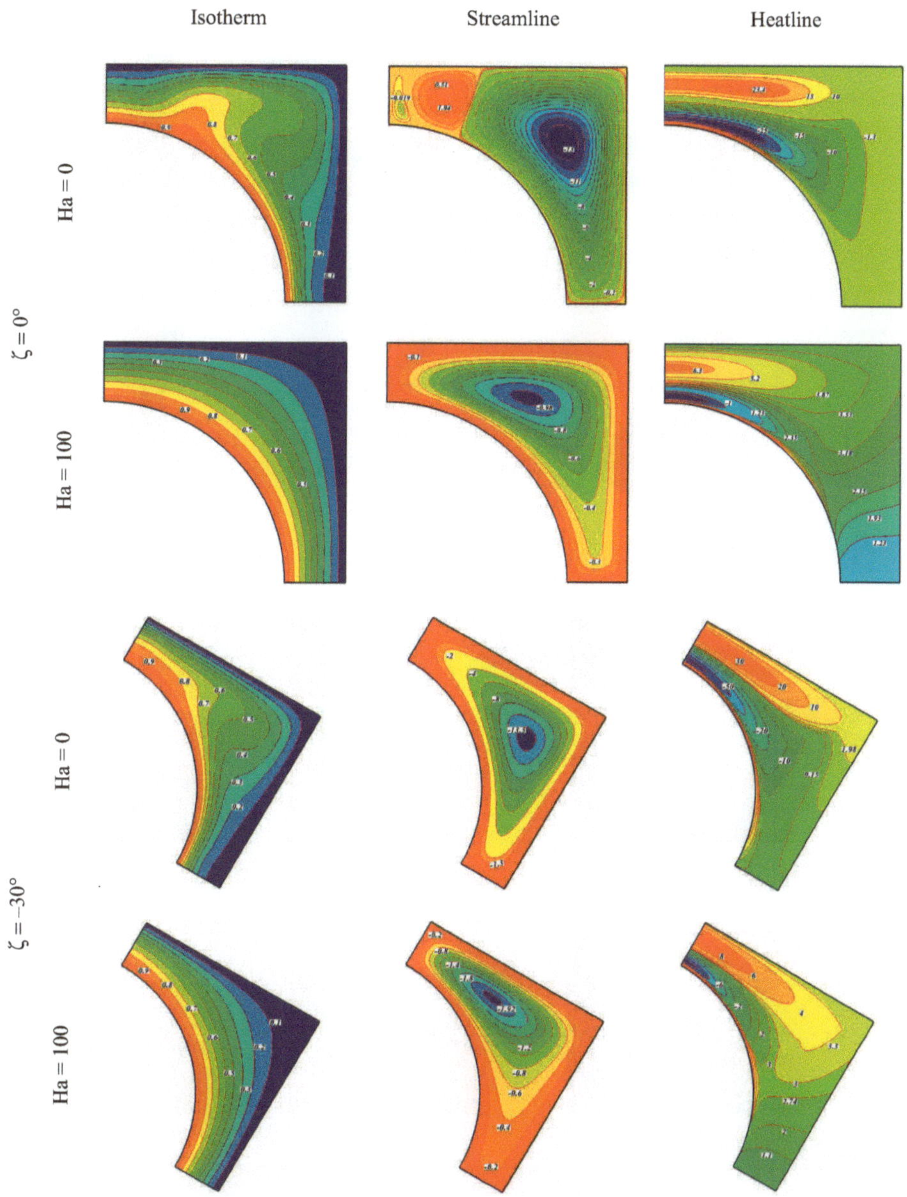

FIGURE 2.47 Comparison of the Isotherms, Streamlines, and Heatlines for Different Values of Hartmann Number and Inclination Angle at Ra = 10^5, $\phi = 0.04$, and Pr = 6.2

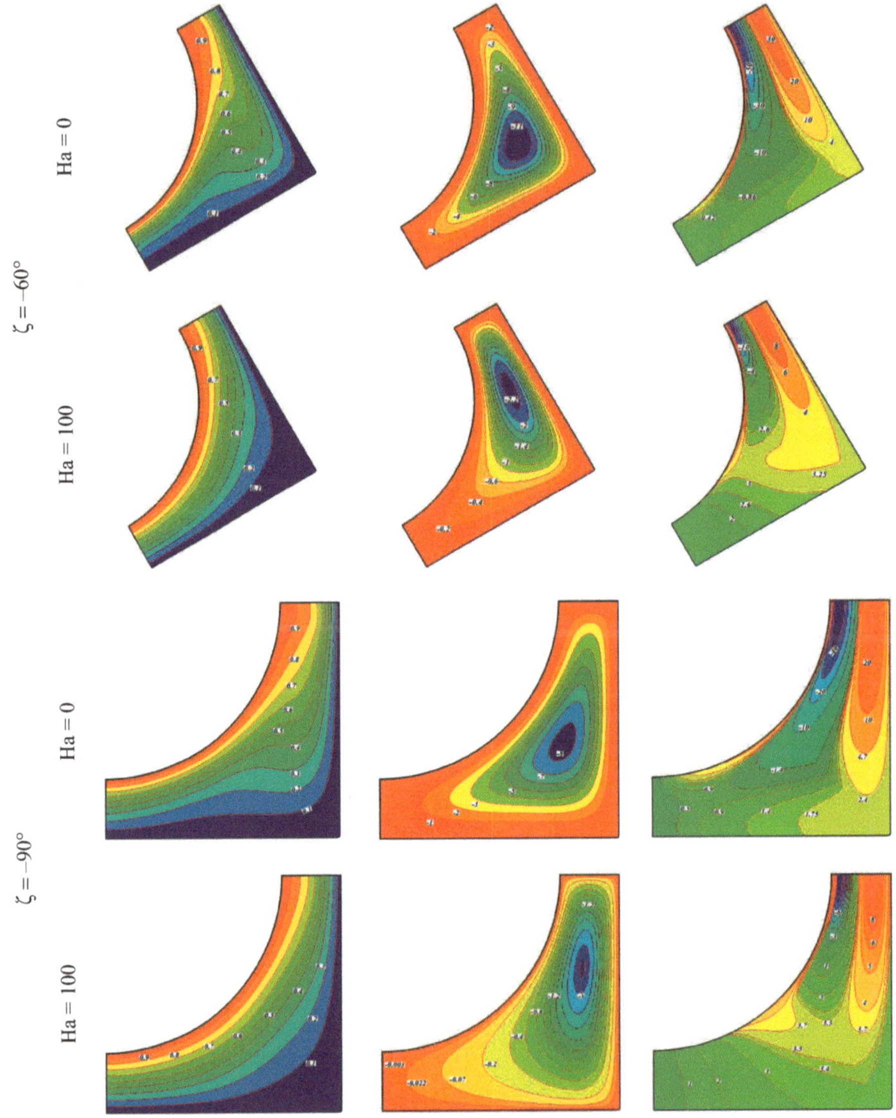

FIGURE 2.47 ***(Continued)***

the presence of thermal plume. In all cases except for $Ra = 10^5$, $\gamma = -90°$, one minimum point exists in the local Nusselt number profile, which occurs in lower values of $|\gamma|$ with an increase of the Hartmann number. The effects of the Rayleigh number, Hartmann number, and inclination angle on the average Nusselt number is shown in Fig. 2.49. The Nusselt number is an increasing function of the Rayleigh number, but it is a decreasing function of the Hartmann number and inclined angle. Also, it

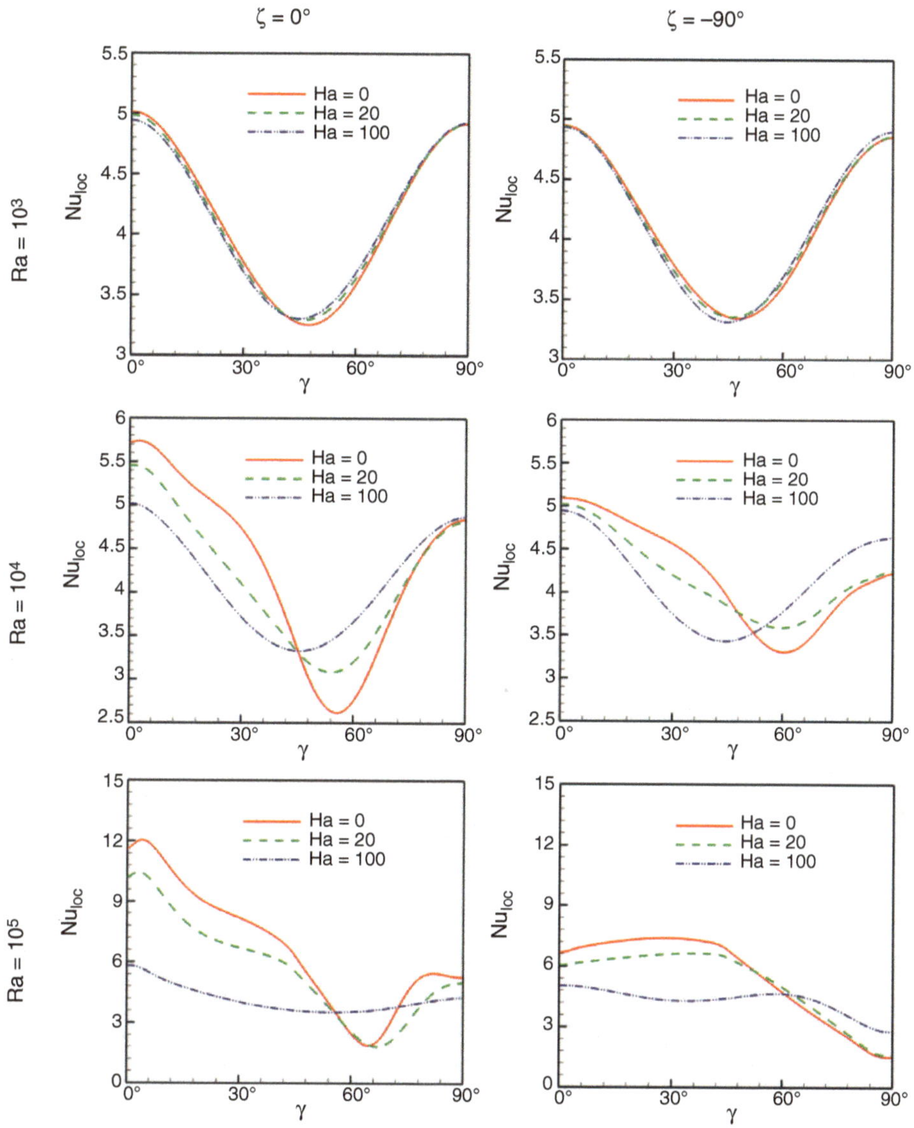

FIGURE 2.48 Effects of the Hartmann Number, Rayleigh Number, and Inclination Angle for Cu–water Nanofluid on the Local Nusselt Number

can be found that the effect of the Hartmann number on the Nusselt number is more pronounced at $\zeta = 0°$.

The heat transfer enhancement ratio caused by the addition of nanoparticles for different values of ζ, Ha, and Ra is shown in Fig. 2.50. In general, it can be found that the effect of nanoparticles is more pronounced at a low Rayleigh number and

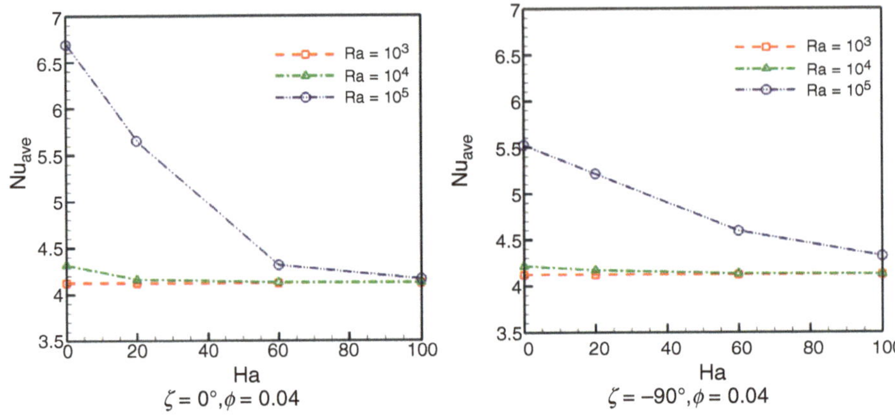

FIGURE 2.49 Effects of the Rayleigh Number, Hartmann Number, and Inclination Angle on Nu_{ave}

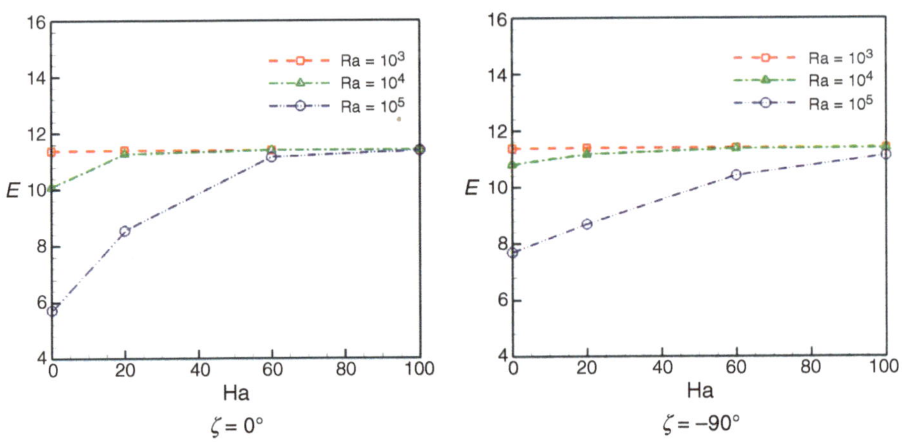

FIGURE 2.50 Effects of the Rayleigh Number, Hartmann Number, and Inclination Angle on the Ratio of Heat Transfer Enhancement Caused by the Addition of Nanoparticles

high values of Hartmann number because of the greater enhancement rate. This observation can be explained by noting that at a low Rayleigh number, the heat transfer is dominant by conduction. Therefore, the addition of high thermal conductivity nanoparticles will increase the conduction and make the enhancement more effective. The inclination angle has no significant effect on the rate on enhancement when $Ra = 10^3$, 10^4 but E increases with increase of $|\zeta|$ when $Ra = 10^5$. Finally, the corresponding polynomial representation of such model for each of Nusselt number and rate of enhancement are presented as follows:

$$\begin{aligned} \mathrm{Nu} &= a_{13} + a_{23}Y_1 + a_{33}Y_2 + a_{43}Y_1^2 + a_{53}Y_2^2 + a_{63}Y_1Y_2 \\ Y_1 &= a_{11} + a_{21}\zeta + a_{31}\phi + a_{41}\zeta^2 + a_{51}\phi^2 + a_{61}\zeta\,\phi \\ Y_2 &= a_{12} + a_{22}\,\mathrm{Ra}^* + a_{32}\,\mathrm{Ha}^* + a_{42}\,\mathrm{Ra}^{*2} + a_{52}\,\mathrm{Ha}^{*2} + a_{62}\,\mathrm{Ra}^*\,\mathrm{Ha}^* \end{aligned} \tag{2.41}$$

$$\begin{aligned} E &= b_{13} + b_{23}Y_1 + b_{33}Y_2 + b_{43}Y_1^2 + b_{53}Y_2^2 + b_{63}Y_1Y_2 \\ Y_1 &= b_{11} + b_{21}\zeta + b_{31}\,\mathrm{Ha}^* + b_{41}\zeta^2 + b_{51}\,\mathrm{Ha}^{*2} + b_{61}\zeta\,\mathrm{Ha}^* \\ Y_2 &= b_{12} + b_{22}\,\mathrm{Ra}^* + b_{32}\zeta + b_{42}\,\mathrm{Ra}^{*2} + b_{52}\zeta^2 + b_{62}\,\mathrm{Ra}^*\,\zeta \end{aligned} \tag{2.42}$$

where $\mathrm{Ra}^* = \log(\mathrm{Ra})$ and $\mathrm{Ha}^* = \mathrm{Ha}/100$. Also, a_{ij} and b_{ij} can be found in Tables 2.7 and 2.8.

2.5.9 APPLICATION OF CVFEM FOR THE COMBINED EFFECT OF FHD AND MHD CONSIDERING THERMAL RADIATION

2.5.9.1 Problem definition

The schematic diagram and the mesh of the semi-annulus enclosure used in the present CVFEM program are shown in Fig. 2.51. The inner wall is under constant heat flux (q''), and the outer wall is maintained at constant temperature (T_c); the two other walls are thermally insulated. In this study, γ equals 45°. The contours of the magnetic field strength are shown in Fig. 2.52. In this study, the magnetic source is located at $(-0.05\ \text{cols}, 0.5\ \text{rows})$.

Table 2.7 Constant Coefficient for Using Eq. (2.41)

a_{ij}	$i = 1$	$i = 2$	$i = 3$	$i = 4$	$i = 5$	$i = 6$
$j = 1$	4.080931	−0.10849	10.44936	−0.10891	0.417974	−0.03188
$j = 2$	7.809771	−2.67057	2.414376	0.452514	0.790997	−0.97843
$j = 3$	8.465516	0.484181	−4.13987	−0.01217	0.4827	0.145098

Table 2.8 Constant Coefficient for Using Eq. (2.42)

b_{ij}	$i = 1$	$i = 2$	$i = 3$	$i = 4$	$i = 5$	$i = 6$
$j = 1$	9.333563	−0.00506	4.537482	0.195157	−2.33844	0.508573
$j = 2$	3.204931	5.085197	0.530967	−0.77932	0.195157	−0.07679
$j = 3$	−134.848	14.51322	11.03078	−0.10825	0.060356	−1.0671

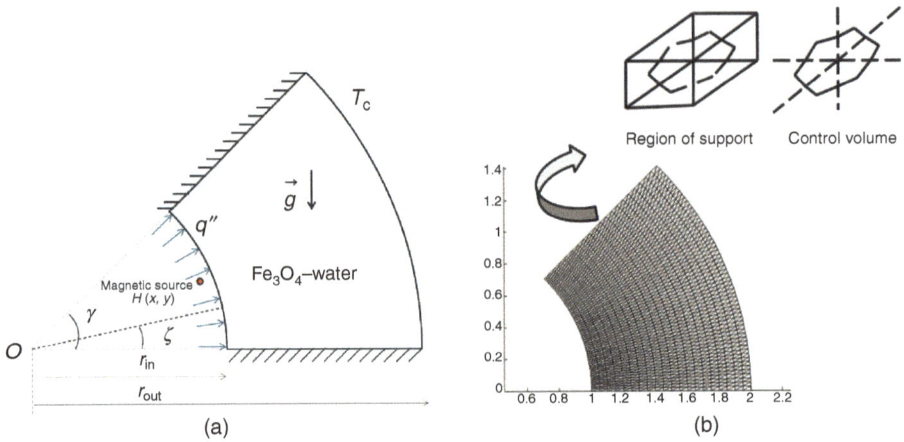

FIGURE 2.51

(a) Geometry and the boundary conditions. (b) The mesh of enclosure considered in this work.

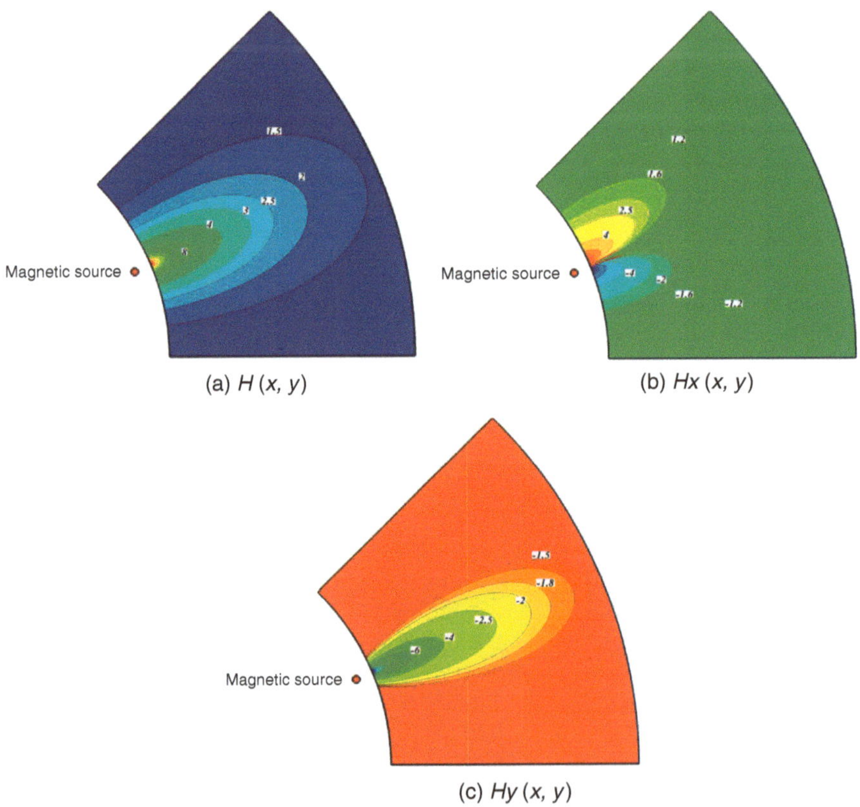

FIGURE 2.52 Contours of (a) the Magnetic Field Strength *H*; (b) the Magnetic Field Intensity Component in *x* Direction *Hx*; (c) the Magnetic Field Intensity Component in *y* Direction *Hy*

2.5.9.2 Effects of active parameters

Ferrofluid (Fe_3O_4–water) flow and heat transfer in a semi-annulus enclosure in presence of an external magnetic field is investigated. CVFEM is used to solve the governing equations. The mathematical model used for the formulation of the problem is consistent with the principles of FHD and MHD. Calculations are made for various values of volume fraction of nanoparticles ($\phi = 0\%$ and 4%), Rayleigh numbers ($Ra = 10^3$, 10^4, and 10^5), magnetic number arising from FHD ($Mn_F = 0$, 20, 60, and 100), radiation parameter ($Rd = 0.0$ and 0.02), and Hartmann number arising from MHD ($Ha = 0$, 2, 6, and 10). In all calculations, the Pr, ε_1, and Ec are set to 6.8, 0.0, and 1e-5, respectively.

Fig. 2.53 shows the comparison of the isotherm and streamlines between nanofluid and pure fluid. The velocity components of nanofluid are increased because of an increase in the energy transport in the fluid with the increase of the volume fraction. Thus, the absolute values of stream functions indicate that the strength of the flow increases with increasing the volume fraction of the nanofluid. Also, it can be found that the thermal boundary layer thickness increases with an increase of the volume fraction of the nanofluid.

Fig. 2.54 shows the effect of Hartmann number on isotherm and streamlines. As the Hartmann number increases, the absolute value of maximum stream function decreases. Also, this figure shows that streamlines become more disturbed near the magnetic source. Isotherm and streamline contours for different values of Rayleigh number, Hartmann number, radiation parameter, and magnetic number are shown in Figs 2.55 and 2.56.

When $Ra = 10^3$, the heat transfer in the enclosure is mainly dominated by the conduction mode. At $Mn_F = 0$, $Ha = 0$, the streamlines show one rotating eddy.

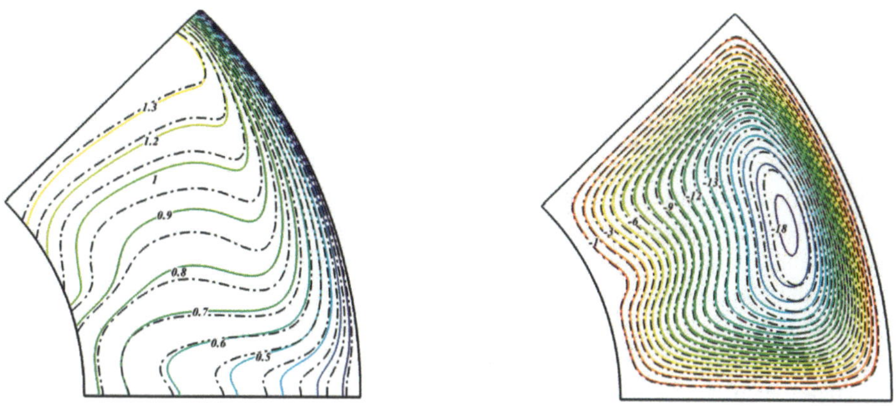

FIGURE 2.53 Comparison of the Isotherm and Streamline Between Nanofluid ($\phi = 0.04$) (—) and Pure Fluid ($\phi = 0$) (– – –) When $Ra = 10^5$, $Mn_F = 100$, $Ha = 10$, and $Pr = 6.8$

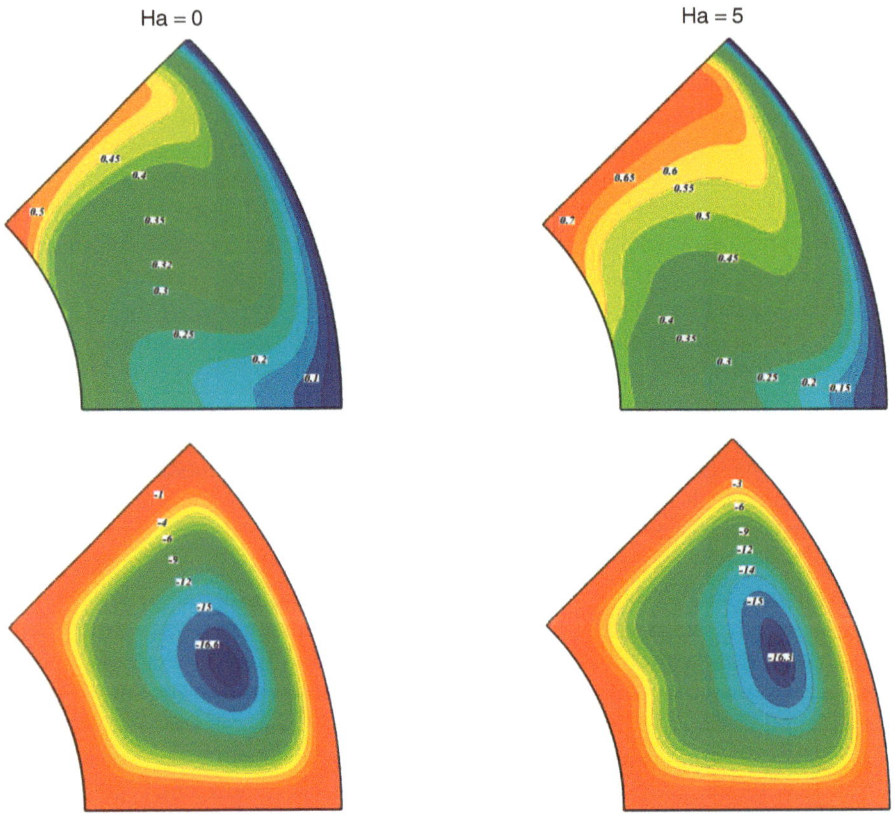

FIGURE 2.54 Effect of Hartmann Number on Isotherm and Streamlines When Ra = 10^5, Mn = 100, $\phi = 0.04$, and Rd = 0.02

In the presence of a magnetic field, the vortex center moves upward. As the Rayleigh number increases, the role of convection in heat transfer becomes more significant, and consequently, the thermal boundary layer on the surface of the inner wall becomes thinner. In addition, a plume starts to appear on the top of the inner circular wall at $\zeta = 45°$. The isotherms and streamlines become denser near the magnetic source ($\zeta = 22.5°$). Also, this figure shows that the absolute value of the maximum stream function increases with an augment of the radiation parameter. The center of the vortex moves downward left by increasing the radiation parameter.

Fig. 2.57 depicts the effects of the magnetic number, Hartmann number, radiation parameter, and Rayleigh number on the local Nusselt number. The local Nusselt number increases with an increase of the Rayleigh number, and it decreases with an

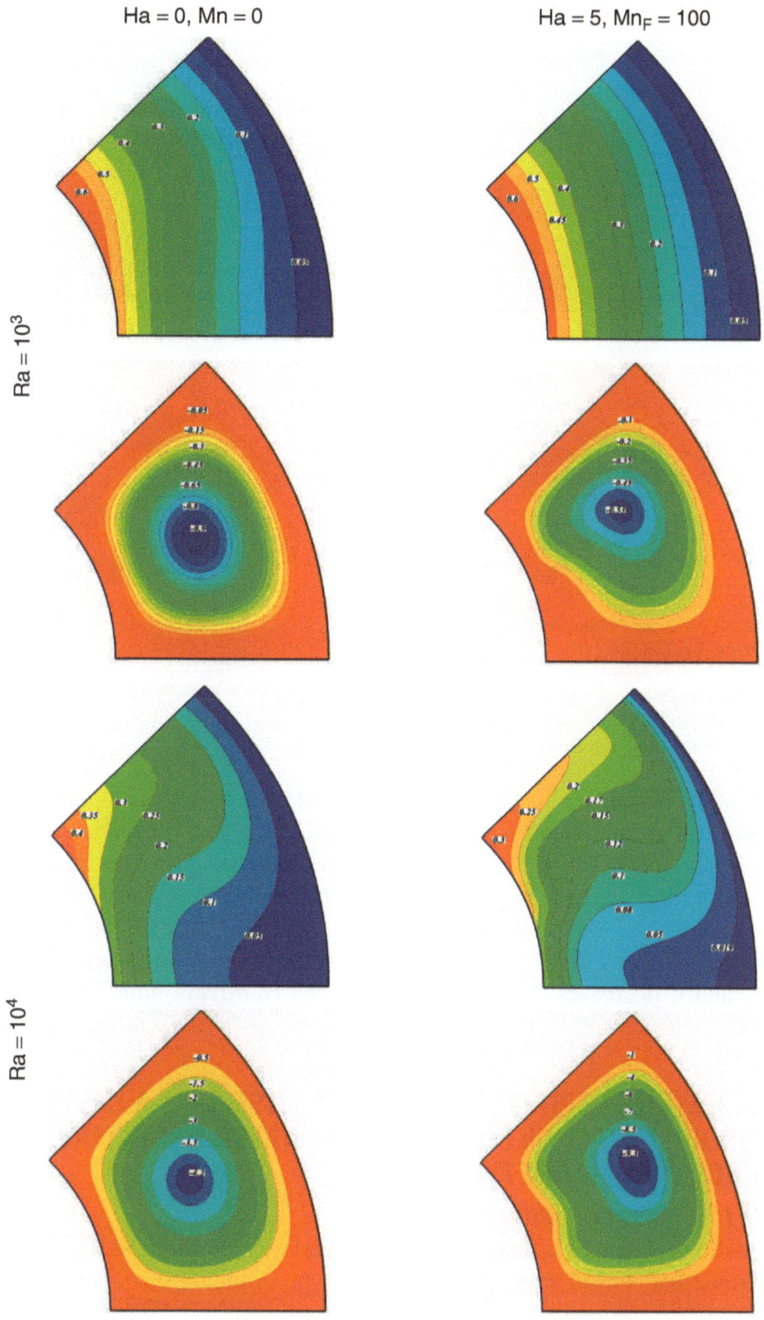

FIGURE 2.55 Isotherm (Up) and Streamlines (Down) Contours for Different Values of Rayleigh Number, Hartmann Number and Magnetic Number When Rd = 0.0, $\phi = 0.04$, and Pr = 6.8

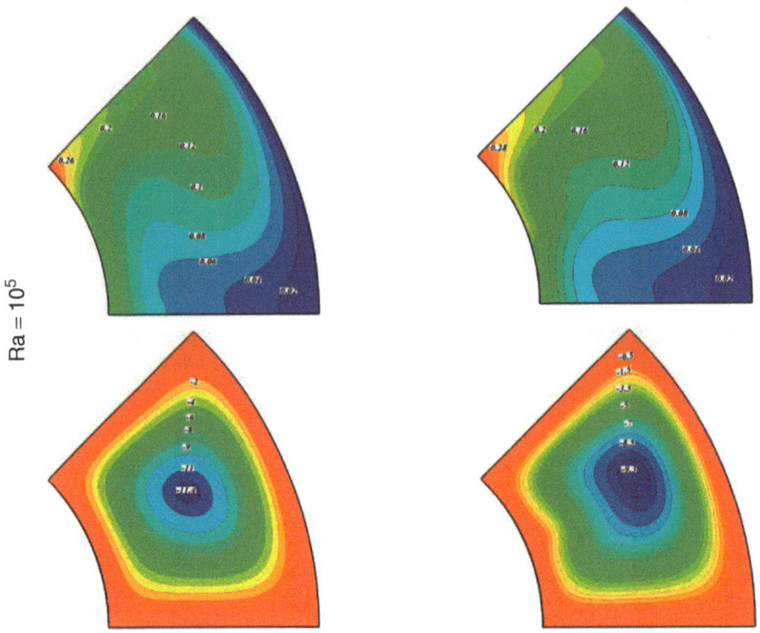

FIGURE 2.55 (*Continued*)

increase of the Hartmann number and radiation parameter. The local minimum point at $\zeta = 22.5°$ in these profiles is related to the existence of a magnetic source at this region. The effects of the volume fraction of nanofluid, magnetic number, Hartmann number, radiation parameter, and Rayleigh number on the average Nusselt number are shown in Fig. 2.58 and Table 2.9. The average Nusselt number increases with an increase of volume fraction of the nanofluid, the magnetic number, and Rayleigh number, and it decreases with an increase of the Hartmann number and radiation parameter.

2.5.10 NATURAL CONVECTION HEAT TRANSFER IN A CAVITY WITH A SINUSOIDAL WALL FILLED WITH CuO–WATER NANOFLUID IN THE PRESENCE OF A MAGNETIC FIELD

2.5.10.1 Problem definition

A schematic of the problem and the related boundary conditions as well as the mesh of enclosure that is used in the present CVFEM program are shown in Fig. 2.59 [8].

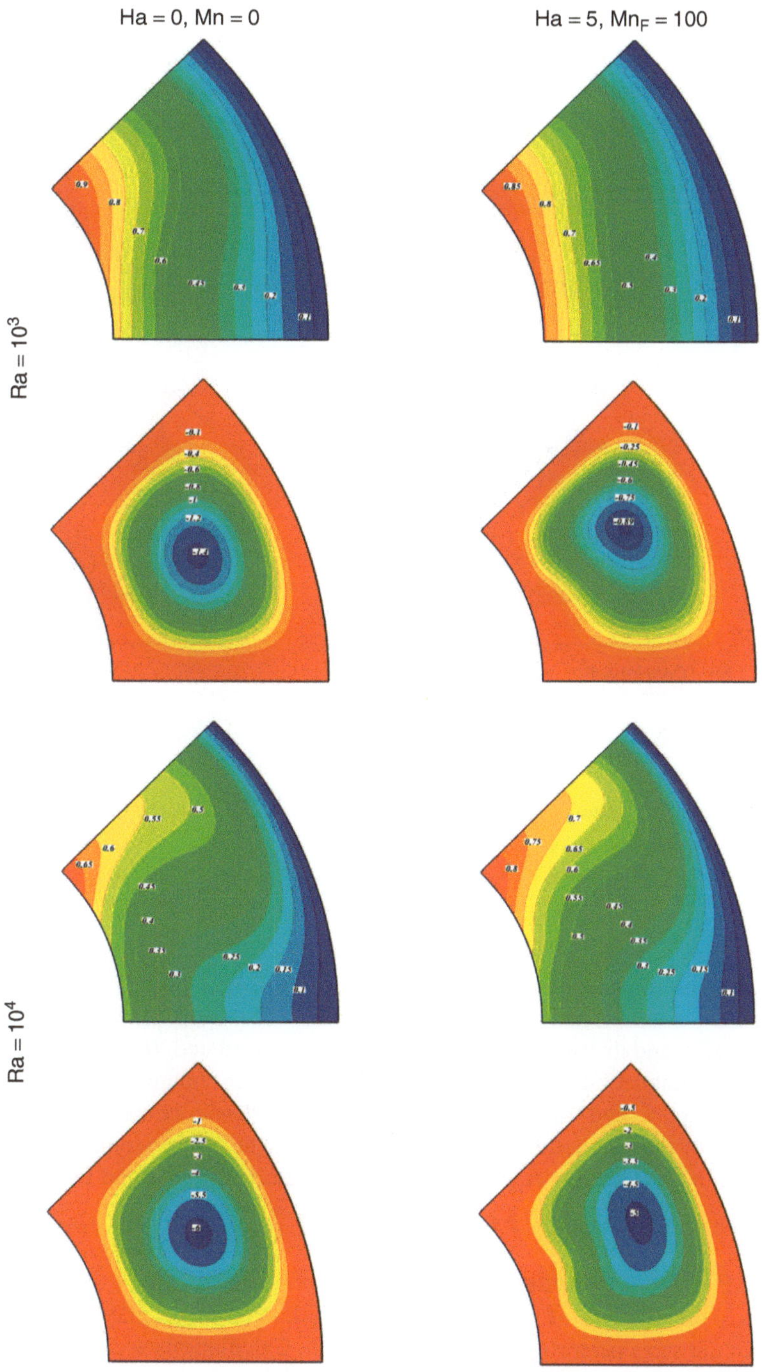

FIGURE 2.56 Isotherm (Up) and Streamline (Down) Contours for Different Values of Rayleigh Number, Hartmann Number and Magnetic Number When Rd = 0.02, ϕ = 0.04, and Pr = 6.8

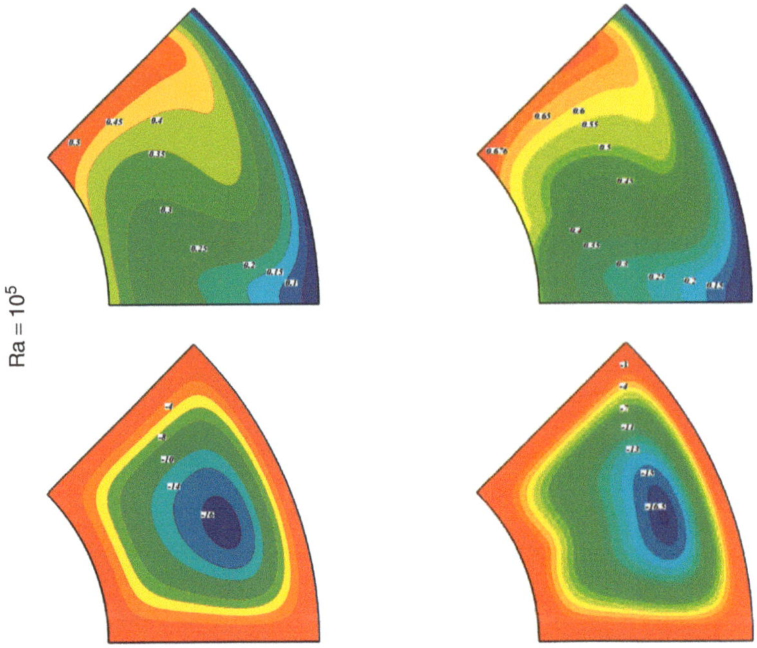

FIGURE 2.56 (*Continued*)

The enclosure has a width-to-height aspect ratio of 2. The shape of the left sinusoidal wall profile is assumed to be the following pattern:

$$\frac{x}{H}=1-\left\{a\left(1+\sin\left(\pi\frac{y}{H}-\pi/2\right)\right)\right\} \tag{2.43}$$

where a is the dimensionless amplitude of the sinusoidal wall. The left sinusoidal wall is under constant heat flux (q''), the right flat wall is maintained at constant temperatures (T_c), and the two other walls are thermally insulated. It is also assumed that the uniform magnetic field ($\vec{B}=B_x\vec{e}_x+B_y\vec{e}_y$) of constant magnitude $B=\sqrt{B_x^2+B_y^2}$ is applied, where $\vec{e}_x$ and $\vec{e}_y$ are unit vectors in the Cartesian coordinate system. The orientation of the magnetic field forms an angle γ with the horizontal axis such that $\gamma=B_x/B_y$. In this study, γ equals to zero. The electric current J and the electromagnetic force F are defined by $J=\sigma\left(\vec{V}\times\vec{B}\right)$ and $F=\sigma\left(\vec{V}\times\vec{B}\right)\times\vec{B}$, respectively.

2.5.10.2 Effects of active parameters

Numerical simulations of natural convection in an enclosure filled with CuO–water nanofluid were performed using the CVFEM. Calculations are made for various

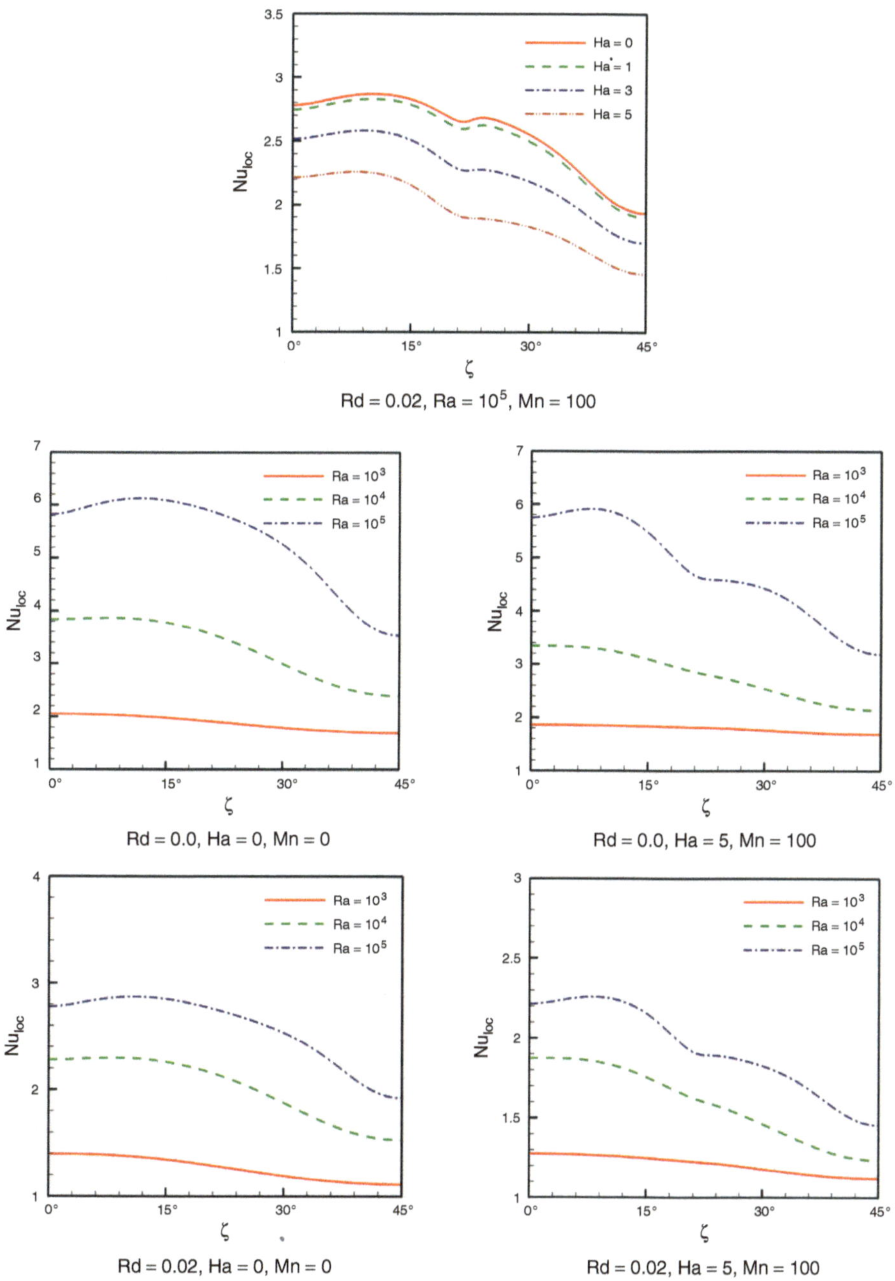

FIGURE 2.57 Effects of Magnetic Number, Hartmann Number, Radiation Parameter, and Rayleigh Number on the Local Nusselt Number Nu_{loc} Along the Hot Wall When $\phi = 0.04$

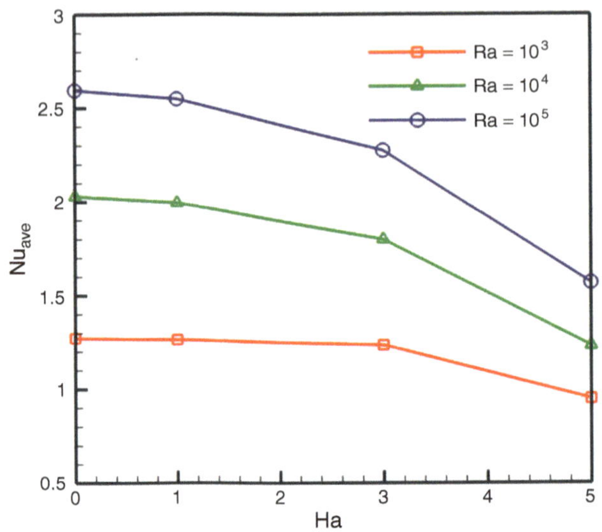

FIGURE 2.58 Effects of the Hartmann Number and Rayleigh Number on the Average Nusselt Number Nu_{ave} Along the Hot Wall When $\phi = 0.04$

values of volume fraction of nanoparticles (ϕ = 0 to 4%), Rayleigh number (Ra = 10^3, 10^4, and 10^5), dimensionless amplitude of the sinusoidal wall (a = 0.1, 0.2, and 0.3), and Hartmann number (Ha = 0–100) at a constant Prandtl number (Pr = 6.2).

Fig. 2.60 shows the effects of dimensionless amplitude of the sinusoidal wall (a) and Rayleigh number (Ra) in the absence of a magnetic field. At Ra = 10^3 for all amplitude of the sinusoidal wall, the isotherms are nearly smooth curves and nearly parallel to each other; this pattern is the characteristic of a conduction-dominant mechanism of heat transfer at low Rayleigh numbers. The space to accelerate the flow inside the cavity decreases with an increase of amplitude of the sinusoidal wall. At a = 0.3, one counter clockwise and two clockwise eddies are observed. By increasing a up to 0.3, another tiny clockwise vorticity appears. In general, as the Rayleigh number increases, the buoyancy-driven circulations inside the enclosure become stronger, as it is clear from more distortion appears in the isotherms.

Isotherm (up) and streamline (down) contours for different values of a and Ra in the presence of a magnetic field are shown in Fig. 2.61. When the magnetic field is imposed on the enclosure, the velocity field is suppressed owing to the retarding

Table 2.9 Effects of Hartmann Number, Radiation Parameter, Nanoparticle Volume Fraction, and Magnetic Number on Average Nusselt Number

		$Mn_F = 0, Ha = 0$		$Mn_F = 100, Ha = 5$		$Mn_F = 100, Ha = 0$		$Mn_F = 0, Ha = 5$	
		Rd		Rd		Rd		Rd	
Ra	ϕ	0	0.02	0	0.02	0	0.02	0	0.02
10^3	0	1.611611	0.995565	1.519592	0.928174	1.711999	1.010865	1.510564	0.953924
10^3	0.04	1.890241	1.270623	1.801949	1.215039	1.972767	1.285426	1.796937	1.234266
10^4	0	2.906977	1.609061	2.477273	1.244099	2.934773	1.636442	2.459938	1.226934
10^4	0.04	3.312842	2.022216	2.802561	1.596926	3.34569	2.048861	2.784897	1.582475
10^5	0	4.799285	2.103907	4.316693	1.570677	4.775208	2.106432	4.331907	1.564888
10^5	0.04	5.372904	2.595023	4.789146	1.942158	5.345175	2.588536	4.807063	1.940414

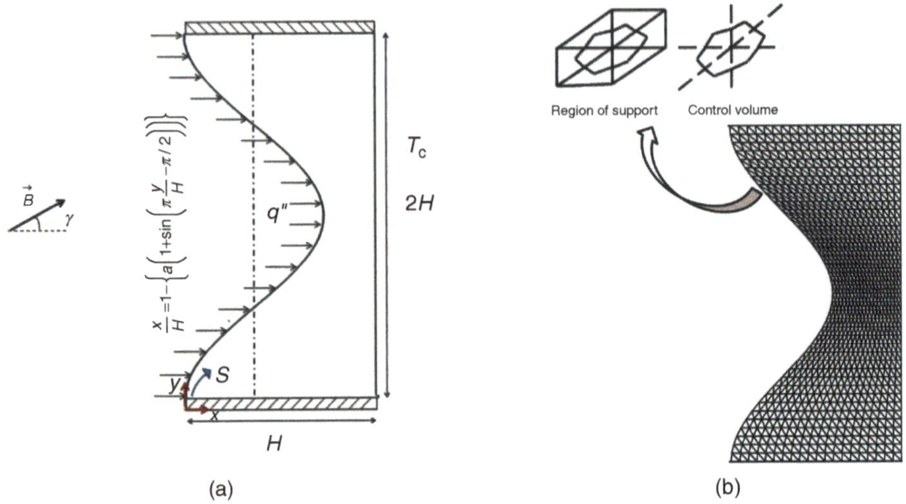

FIGURE 2.59

(a) Geometry and the boundary conditions with (b) the mesh of enclosure considered in this work; (c) A sample triangular element and its corresponding control volume.

effect of the Lorenz force. So, the intensity of convection weakens significantly. An increasing Hartmann number leads to stretch eddies along a vertical line. Also, the magnetic field makes the isotherms parallel to each other because of domination of the conduction mode of heat transfer. The corresponding polynomial representation of the Nusselt number is as follows:

$$\begin{aligned}
\mathrm{Nu} &= a_{13}+a_{23}Y_1+a_{33}Y_2+a_{43}Y_1^2+a_{53}Y_2^2+a_{63}Y_1Y_2 \\
Y_1 &= a_{11}+a_{21}\,\mathrm{Ha}^*+a_{31}\phi+a_{41}\,\mathrm{Ha}^{*2}+a_{51}\phi^2+a_{61}\,\mathrm{Ha}^*\,\phi \\
Y_2 &= a_{12}+a_{22}\,\mathrm{Ra}^*+a_{32}a+a_{42}\,\mathrm{Ra}^{*2}+a_{52}a^2+a_{62}\,\mathrm{Ra}^*\,a
\end{aligned} \tag{2.44}$$

where $\mathrm{Ha}^*=\mathrm{Ha}/100$ and $\mathrm{Ra}^*=\log(\mathrm{Ra})$. Also, a_{ij} can be found in Table 2.10; for example, a_{21} equals (−6.49034).

Figs 2.62 and 2.63 show the effects of active parameters on the average Nusselt number. Increasing the Rayleigh number is associated with an increase in the heat transfer and the Nusselt number. This is caused by stronger convective heat transfer for the higher Rayleigh number. Although the thermal boundary layer thickness increases with an increase of volume fraction of the nanoparticle, the Nusselt number increases with an augment of ϕ. As the volume fraction of the nanoparticle increases, thermal diffusivity increases. Therefore, the high values of thermal diffusivity cause the boundary thickness to increase and accordingly decrease $1/\theta$. The Nusselt number is a function of $1/\theta$ and k_{nf}/k_f. Because the reduction in $1/\theta$

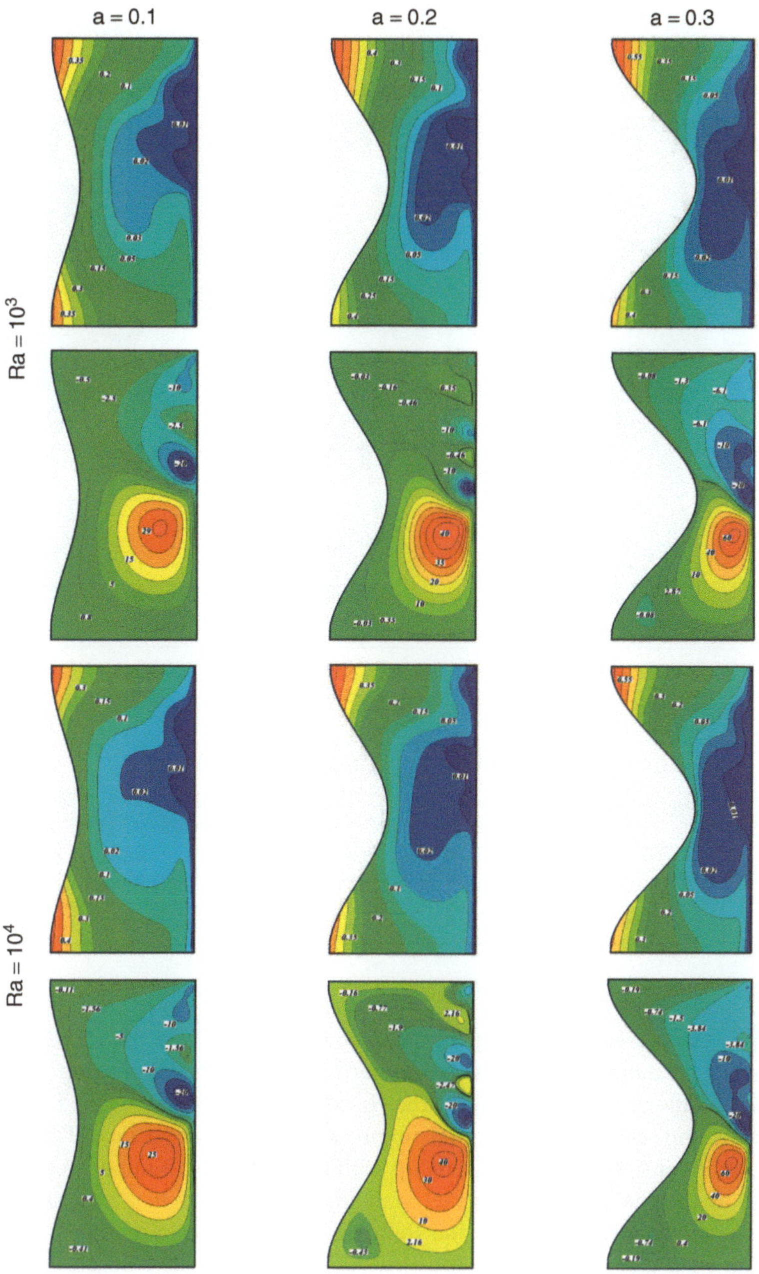

FIGURE 2.60 Isotherm (Up) and Streamline (Down) Contours for Different Values of *a* and Ra at Ha = 0, $\phi = 0.04$, and Pr = 6.2

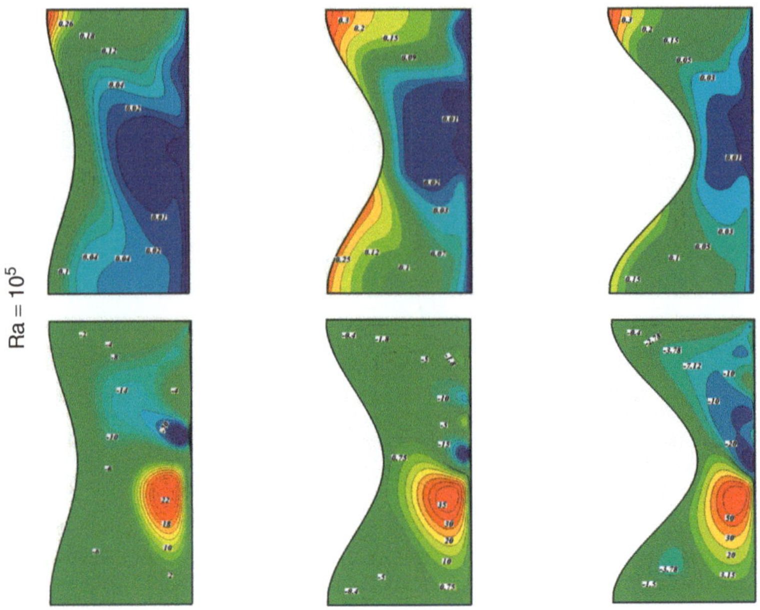

FIGURE 2.60 (*Continued*)

Table 2.10 Constant Coefficient for Using Eq. (2.44)

a_{ij}	$i = 1$	$i = 2$	$i = 3$	$i = 4$	$i = 5$	$i = 6$
$j = 1$	4.707421	−6.49034	15.71694	3.436246	0.628678	−6.63023
$j = 2$	5.70953	−2.38534	0.866596	0.402887	28.59651	−1.4327
$j = 3$	−0.45209	−0.09317	0.376291	0.017439	−0.05352	0.300014

caused by the presence of nanoparticles is much smaller than the thermal conductivity ratio, an augment in Nusselt number takes place by increasing the volume fraction of nanoparticles. Increasing the Hartmann number causes the Lorenz force to increase and leads to a substantial suppression of the convection. Therefore, the Nusselt number has reveres relationship with the Hartmann number. As the dimensionless amplitude of the sinusoidal wall increases, space to accelerate the flow inside the cavity decreases. So, the thermal boundary layer thickness decreases, and in turn $1/\theta$ increases. This is the reason of increasing the Nusselt number with enhance of a.

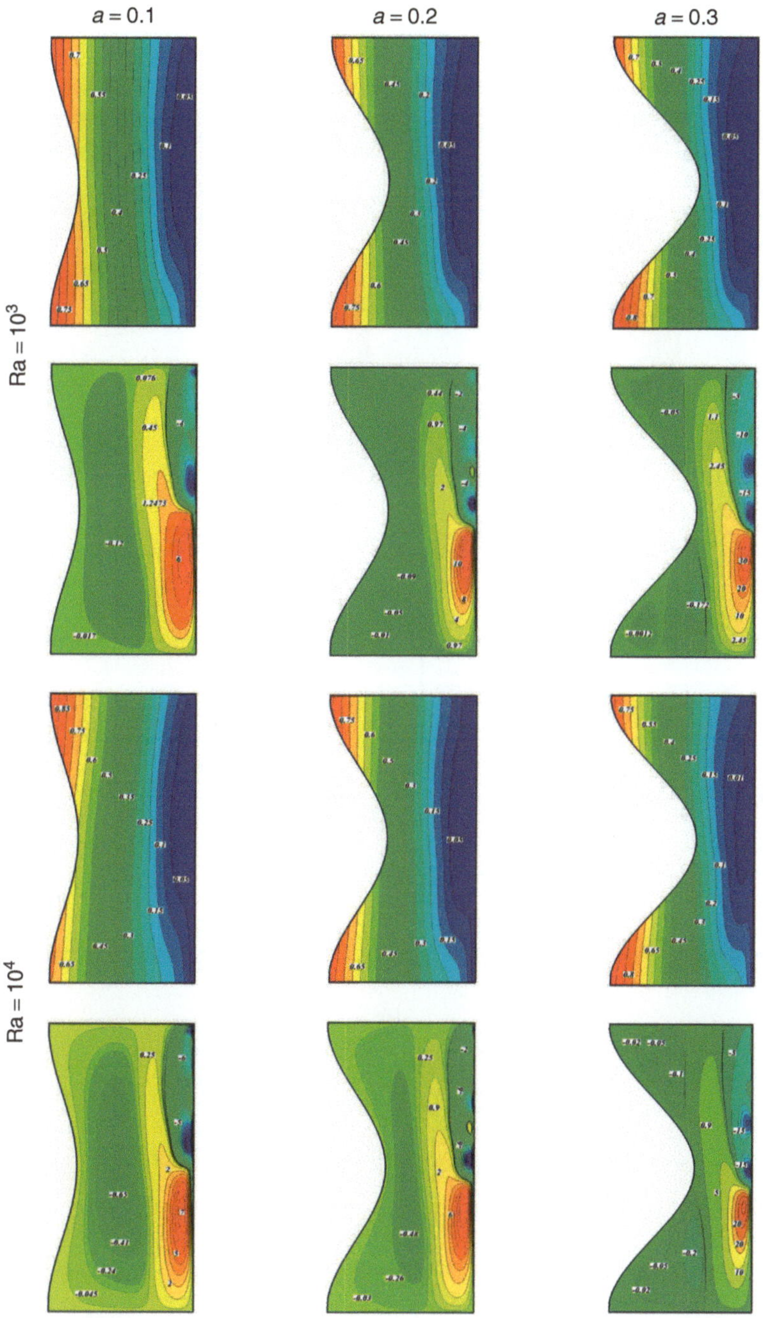

FIGURE 2.61 Isotherm (Up) and Streamline (Down) Contours for Different Values of *a* and Ra at Ha = 100, $\phi = 0.04$, and Pr = 6.2

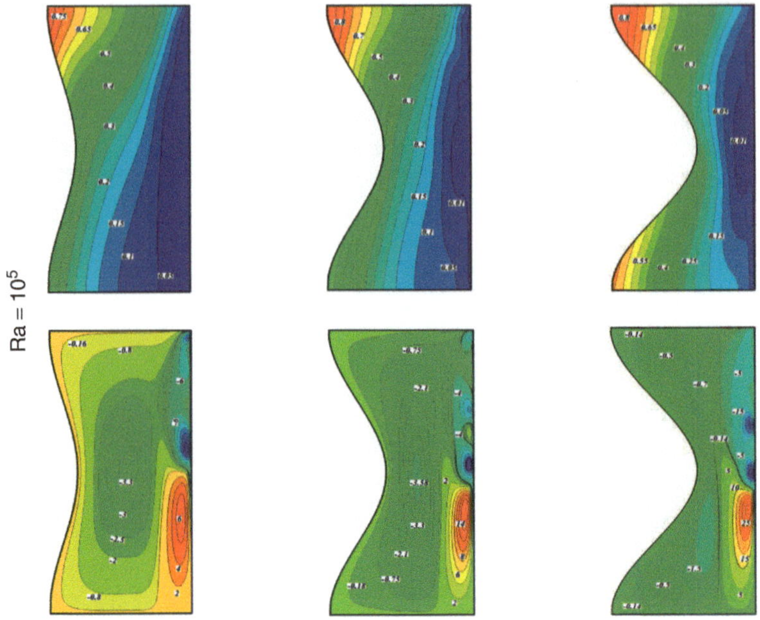

FIGURE 2.61 (***Continued***)

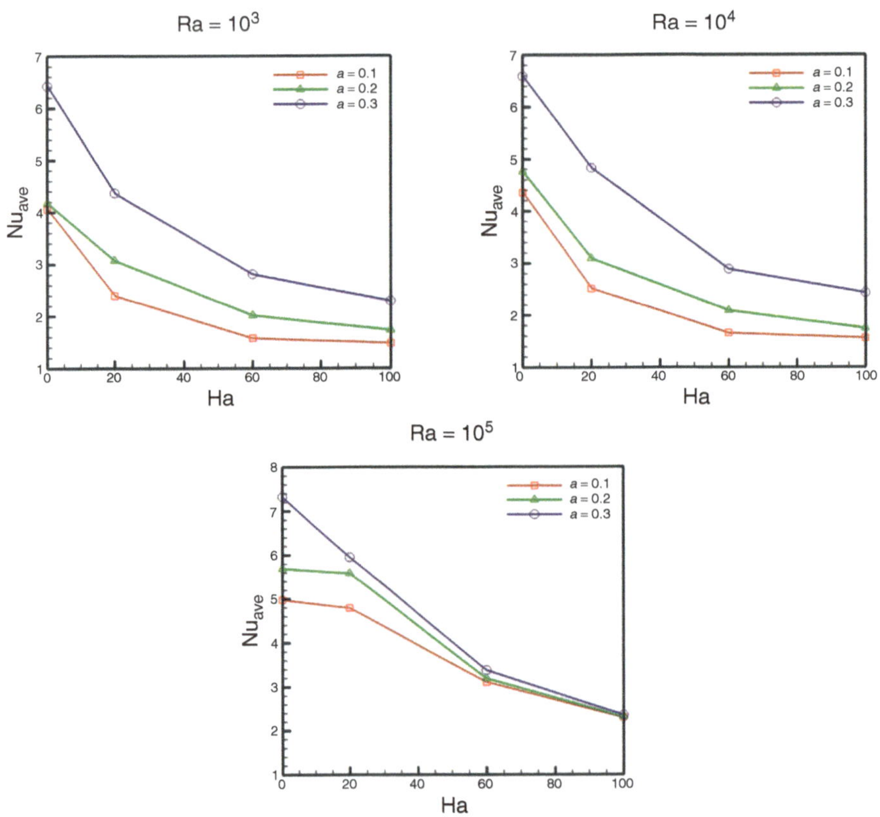

FIGURE 2.62 Variation of Average Nusselt Number Along the Wall With Constant Heat Flux for Different Values of a and Ra at Pr = 6.2 and $\phi = 0.04$

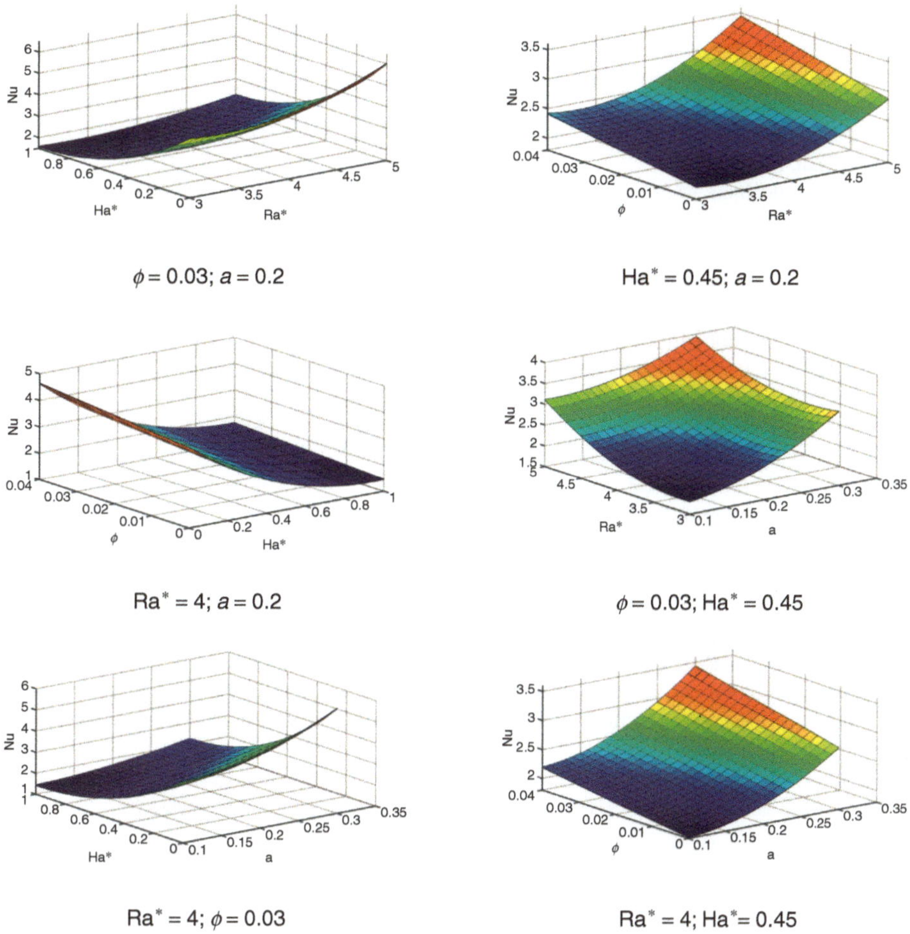

FIGURE 2.63 Variation of Nu for Various Active Parameters

REFERENCES

[1] Baliga BR. Control-volume finite element methods for fluid flow and heat transfer. In: Minkowycz WJ, Sparrow EM, editors. Advances in numerical heat transfer, vol. 1. New York: Taylor & Francis; 1996. p. 97–136.

[2] Baliga BR, Patankar SV. A control volume finite-element method for two-dimensional fluid flow and heat transfer. Numer Heat Transfer 1983;6:245–61.

[3] Soleimani S, Sheikholeslami M, Ganji DD, Gorji-Bandpay M. Natural convection heat transfer in a nanofluid filled semi-annulus enclosure. Int Commun Heat Mass Transfer 2012;39:565–74.

[4] Sheikholeslami M, Gorji-Bandpy M, Pop I, Soleimani S. Numerical study of natural convection between a circular enclosure and a sinusoidal cylinder using control volume based finite element method. Int J Therm Sci 2013;72:147–58.

[5] Sheikholeslami M, Hashim I, Soleimani S. Numerical investigation of the effect of magnetic field on natural convection in a curved-shape enclosure. Math Probl Eng 2013;2013. Article ID 831725.

[6] Sheikholeslami M, Gorji-Bandpy M, Soleimani S. Two phase simulation of nanofluid flow and heat transfer using heatline analysis. Int Commun Heat Mass Transfer 2013;47:73–81.

[7] Sheikholeslami M, Gorji-Bandpy M, Ganji DD, Soleimani S. Thermal management for free convection of nanofluid using two phase model. J Mol Liq 2014;194:179–87.

[8] Sheikholeslami M, Gorji-Bandpy M, Ganji DD, Soleimani S. Natural convection heat transfer in a cavity with sinusoidal wall filled with CuO-water nanofluid in presence of magnetic field. J Taiwan Inst Chem Eng 2014;45:40–9.

[9] Sheikholeslami M, Gorji-Bandpy M, Ganji DD, Soleimani S. MHD natural convection in a nanofluid filled inclined enclosure with sinusoidal wall using CVFEM. Neural Comput Applic 2014;24:873–82.

[10] Sheikholeslami M, Gorji Bandpy M, Ellahi R, Hassan M, Soleimani S. Effects of MHD on Cu-water nanofluid flow and heat transfer by means of CVFEM. J Magn Magn Mater 2014;349:188–200.

[11] Sheikholeslami M, Ganji DD, Gorji-Bandpy M, Soleimani S. Magnetic field effect on nanofluid flow and heat transfer using KKL model. J Taiwan Inst Chem Eng 2014;45:795–807.

[12] Sheikholeslami M, Gorji-Bandpy M, Ganji DD, Soleimani S. Natural convection heat transfer in a nanofluid filled inclined L-shaped enclosure, IJST. Trans Mech Eng 2014;38:217–26.

[13] Sheikholeslami M, Gorji-Bandpy M, Ganji DD, Soleimani S. Heat flux boundary condition for nanofluid filled enclosure in presence of magnetic field. J Mol Liq 2014;193: 174–84.

[14] Sheikholeslami M, Gorji-Bandpy M, Ganji DD, Rana P, Soleimani S. Magnetohydrodynamic free convection of Al_2O_3-water nanofluid considering thermophoresis and Brownian motion effects. Comput Fluids 2014;94:147–60.

[15] Sheikholeslami M, Gorji-Bandpy M, Ganji DD, Soleimani S. Effect of a magnetic field on natural convection in an inclined half-annulus enclosure filled with Cu–water nanofluid using CVFEM. Adv Powder Technol 2013;24:980–91.

[16] Sheikholeslami M, Ganji DD, Rashidi MM. Ferrofluid flow and heat transfer in a semi annulus enclosure in the presence of magnetic source considering thermal radiation. J Taiwan Inst Chem Eng 2015;47:6–17.

[17] Khanafer K, Vafai K, Lightstone M. Buoyancy-driven heat transfer enhancement in a two-dimensional enclosure utilizing nanofluids. Int J Heat Mass Transfer 2003;46: 3639–53.

[18] De Vahl Davis G. Natural convection of air in a square cavity, a benchmark numerical solution. Int J Numer Methods Fluids 1962;3:249–64.

[19] Rudraiah N, Barron RM, Venkatachalappa M, Subbaraya CK. Effect of a magnetic field on free convection in a rectangular enclosure. Int J Engrg Sci 1995;33(8):1075–84.

[20] Sharif MAR, Mohammad TR. Natural convection in cavities with constant flux heating at the bottom wall and isothermal cooling from the sidewalls. Int J Therm Sci 2005;44: 865–78.

[21] Kuehn TH, Goldstein RJ. An experimental and theoretical study of natural convection in the annulus between horizontal concentric cylinders. J Fluid Mech 1976;74:695–719.
[22] Labonia G, Guj G. Natural convection in a horizontal concentric cylindrical annulus: oscillatory flow and transition to chaos. J Fluid Mech 1998;375:179–202.
[23] Sheikholeslami M, Gorji-Bandpy M, Ganji DD, Soleimani S, Seyyedi SM. Natural convection of nanofluids in an enclosure between a circular and a sinusoidal cylinder in the presence of magnetic field. Int Commun Heat Mass Transfer 2012;39:1435–43.
[24] Sheikholeslami M, Rashidi MM, Ganji DD. Effect of non-uniform magnetic field on forced convection heat transfer of Fe_3O_4–water nanofluid. Comput Methods Appl Mech Engrg 2015;294:299–312.
[25] Sheikholeslami M, Rashidi MM. Effect of space dependent magnetic field on free convection of Fe_3O_4–water nanofluid. J Taiwan Inst Chem Eng 2015;56:6–15.
[26] Sheikholeslami M, Sheykholeslami FB, Khoshhal S, Mola-Abasi H, Ganji DD, Rokni Houman B. Effect of magnetic field on Cu–water nanofluid heat transfer using GMDH-type neural network. Neural Comput Applic 2014;25:171–8.
[27] Sheikholeslami M, Ganji DD. Ferrohydrodynamic and magnetohydrodynamic effects on ferrofluid flow and convective heat transfer. Energy 2014;75:400–10.
[28] Kandelousi MS. Effect of spatially variable magnetic field on ferrofluid flow and heat transfer considering constant heat flux boundary condition. Eur Phys J Plus 2014;1:129–248.
[29] Sheikholeslami M, Rashidi MM. Ferrofluid heat transfer treatment in the presence of variable magnetic field. Eur Phys J Plus 2015;130:115.

CHAPTER

New semianalytical methods: application for MHD nanofluid hydrothermal behavior

3

CHAPTER OUTLINE

3.1 INTRODUCTION

Most engineering problems, especially some heat transfer equations, are nonlinear; therefore, some of them are solved using numerical solutions, and some are solved using different analytic methods, such as the perturbation method, homotopy perturbation method (HPM), and variational iteration method introduced by He. Therefore, many different methods have recently introduced some ways to eliminate the small parameter. One of the semiexact methods that does not need small parameters is the HPM. HPM was proposed first by He in 1998 and was further developed and improved by He [1]. The method yields a very rapid convergence of the solution series in the most cases. The method yields a very rapid convergence of the solution series in the most cases. The HPM proved its capability to solve a large class of nonlinear problems efficiently, accurately, and easily with approximations convergency very rapidly to solution. Usually, few iterations lead to a highly accurate solution. This method is used for much research in engineering sciences. He's HPM was applied to obtain approximate analytical solutions for the motion of a spherical particle in a plane Couette flow by Jalal et al. [2]. Hydromagnetic flow between two horizontal plates in a rotating system, in which the lower plate was a stretching sheet and the upper is a porous solid plate, has been analyzed by Sheikholeslami et al. [3]. They reported that increasing the magnetic parameter or viscosity parameter leads to decreasing Nu; with increasing the rotation parameter, blowing velocity parameter, and

Prandtl number, the Nusselt number increases. The three-dimensional problem of steady fluid deposition on an inclined rotating disk was studied by Sheikholeslami et al. [4]. They concluded that by increasing normalized thickness, the Nusselt number increases; however, this trend is more noticeable with greater Prandtl numbers. With low Prandtl numbers, Nusselt number changes are not sensible; in larger Prandtl numbers, the changes in the Nusselt number are more noticeable.

Heat transfer of a nanofluid flow that was squeezed between parallel plates was investigated analytically using HPM by Sheikholeslami and Ganji [5]. They reported that the Nusselt number has a direct relationship with the nanoparticle volume fraction, squeeze number, and Eckert number when two plates are separated, but it has a reverse relationship with the squeeze number when two plates are squeezed.

Sheikholeslami et al. [6] studied the problem of laminar nanofluid flow in a semiporous channel. They found that that the velocity boundary layer thickness decreases with increasing Reynolds number and nanoparticle volume fraction, and it increases while the Hartmann number increases. Ashorynejad et al. [7] studied the flow and heat transfer of a nanofluid over a stretching cylinder in the presence of a magnetic field. They found that choosing copper (for a small magnetic parameter) and alumina (for large values of magnetic parameter) leads to the highest cooling performance for this problem.

Sheikholeslami et al. [8] investigated two-phase unsteady nanofluid flow and heat transfer between parallel plates in the presence of a time-dependent magnetic field. They found that the Nusselt number is an increasing function of the Hartmann number, Eckert number, and Schmidt number, but it is a decreasing function of the squeeze number. Micropolar fluid flow and heat transfer in a permeable channel has been investigated by Sheikholeslami et al. [9]. They showed that the Reynolds number and Peclet number have direct relationship with the Nusselt number and Sherwood number.

One of the powerful methods that is stronger for solving nonlinear problems without depending on the small parameter is the optimal homotopy asymptotic method (OHAM). OHAM was considered for the solution of strongly nonlinear differential equations arising in heat transfer by Haq et al. [10]. They showed that this method provides a convenient way to optimally control convergence of the approximation series and adjust convergence regions. Hashmi et al. [11] used OHAM for finding the approximate solutions of a class of Volterra integral equations with weakly singular kernels. They showed that OHAM is a reliable and efficient technique for finding the solutions of weakly singular Volterra integral equations. Sheikholeslami et al. [12] analyzed the problem of laminar viscous flow in a semiporous channel in the presence of a transverse magnetic field. Their results provide further support that OHAM is a powerful tool for solving nonlinear differential equations.

The use of the least square method in minimizing the residual of OHAM might increase the CPU execution time of the algorithm. In this work, we shall instead adopt the Galerkin technique for minimizing the residual of OHAM. We shall develop and apply the so-called Galerkin optimal homotopy asymptotic method (GOHAM) to study the unsteady two-dimensional squeezing nanofluid flow between infinite parallel plates. To the best of our knowledge, such a problem has never been studied before. Sheikholeslami and Ganji [13] studied the magnetohydrodynamic (MHD) flow in a permeable channel filled with nanofluid. They showed that that velocity boundary layer thickness decreases with an increase of the Reynolds number and nanoparticle volume fraction, and it increases as the Hartmann number increases. The problem of laminar viscous flow in a semiporous channel in the presence of transverse magnetic field was studied by Sheikholeslami et al. [14]. They showed that OHAM was a powerful approach for solving nonlinear differential equations such as this problem. Also they investigated the effects of some important parameters to evaluate how these parameters effect on this fluid behavior. Sheikholeslami et al. [15] applied GMDH to investigate heat transfer of Cu–water nanofluid over a stretching cylinder in the presence of a magnetic field. Their results indicate that GMDH-type NN in comparison with a fourth-order Runge–Kutta integration scheme. Sheikholeslami et al. [16] analyzed the MHD nanofluid flow and heat transfer between two horizontal plates in a rotating system. Their results indicated that for both suction and injection, the Nusselt number has a direct relationship with the nanoparticle volume fraction.

Another semiexact method that does not need small parameters is the differential transformation method (DTM). DTM can overcome the foregoing restrictions and limitations of perturbation methods. This method constructs an analytical solution in the form of a polynomial. It is different from the traditional higher order Taylor series method. The Taylor series method is computationally expensive for large orders. The DTM is an alternative procedure for obtaining an analytic Taylor series solution of differential equations. The main advantage of this method is that it can be applied directly to nonlinear differential equations without requiring linearization or discretization, and therefore, it is not affected by errors associated with discretization. The concept of DTM was first introduced by Zhou [17], who solved linear and nonlinear problems in electrical circuits. Chen and Ho [18] developed this method for partial differential equations, and Ayaz [19] applied it to the system of differential equations; this method is very powerful [20].

Jang et al. [21] applied the two-dimensional DTM to the solution of partial differential equations. Finally, Hassan [22] adapted the DTM to solve some problems. This method was successfully applied to various application problems [23–31]. All of these successful applications verified the validity, effectiveness, and flexibility of DTM. Natural convection of a non-Newtonian copper–water nanofluid between two infinite parallel vertical flat plates was investigated by Domairry et al. [32]. They concluded that as the nanoparticle volume fraction increases, the momentum boundary layer thickness increases, and the thermal boundary layer thickness decreases. Sheikholeslami and Ganji [33] applied DTM to solve the problem of nanofluid flow and heat transfer between parallel plates considering Brownian motion. They concluded whereas that the Nusselt number increases with an augment of nanoparticle volume fraction and Hartmann number, it decreases with an increase of the squeeze number. Hatami et al. [34] used multistep DTM to solve the problem of motion of a spherical particle in plane Couette fluid. Chen and Ju [35] used the hybrid method to solve transient advective–dispersive transport problems. Peng and Chen [36] performed a hybrid numerical technique that combines the differential transformation and finite difference method (DTM–FDM) to investigate the annular fin with temperature-dependent thermal conductivity. They showed that the convective heat transfer plays a dominant role for heat dissipation under the convection–radiation condition. Mosayebidorcheh et al. [37] applied hybrid DTM–FDM to simulate turbulent MHD Couette nanofluid flow and heat transfer. They showed that velocity profiles increase with increase of Hall parameter and thus the skin friction coefficient increases with an increase of the Hall parameter.

The Adomian decomposition method (ADM) is used to solve a wide range of physical problems. One of the semiexact methods that does not need linearization or discretization is ADM, and several modification have improved its ability [38]. An advantage of this method is that it can provide analytical approximation or an approximated solution to a rather wide class of nonlinear (and stochastic) equations without linearization, perturbation, closure approximation, or discretization methods. Unlike the common methods (i.e., weak nonlinearity and small perturbation that change the physics of the problem because of simplification), ADM gives the approximated solution of the problem without any simplification. Thus, its results are more realistic. ADM's abilities have attracted many authors to use this method for solving fluid dynamic problems. Jafari et al. [39] presented a modified ADM to solve a system of nonlinear equations, which yielded series of solution with faster accelerated convergence than the series obtained by the standard ADM. Allen and Syam [40] investigated nonhomogeneous and classical Blasius equations. Hashim [36] used ADM to solve boundary value problem (BVPs) for fourth-order integro-differential equations and the Blasius equation [41]. Kechil et al. [42] presented a nonperturbative solution of free-convective boundary-layer equation by ADM. Arslanturk [43] inspected the fins' efficiency of convective straight fins with temperature-dependent thermal conductivity using the decomposition method. Sheikholeslami et al. [44] studied the effects of magnetic field and nanoparticle on the Jeffery-Hamel flow using ADM. They show that increasing Hartmann number will lead to backflow reduction. In greater angles or Reynolds numbers, a high Hartmann number is needed for reduction of backflow. Also, the results show that the momentum boundary layer thickness increases as the nanoparticle volume fraction increases. Squeezing unsteady nanofluid flow and heat transfer has been studied by Sheikholeslami et al. [45]. They showed that for the case in which two plates are moving together, the Nusselt number increases with an increase of nanoparticle volume fraction and Eckert number, and it decreases with growth of the squeeze number.

The homotopy analysis method (HAM) was first devised in 1992 by Liao Shijun of Shanghai Jiaotong University in his PhD dissertation [46] and further modified in 1997 to introduced a nonzero auxiliary parameter, referred to as the *convergence-control parameter*, c0, to construct a homotopy on a differential system in general form [47]. The convergence-control parameter is a nonphysical variable that provides a simple way to verify and enforce convergence of a solution series. The capability of the HAM to naturally show convergence of the series solution is unusual in analytical and semianalytic approaches to nonlinear partial differential equations. Sheikholeslami et al. [48] investigated the boundary layer flow of nanofluid over a permeable stretching wall. Their results indicated that whereas an increase in the nanoparticle volume fraction decreases the momentum boundary layer thickness and entropy generation rate, the thermal boundary layer thickness increases. Sheikholeslami et al. [49] investigated the flow of nanofluid and heat transfer characteristics between two horizontal plates in a rotating system. Their results show that for suction and injection, the heat transfer rate at the surface increases with increasing the nanoparticle volume fraction, Reynolds number, and injection/suction parameter, and it decreases with power of the rotation parameter. This method was applied in several scientific problems [50–56].

3.2 NEW SEMIANALYTICAL METHODS: BASIC IDEA AND SAMPLE CODES

In this section, the basic idea of different methods is presented. To show the application of these methods to solve the nonlinear problem, the following equation with its boundary conditions is solved via these methods. Sample Maple codes are provided in the appendix.

$$f'''(\eta)+2\alpha\,\mathrm{Re}\,f(\eta)f'(\eta)+(4-\mathrm{Ha})\alpha^2 f'(\eta)=0 \tag{3.1}$$

$$f(0)=1,\ f'(0)=0,\ f(1)=0 \tag{3.2}$$

3.2.1 HOMOTOPY PERTURBATION METHOD

3.2.1.1 Basic idea of HPM

To illustrate the basic ideas of this method, we consider the following Eq. [8]:

$$A(u)-f(r)=0 \quad r\in\Omega \tag{3.3}$$

with the boundary condition of:

$$B\left(u,\frac{\partial u}{\partial n}\right)=0 \quad r\in\Gamma, \tag{3.4}$$

where A is a general differential operator, B is a boundary operator, $f(r)$ is a known analytical function, and Γ is the boundary of the domain Ω.

A can be divided into two parts, which are L and N, where L is linear and N is nonlinear. Eq. (3.3) can therefore be rewritten as follows:

$$L(u)+N(u)-f(r)=0, \quad r\in\Omega \tag{3.5}$$

The homotopy perturbation structure is shown as follows:

$$H(v,p)=(1-p)\left[L(v)-L(u_0)\right]+p\left[A(v)-f(r)\right]=0 \tag{3.6}$$

where,

$$v(r,p):\Omega\times[0,1]\to R \tag{3.7}$$

In Eq. (3.6), $p \in [0,1]$ is an embedding parameter, and u_0 is the first approximation that satisfies the boundary condition. We can assume that the solution of Eq. (3.6) can be written as a power series in p, as following:

$$v = v_0 + pv_1 + p^2v_2 + ... \tag{3.8}$$

and the best approximation for solution is:

$$u = \lim_{p\to 1} v = v_0 + v_1 + v_2 + ... \tag{3.9}$$

3.2.1.2 Application of HPM in an example providing Maple code

Using HPM, suppose the solution of Eq. (3.1) has the form:

$$(1-p)(f''' - f_0''') + H_1(P)(f'''(\eta) + 2\alpha \operatorname{Re} f(\eta)f'(\eta) + (4-\mathrm{Ha})\alpha^2 f'(\eta)) = 0 \tag{3.10}$$

We consider f as follows:

$$f(\eta) = f_0(\eta) + f_1(\eta) + ... = \sum_{i=0}^{n} f_i(\eta) \tag{3.11}$$

with substituting f from Eq. (3.11) to Eq. (3.10) and some simplification and rearranging based on powers of p terms, we have:

$$\begin{aligned} p^0:\ & f''' = 0, \\ & f_0(0) = 1,\ f_0'(0) = 0,\ f_0(1) = 0. \end{aligned} \tag{3.12}$$

$$\begin{aligned} & p^1: \\ & f_1''' + 2\alpha \operatorname{Re} f_0' f_0 - f_0''' - \alpha^2 \mathrm{Ha}\, f_0' + 4\alpha^2 f_0' = 0, \\ & f_1(0) = 0,\ f_1'(0) = 0,\ f_1(1) = 0. \end{aligned} \tag{3.13}$$

Solving Eqs. (3.12) and (3.13) with boundary conditions, we have

$$f_0(\eta) = -\eta^2 + 1, \tag{3.14}$$

$$\begin{aligned} f_1(\eta) = (&-0.3333\alpha \operatorname{Re}\eta^6 - 0.0833\alpha^2 \mathrm{Ha}\,\eta^4 + 0.3333\alpha^2\eta^4 \\ &+ 0.1667\,\alpha \operatorname{Re}\eta^4 - 0.1333\alpha \operatorname{Re}\eta^2 \\ &+ 0.0833\,\alpha\, \mathrm{Ha}\,\eta^2 - 0.3333\alpha^2\eta^2) \end{aligned} \tag{3.15}$$

The solution of this equation, when $p \to 1$, will be as follows:

$$f(\eta) = f_0(y) + f_1(y) + f_2(y) + ... \tag{3.16}$$

Appendix A3 presents sample code for this method.

3.2.2 GALERKIN OPTIMAL HOMOTOPY ANALYSIS METHOD

3.2.2.1 Basic idea of GOHAM

The following differential equation is considered:

$$L(u(t))+N(u(t))+g(t)=0, \quad B(u)=0, \tag{3.17}$$

where L is a linear operator, τ is an independent variable, $u(t)$ is an unknown function, $g(t)$ is a known function, $N(u(t))$ is a nonlinear operator, and B is a boundary operator. By means of OHAM, one first constructs a set of equations [13]:

$$\begin{aligned}&(1-p)[L(\phi(\tau,p))+g(\tau)]-H(p)[L(\phi(\tau,p))+g(\tau)+N(\phi(\tau,p))]=0\\&\mathrm{B}(\phi(\tau,p))=0,\end{aligned} \tag{3.18}$$

where $p \in [0,1]$ is an embedding parameter, $H(p)$ denotes a nonzero auxiliary function for $p \neq 0$ and $H(0) = 0$, and $\phi(\tau, p)$ is an unknown function. Obviously, when $p = 0$ and $p = 1$, it holds that:

$$\phi(\tau,0)=u_0(\tau), \quad \phi(\tau,1)=u(\tau). \tag{3.19}$$

Thus, as p increases from 0 to 1, the solution $\phi(\tau, p)$ varies from $u_0(\tau)$ to the solution $u(\tau)$, where $u_0(\tau)$ is obtained from Eq. (3.18) for $p = 0$:

$$L(u_0(\tau))+g(\tau)=0, \quad B(u_0)=0. \tag{3.20}$$

We choose the auxiliary function $H(p)$ in the form:

$$H(p)=pC_1+p_2C_2+\ldots \tag{3.21}$$

where C_1, C_2, and so on are constants that can be determined later.

Expanding $\phi(\tau, p)$ in a series with respect to p, one has:

$$\phi(\tau,p,C_i)=u_0(\tau)+\sum_{k\geq 1}u_k(\tau,C_i)p_k, \quad i=1,2,\ldots \tag{3.22}$$

Substituting Eq. (3.22) into Eq. (3.18), collecting the same powers of p, and equating each coefficient of p to zero, we obtain set of differential equation with boundary conditions. Solving differential equations by boundary conditions $u_0(\tau)$, $u_1(\tau, C_1)$, $u_2(\tau, C_2)$,... are obtained. Generally speaking, the solution of Eq. (3.17) can be determined approximately in the form:

$$\tilde{u}^{(m)}=u_0(\tau)+\sum_{k=1}^{m}u_k(\tau,C_i). \tag{3.23}$$

Note that the last coefficient C_{m} can be a function of τ. Substituting Eq. (3.18) into Eq. (3.17), the following residual results:

$$R(\tau,C_i)=L(\tilde{u}^{(m)}(\tau,C_i))+g(\tau)+N(\tilde{u}^{(m)}(\tau,C_i)). \tag{3.24}$$

If $R(\tau, C_i) = 0$, then $\tilde{u}^{(m)}(\tau, C_i)$ happens to be the exact solution. Generally, such a case will not arise for nonlinear problems, but we can minimize the functional by the Galerkin method:

$$w_i = \frac{\partial R(\tau, C_1, C_2, ..., C_m)}{\partial C_i}, \quad i = 1, 2, ..., m, \tag{3.25}$$

The unknown constants $C_i(i = 1,2,\ldots,m)$ can be identified from the conditions:

$$J(C_1, C_2) = \int_a^b w_i \,.\, R(\tau, C_1, C_2, ..., C_m)\, d\tau = 0, \tag{3.26}$$

where a and b are two values, depending on the given problem. With these constants, the approximate solution (of order m) (Eq. (3.23)) is well determined. It can be observed that the method proposed in this work generalizes these two methods using the special (more general) auxiliary function $H(p)$.

3.2.2.2 Application of GOHAM in an example providing Maple code

In this section, GOHAM is applied to nonlinear ordinary differential Eq. (3.1) According to GOHAM, applying Eq. (3.18) to Eq. (3.1):

$$\begin{aligned}&(1-p)(f'''(\eta) - f_0'''(\eta)) + H_1(p)(f'''(\eta)\\&+2\alpha \operatorname{Re} f(\eta) f'(\eta) + (4 - \mathrm{Ha})\alpha^2 f'(\eta)) = 0\end{aligned} \tag{3.27}$$

where primes denote differentiation with respect to η.

We consider f and $H_1(p)$ as follows:

$$f = f_0 + p f_1 + p^2 f_2, \quad H_1(p) = pC_1 + p^2 C_2. \tag{3.28}$$

Substituting f and $H_1(p)$ from Eq. (3.28) into Eq. (3.27) and using some simplification and rearranging based on powers of p-terms, we have:

$$\begin{aligned}p^0 :\ & f''' = 0,\\ & f_0(0) = 1,\ f_0'(0) = 0,\ f_0(1) = 0.\end{aligned} \tag{3.29}$$

$$\begin{aligned}p^1 :&\\ & f_1''' + C_1 f_0''' + 2C_1 \alpha \operatorname{Re} f_0' f_0 - f_0''' - C_1 \alpha^2 \mathrm{Ha}\, f_0' + 4C_1 \alpha^2 f_0' = 0,\\ & f_1(0) = 0,\ f_1'(0) = 0,\ f_1(1) = 0.\end{aligned} \tag{3.30}$$

Solving Eqs. (3.29) and (3.30) with boundary conditions:

$$f_0(\eta) = -\eta^2 + 1, \tag{3.31}$$

$$\begin{aligned}f_1(\eta) = C_1\, (&-0.3333\alpha \operatorname{Re} \eta^6 - 0.0833\alpha^2\, \mathrm{Ha}\, \eta^4 + 0.3333\alpha^2 \eta^4\\ &+ 0.1667\,\alpha \operatorname{Re} \eta^4 - 0.1333\alpha \operatorname{Re} \eta^2\\ &+ 0.0833\,\alpha\, \mathrm{Ha}\, \eta^2 - 0.3333\alpha^2\, \eta^2)\end{aligned} \tag{3.32}$$

The term $f_2(\eta)$ is so large that is mentioned graphically. Therefore, the final expression for $f(\eta)$ is:

$$f(\eta) = f_0(\eta) + f_1(\eta) + f_2(\eta) \tag{3.33}$$

From Eq. (3.24), by substituting $f(\eta)$ into Eq. (3.27), $R(\eta, C_{11}, C_{12})$ are obtained, and J is obtained in the following manner:

$$J(C_{11}, C_{12}) = \int_a^b w_i \,.\, R(\tau, C_1, C_2, \ldots, C_m)\, d\tau = 0 \tag{3.34}$$

The constants C_1, C_2 are obtained from Eq. (3.34). By constants C_1 and C_2 into Eq. (3.33), an expression for $f(\eta)$ is obtained. Appendix A4 presents sample code for this method.

3.2.3 DIFFERENTIAL TRANSFORMATION METHOD

3.2.3.1 Basic idea of DTM

We suppose $\chi(\tau)$ to be an analytic function in a domain D, and $\tau = \tau_i$ represent any point in D. The function $\chi(\tau)$ is then represented by one power series whose center is located at τ_i. The Taylor series expansion function of $\chi(\tau)$ is of the form [33]:

$$\chi(\tau) = \sum_{k=0}^{\infty} \frac{(\tau - \tau_i)^k}{k!} \left[\frac{d^k \chi(\tau)}{dt^k} \right]_{\tau=\tau_i} \quad \forall \tau \in D \tag{3.35}$$

The particular case of Eq. (3.35) when $\tau = \tau_i$ is referred to as the Maclaurin series of $\chi(\tau)$ and is expressed as:

$$\chi(\tau) = \sum_{k=0}^{\infty} \frac{\tau^k}{k!} \left[\frac{d^k \chi(\tau)}{dt^k} \right]_{\tau=0} \quad \forall \tau \in D \tag{3.36}$$

As explained in the differential transformation of the function, $\chi(\tau)$ is defined as follows:

$$\chi(k) = \sum_{k=0}^{\infty} \frac{H^k}{k!} \left[\frac{d^k \chi(\tau)}{dt^k} \right]_{\tau=0} \tag{3.37}$$

where $\chi(\tau)$ is the original function and $\chi(\tau)$ is the transformed function. The differential spectrum of $\chi(\tau)$ is confined within the interval $\tau \in [0,H]$, where H is a constant. The differential inverse transform of $\chi(\tau)$ is defined as follows:

$$\chi(\tau) = \sum_{k=0}^{\infty} \left(\frac{\tau}{H} \right)^k \chi(k) \tag{3.38}$$

It is clear that the concept of the differential transformation is based on the Taylor series expansion. The values of function $\chi(\tau)$ at values of argument k are referred to as discrete, that is, $\chi(0)$ is known as the zero discrete, $\chi(1)$ as the first discrete, and so on. The more discrete available, the more precisely it is possible to restore the unknown function. The function $\chi(\tau)$ consists of the T-function $\chi(k)$, and its value is given by the sum of the T-function with $(\tau/H)^k$ as its coefficient. In real applications, at the right choice of constant H, the larger values of argument k, the discrete of spectrum reduces rapidly. Mathematical operations performed by DTM are listed in Table 3.1.

3.2.3.2 Application of DTM in an example providing Maple code

Now we apply DTM into Eq. (3.1). Taking the differential transform of Eq. (3.37) with respect to χ and considering $H = 1$ gives:

$$(k+1)(k+2)(k+3)F(k+3)$$
$$+2\alpha \operatorname{Re}\sum_{r=0}^{k}[(k-r+1)F(r)F(k-r+1)]+(4-\mathrm{Ha})\alpha^2(k+1)F(k+1)=0 \tag{3.39}$$

$$F(0)=1, F(1)=0, F(2)=\beta, \tag{3.40}$$

where $F(k)$ is the differential transformation of $F(\chi)$ and β is a constant that can be obtained through boundary condition, Eq. (3.2).

$$f(1)=0 \quad \text{or} \quad \sum_{k=0}^{N}F(k)=0 \tag{3.41}$$

Table 3.1 Some of the Basic Operations of the Differential Transformation Method

Original Function	Transformed Function
$f(\eta)=\alpha g(\eta)\pm\beta h(\eta)$	$F[k]=\alpha G[k]\pm\beta H[k]$
$f(\eta)=\dfrac{d^n g(\eta)}{d\eta^n}$	$F[k]=\dfrac{(k+n)!}{k!}G[k+n]$
$f(\eta)=g(\eta)h(\eta)$	$F[k]=\sum_{m-0}^{k}F[m]H[k-m]$
$f(\tau)=\sin(\varpi\eta+\alpha)$	$F[k]=\dfrac{\varpi^k}{k!}\sin\left(\dfrac{\pi k}{2}+\alpha\right)$
$f(\tau)=\cos(\varpi\eta+\alpha)$	$F[k]=\dfrac{\varpi^k}{k!}\cos\left(\dfrac{\pi k}{2}+\alpha\right)$
$f(\eta)=e^{\lambda\eta}$	$F[k]=\dfrac{\lambda^k}{k!}$
$F(\eta)=(1+\eta)^m$	$F[k]=\dfrac{m(m-1)\ldots(m-k+1)}{k!}$
$f(\eta)=\eta^m$	$F[k]=\delta(k-m)=\begin{cases}1, k=m\\ 0, k\neq m\end{cases}$

This problem can be solved for different values of H:

$$F(0)=1, F(1)=0, F(2)=\beta, F(3)=0$$
$$F(4)=\frac{-1}{6}\alpha\,\mathrm{Re}\,\beta-\frac{1}{3}\alpha^2\beta+\frac{1}{12}\alpha^2\beta\,\mathrm{Ha} \tag{3.42}$$
$$F(5)=0$$

The above process is continuous. Substituting Eq. (3.42) into the main equation based on DTM, it can be obtained that the closed form of the solutions is:

$$F(\eta)=1+\beta\eta^2-\left(\frac{-1}{6}\alpha\,\mathrm{Re}\,\beta-\frac{1}{3}\alpha^2\beta+\frac{1}{12}\alpha^2\beta\right)\eta^4+\cdots \tag{3.43}$$

To obtain the value of β, we substitute the boundary condition from Eq. (3.2) into Eq. (3.43) in point $\eta=1$. By substituting the obtained β into Eq. (3.43), we can find the expressions of $F(\eta)$.

Appendix A5 presents sample code for this method.

3.2.4 ADOMIAN DECOMPOSITION METHOD

3.2.4.1 Basic idea of ADM

Consider equation $F\,u(t)=g(t)$, where F represents a general nonlinear ordinary or partial differential operator including both linear and nonlinear terms. The linear terms are decomposed into $L+R$, where L is easily invertible (usually the highest order derivative) and R is the remained of the linear operator. Thus, the equation can be written as [44]:

$$\mathrm{Lu}+\mathrm{Nu}+\mathrm{Ru}=g \tag{3.44}$$

where Nu indicates the nonlinear terms. By solving this equation for Lu, since L is invertible, we can write:

$$L^{-1}Lu=L^{-1}g-L^{-1}Ru-L^{-1}Nu \tag{3.45}$$

If L is a second-order operator, L^{-1} is a twofold indefinite integral. By solving Eq. (3.44), we have:

$$u=A+Bt+L^{-1}g-L^{-1}Ru-L^{-1}Nu \tag{3.46}$$

Where A and B are constants of integration and can be found from the boundary or initial conditions. Adomian method assumes the solution u can be expanded into infinite series as:

$$u=\sum_{n=0}^{\infty}u_n \tag{3.47}$$

Also, the nonlinear term Nu will be written as:

$$\mathrm{Nu}=\sum_{n=0}^{\infty}A_n \tag{3.48}$$

Where A_n are the special Adomian polynomials. By specified A_n, next component of u can be determined:

$$u_{n+1} = L^{-1}\sum_{n=0}^{n} A_n \tag{3.49}$$

Finally, after some iteration and getting sufficient accuracy, the solution can be expressed by Eq. (3.46). In Eq. (3.47), the Adomian polynomials can be generated by several means. Here we used the following recursive formulation:

$$A_n = \frac{1}{n!}\left[\frac{d^n}{d\lambda^n}\left[N\left(\sum_{i=0}^{n}\lambda^i u_i\right)\right]\right]_{\lambda=0}, \qquad n = 0,1,2,3,... \tag{3.50}$$

Since the method does not resort to linearization or assumption of weak nonlinearity, the solution generated is in general more realistic than those achieved by simplifying the model of the physical problem.

3.2.4.2 Application of ADM in an example providing Maple code

According to Eq. (3.33), Eq. (3.1) must be written as following:

$$Lf = -2\,\mathrm{Re}\,\alpha\, f\, f' - (4 - \mathrm{Ha})\,\alpha^2 f' \tag{3.51}$$

where the differential operator L is given by $L = d^3/d\eta^3$. Assume the inverse of the operator L exists and it can be integrated from 0 to η, that is, $L^{-1} = \int_0^{\eta}\int_0^{\eta}\int_0^{\eta}(\bullet)\,d\eta\, d\eta\, d\eta$.

Operating with L^{-1} on Eq. (3.51) and after exerting boundary condition on it, we have:

$$f(\eta) = f(0) + f'(0)\,\eta + f''(0)\frac{\eta^2}{2} + L^{-1}(Nu) \tag{3.52}$$

Where

$$\mathrm{Nu} = -2\,\mathrm{Re}\,\alpha\, f(\eta)\, f'(\eta) - (4 - \mathrm{Ha}) f'(\eta) \tag{3.53}$$

ADM introduced the following expression:

$$f(\eta) = \sum_{m=0}^{\infty} f_m(\eta) \tag{3.54}$$

$$f(\eta) = \sum_{m=0}^{\infty} f_m = f_0 + L^{-1}(\mathrm{Nu}) \tag{3.55}$$

To determine the components of $f_m(\eta)$ the $f_0(\eta)$ is defined by applying the boundary condition of Eq. (3.1), and by assuming $f''(0) = \beta$.

$$f_0(\eta) = 1 + \beta\frac{\eta^2}{2} \tag{3.56}$$

$$f_1(\eta)=\left(\frac{-1}{120}\right)\alpha\,\mathrm{Re}\ \beta^2\eta^6-\frac{1}{4}\left(\frac{1}{3}\alpha\,\mathrm{Re}\,\beta\right.$$
$$\left.+\frac{1}{6}(4-\mathrm{Ha})\alpha^2\beta\right)\eta^4-\frac{1}{120}(4-\mathrm{Ha})\alpha^2\beta\,\eta^5 \tag{3.57}$$
$$-\frac{1}{6}(4-\mathrm{Ha})\alpha^2\eta^3$$

$$f_2(\eta)=\left(\frac{1}{10800}\right)\alpha^2(\mathrm{Re})^2\beta^3\eta^{10}+\frac{1}{7560}\alpha^3\,\mathrm{Re}(4-\mathrm{Ha})\beta^2\eta$$
$$-\frac{1}{8}\left(\frac{1}{21}\alpha\,\mathrm{Re}\left(-\frac{1}{12}\alpha\,\mathrm{Re}\ \beta-\frac{1}{24}(4-\mathrm{Ha})\alpha^2\beta\right)\beta\right.$$
$$-\frac{1}{420}\alpha^2(\mathrm{Re})^2+\frac{1}{42}\alpha^2\,\mathrm{Re}\,\beta\left(-\frac{1}{3}\alpha\,\mathrm{Re}\ \beta\right.$$
$$\left.\left.-\frac{1}{6}(4-\mathrm{Ha})\alpha^2\beta\right)-\frac{1}{840}\alpha^3\,\mathrm{Re}.A^*(4-\mathrm{Ha})\beta^2\right)\eta^8$$
$$-\frac{1}{7}\left(-\frac{11}{360}\alpha^3\,\mathrm{Re}(4-\mathrm{Ha})\beta-\frac{1}{720}(4-\mathrm{Ha})^2\alpha^4\beta\right)\eta^7$$
$$-\frac{1}{6}\left(\frac{1}{10}\alpha\,\mathrm{Re}\left(-\frac{1}{3}\alpha\,\mathrm{Re}\,\beta-\frac{1}{6}(4-\mathrm{Ha})\alpha^2\beta\right)\right. \tag{3.58}$$
$$\left.+\frac{1}{20}(4-\mathrm{Ha})\alpha^2\left(-\frac{1}{3}\alpha\,\mathrm{Re}\,\beta-\frac{1}{6}(4-\mathrm{Ha})\alpha^2\beta\right)\right)\eta^6$$
$$-\frac{1}{5}\left(-\frac{1}{12}\alpha^3\,\mathrm{Re}(4-\mathrm{Ha})-\frac{1}{720}(4-\mathrm{Ha})^2\alpha^4\right)\eta^5$$
$$+\frac{1}{40320}(4-\mathrm{Ha})^2\alpha^4\beta\eta^8-\frac{1}{120}(4-\mathrm{Ha})\alpha^2\left(-\frac{1}{12}\alpha\,\mathrm{Re}\,\beta\right.$$
$$\left.-\frac{1}{24}(4-\mathrm{Ha})\alpha^2\beta\right)\eta^7+\frac{1}{720}(4-\mathrm{Ha})^2\alpha^4\eta^6$$

$f_3(\eta), f_4(\eta),\ldots$ can be determined in similar way from Eq. (3.55).

Using $f(\eta)=\sum_{m=0}^{\infty} f_m(\eta)=f_0(\eta)+f_1(\eta)+f_2(\eta)+f_3(\eta)+\ldots.$, thus

$$f(\eta)=1+\beta\frac{\eta^2}{2}+\left(\frac{-1}{120}\right)\alpha\,\mathrm{Re}.A^*(1-\phi)^{2.5}\ \beta^2\eta^6-\frac{1}{4}\left(\frac{1}{3}\alpha\,\mathrm{Re}.A^*(1-\phi)^{2.5}\ \beta\right.$$
$$\left.+\frac{1}{6}(4-(1-\phi)^{1.25}\ \mathrm{Ha})\alpha^2\ \beta\right)\eta^4-\frac{1}{120}(4-(1-\phi)^{1.25}\ \mathrm{Ha})\alpha^2\beta\,\eta^5 \tag{3.59}$$
$$-\frac{1}{6}(4-\mathrm{Ha})\alpha^2\eta^3+\ldots$$

According to Eq. (3.59), the accuracy of ADM solution increases by increasing the number of solution terms (m). For the complete solution of Eq. (3.27), β should be determined, with boundary condition of $f(\eta)=0$. Appendix A6 presents sample code for this method.

3.2.5 HOMOTOPY ANALYSIS METHOD

3.2.5.1 Basic idea of HAM

Let us assume the following nonlinear differential equation in the form of:

$$N[u(\tau)]=0, \tag{3.60}$$

where N is a nonlinear operator, τ is an independent variable, and $u(\tau)$ is the solution of the equation. We define the function $\phi(\tau, p)$ as follows:

$$\lim_{p\to 0} \phi(\tau, p)=u_0(\tau) \tag{3.61}$$

where $p \in [0,1]$ and $u_0(\tau)$ is the initial guess that satisfies the initial or boundary condition,

$$\lim_{p\to 1} \phi(\tau, p)=u_0(\tau) \tag{3.62}$$

And by using the generalized homotopy method, Liao's so-called zeroth-order deformation equation will be [48]:

$$(1-p)L[\phi(\tau, p)-u_0(\tau)]=p\hbar H(\tau)\,N[\phi(\tau, p)] \tag{3.63}$$

where $\hbar$ is the auxiliary parameter that helps us increase the results convergence, $H(\tau)$ is the auxiliary function, and L is the linear operator. It should be noted that there is a great freedom to choose the auxiliary parameter $\hbar$, the auxiliary function $H(\tau)$, the initial guess $u_0(\tau)$, and the auxiliary linear operator L. This freedom plays an important role in establishing the keystone of validity and flexibility of HAM as shown in this chapter. Thus, when p increases from 0 to 1, the solution $\phi(\tau, p)$ changes between the initial guess $u_0(\tau)$ and the solution $u(\tau)$. The Taylor series expansion of $\phi(\tau, p)$ with respect to p is:

$$\phi(\tau, p)=u_0(\tau)+\sum_{m=1}^{\infty} u_m(\tau)p^m \tag{3.64}$$

and

$$u_0^{[m]}(\tau)=\left.\frac{\partial^m \phi(\tau, p)}{\partial p^m}\right|_{p=0} \tag{3.65}$$

where $u_0^{[m]}(\tau)$ for brevity is called the mth order of deformation derivation, which reads:

$$u_m(\tau)=\frac{u_0^{[m]}}{m!}=\frac{1}{m!}\left.\frac{\partial^m \phi(\tau, p)}{\partial p^m}\right|_{p=0} \tag{3.66}$$

It's clear that if the auxiliary parameter is $\hbar = -1$ and the auxiliary function is determined to be $H(\tau) = 1$, Eq. (3.63) will be:

$$(1-p)L[\phi(\tau,p)-u_0(\tau)]+p(\tau)N[\phi(\tau,p)]=0 \tag{3.67}$$

This statement is commonly used in the HAM procedure. Indeed, in HAM, we solve the nonlinear differential equation by separating any Taylor expansion term. Now we define the vector of:

$$\overrightarrow{u_m} = \{\overrightarrow{u_1}, \overrightarrow{u_2}, \overrightarrow{u_3}, \ldots, \overrightarrow{u_n}\} \tag{3.68}$$

According to the definition in Eq. (3.66), the governing equation and the corresponding initial condition of $u_m(\tau)$ can be deduced from zeroth-order deformation Eq. (3.60). Differentiating Eq. (3.60) for m times with respect to the embedding parameter p and setting $p = 0$ and finally dividing by $m!$, we will have the so-called mth order deformation equation in the form

$$L[u_m(\tau)-\chi_m u_{m-1}(\tau)] = \hbar H(\tau) R(\vec{u}_{m-1}) \tag{3.69}$$

where

$$R(\vec{u}_{m-1}) = \frac{1}{(m-1)!} \left. \frac{\partial^m \phi(\tau,p)}{\partial p^m} \right|_{p=0} \tag{3.70}$$

and

$$\chi_m = \begin{cases} 0 & m \le 1 \\ 1 & m > 1 \end{cases} \tag{3.71}$$

3.2.5.2 Application of HAM in an example providing Maple code

For HAM solutions of the governing equations, we choose the initial approximations of $f(\eta)$ as follows:

$$f_0(\eta) = 1-\eta^2 \tag{3.72}$$

and the auxiliary linear operators are

$$L(f) = f''' \tag{3.73}$$

These auxiliary linear operators satisfy

$$L(C_3 + C_2\eta + 0.5C_1\eta^2) \tag{3.74}$$

where C_i (i = 1,2,3) are constants. Introducing a nonzero auxiliary parameter $\hbar$, we develop the zeroth-order deformation problems as follows:

$$(1-p)L[f(\eta;p)-f_0(\eta)] = p\hbar N[f(\eta;p)] \tag{3.75}$$

$$f(0;p)=1,\ f'(0;p)=0,\ f(1;p)=0 \tag{3.76}$$

where nonlinear operators N_1 and N_2 are defined as

$$N_1[f(\eta;p),\theta(\eta;p)]=\frac{\partial^3 f(\eta;p)}{\partial\eta^3}+2\alpha \mathrm{Re} f(\eta;p)\frac{\partial f(\eta;p)}{\partial\eta}+(4-\mathrm{Ha})\alpha^2\frac{\partial f(\eta;p)}{\partial\eta} \tag{3.77}$$

For $p = 0$ and $p = 1$, respectively, we have:

$$f(\eta;0)=f_0(\eta) \qquad f(\eta;1)=f(\eta) \tag{3.78}$$

As p increases from 0 to 1, $f(\eta;\ p)$ from $f_0(\eta)$ to $f(\eta)$. By Taylor's theorem, $f(\eta)$ can be expanded in a power series of p as follows:

$$f(\eta;p)=f_0(\eta)+\sum_{m=1}^{\infty}f_m(\eta)p^m \tag{3.78}$$

$$f_m(\tau)=\frac{1}{m!}\frac{\partial^m f(\eta,p)}{\partial p^m} \tag{3.79}$$

in which $\hbar$ are chosen in such a way that these series are convergent at $p = 1$. Convergence of the series (3.78) depends on the auxiliary parameter $\hbar$.

Assume that $\hbar$ is selected such that the series (3.78) is convergent at $p = 1$. Then from Eq. (3.78), we have:

$$f(\eta)=f_0(\eta)+\sum_{m=1}^{\infty}f_m(\eta) \tag{3.80}$$

Differentiating the zeroth-order deformation Eq. (3.75) m times with respect to p and then dividing them by $m!$ and finally setting $p = 0$, we have the following mth-order deformation problem:

$$L_1[f_m(\eta)-\chi_m f_{m-1}(\eta)]=\hbar R_m^f(\eta) \tag{3.81}$$

$$f_m(0;p)=0,\ f_m'(0;p)=0,\ f_m(1;p)=0 \tag{3.82}$$

$$R_m^f(\eta)=f_{m-1}'''+2\alpha \mathrm{Re}\sum_{n=0}^{m-1}f_{m-1-n}f_n'+(4-\mathrm{Ha})\alpha^2 f_{m-1}' \tag{3.83}$$

We use Maple software to obtain the solutions for these equations. Two first deformations of the coupled solutions are presented as follows:

$$f_1(\eta)=-\hbar.\mathrm{Re}.A_1.\exp(-\eta)[1+f_w+\eta+f_w\eta-(1+f_w)\exp(\eta) \tag{3.84}$$

The solutions $f_2(\eta),\ldots$ were too long to be mentioned here; therefore, they are shown graphically. Appendix A7 presents sample code for this method..

3.3 APPLICATION OF SEMIANALYTICAL METHODS FOR NANOFLUID HYDROTHERMAL BEHAVIOR IN THE PRESENCE OF MAGNETIC FIELD

3.3.1 TWO-PHASE UNSTEADY NANOFLUID FLOW AND HEAT TRANSFER BETWEEN PARALLEL PLATES IN THE PRESENCE OF A TIME-DEPENDENT MAGNETIC FIELD

3.3.1.1 Definition of the problem

Heat and mass transfer analysis in unsteady two-dimensional squeezing flow of nanofluid between the infinite parallel plates is considered (Fig. 3.1). The two plates are placed at $\ell(1-\gamma t)^{1/2} = h(t)$. When $\gamma > 0$, the two plates are squeezed until they touch $t = 1/\gamma$, and for $\gamma < 0$, the two plates are separated. The viscous dissipation effect, the generation of heat caused by friction caused by shear in the flow, is retained. Also, it is also assumed that the uniform magnetic field ($\vec{B} = B\vec{e_y}$) is applied, where $\vec{e_y}$ is unit vectors in the Cartesian coordinate system. The electric current J and the electromagnetic force F are defined by $J = \sigma\left(\vec{V} \times \vec{B}\right)$ and $F = \sigma\left(\vec{V} \times \vec{B}\right) \times \vec{B}$, respectively.

The governing equations for mass, momentum, energy, and mass transfer in unsteady two-dimensional flow of nanofluid are [8]

$$\frac{\partial u}{\partial x} + \frac{\partial v}{\partial y} = 0, \tag{3.85}$$

$$\rho_f\left(\frac{\partial u}{\partial t} + u\frac{\partial u}{\partial v} + v\frac{\partial u}{\partial y}\right) = -\frac{\partial p}{\partial x} + \mu\left(\frac{\partial^2 u}{\partial x^2} + \frac{\partial^2 u}{\partial y^2}\right) - \sigma B^2 u, \tag{3.86}$$

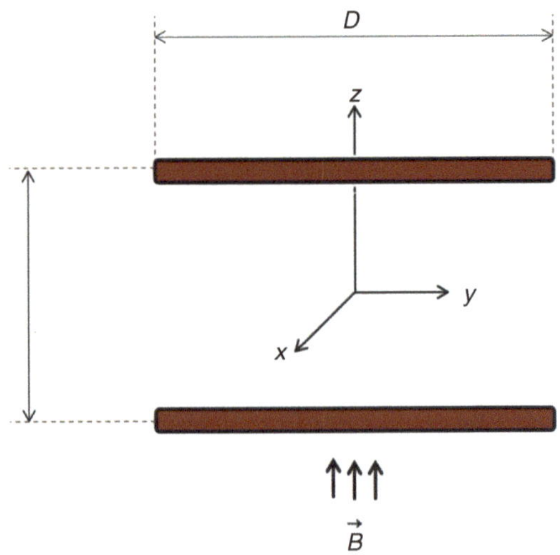

FIGURE 3.1 Geometry of the Problem

$$\rho_f\left(\frac{\partial v}{\partial t}+u\frac{\partial v}{\partial x}+v\frac{\partial v}{\partial y}\right)=-\frac{\partial p}{\partial y}+\mu\left(\frac{\partial^2 v}{\partial x^2}+\frac{\partial^2 v}{\partial y^2}\right), \tag{3.87}$$

$$\begin{aligned}\frac{\partial T}{\partial t}+u\frac{\partial T}{\partial x}+v\frac{\partial T}{\partial y}&=\alpha\left(\frac{\partial^2 T}{\partial x^2}+\frac{\partial^2 T}{\partial y^2}\right)+\frac{\mu}{(\rho c_{\mathrm{p}})_{\mathrm{f}}}\left(4\left(\frac{\partial u}{\partial x}\right)^2\right)\\&+\frac{(\rho c_{\mathrm{P}})_{\mathrm{p}}}{(\rho c_{\mathrm{P}})_{\mathrm{f}}}\left[\begin{array}{l}D_{\mathrm{B}}\left\{\frac{\partial C}{\partial x}\cdot\frac{\partial T}{\partial x}+\frac{\partial C}{\partial y}\cdot\frac{\partial T}{\partial y}\right\}+\\(D_{\mathrm{T}}/T_{\mathrm{c}})\left\{\left(\frac{\partial T}{\partial x}\right)^2+\left(\frac{\partial T}{\partial y}\right)^2\right\}\end{array}\right],\end{aligned} \tag{3.88}$$

$$\frac{\partial C}{\partial t}+u\frac{\partial C}{\partial x}+v\frac{\partial C}{\partial y}=D_{\mathrm{B}}\left(\frac{\partial^2 C}{\partial x^2}+\frac{\partial^2 C}{\partial y^2}\right)+\left(\frac{D_{\mathrm{T}}}{T_{\mathrm{c}}}\right)\left\{\frac{\partial^2 T}{\partial x^2}+\frac{\partial^2 T}{\partial y^2}\right\} \tag{3.89}$$

Here u and v are the velocities in the x and y directions, respectively; T is the temperature; C is the concentration; P is the pressure; ρ_{f} is the base fluid's density; μ is the dynamic viscosity; k is the thermal conductivity; c_{p} is the specific heat of nanofluid; and D_{B} is the diffusion coefficient of the diffusing species. The relevant boundary conditions are

$$\begin{aligned}&C=0,\quad v=v_w=dh/dt,\quad T=T_{\mathrm{H}},C=C_{\mathrm{H}} && at\ y=h(t),\\&v=\partial u/\partial y=\partial T/\partial y=\partial C/\partial y=0 && at\ y=0.\end{aligned} \tag{3.90}$$

We introduce these parameters:

$$\begin{aligned}&\eta=\frac{y}{[\ell(1-\gamma t)^{1/2}]},\qquad u=\frac{\gamma x}{[2(1-\gamma t)]}f'(\eta),\\&v=-\frac{\gamma\ell}{[2(1-\gamma t)^{1/2}]}f(\eta),\quad \theta=\frac{T}{T_{\mathrm{H}}},\\&\phi=\frac{C}{C_{\mathrm{h}}}.\end{aligned} \tag{3.91}$$

Substituting the above variables into Eqs (3.86) and (3.87) and then eliminating the pressure gradient from the resulting equations gives

$$f^{iv}-S(\eta f'''+3f''+f'f''-ff''')-\mathrm{Ha}^2 f''=0, \tag{3.92}$$

Using Eq. (3.91), Eqs (3.88) and (3.89) reduce to the following differential equations:

$$\theta''+\Pr S(f\theta'-\eta\theta')+\Pr\mathrm{Ec}(f''^2)+\mathrm{Nb}\,\phi'\theta'+\mathrm{Nt}\,\theta'^2=0, \tag{3.93}$$

$$\phi''+S.\mathrm{Sc}(f\phi'-\eta\phi')+\frac{\mathrm{Nt}}{\mathrm{Nb}}\theta''=0, \tag{3.94}$$

With these boundary conditions,

$$f(0)=0,\quad f''(0)=0,\quad \theta'(0)=0,$$
$$\phi'(0)=0,\quad f(1)=1,\quad f'(1)=0,$$
$$\theta(1)=\phi(1)=1, \tag{3.95}$$

where S is the squeeze number, Pr is the Prandtl number, Ec is the Eckert number, Sc is the Schmidt number, Ha is Hartman number of nanofluid, Nb is the Brownian motion parameter, and Nt is the thermophoretic parameter, which are defined as:

$$S=\frac{\gamma l^2}{2\mu}\rho_{\mathrm{f}},\quad \Pr=\frac{\mu}{\rho_{\mathrm{f}}\alpha},\quad \mathrm{Ec}=\frac{1}{c_{\mathrm{p}}}\left(\frac{\gamma x}{2(1-\gamma t)}\right)^2$$
$$\mathrm{Sc}=\frac{\mu}{\rho_{\mathrm{f}}D},\quad \mathrm{Ha}=\ell B\sqrt{\frac{\sigma}{\mu}}(1-\gamma t), \tag{3.96}$$
$$\mathrm{Nb}=(\rho c)_{\mathrm{p}}D_{\mathrm{B}}(C_{\mathrm{h}})/((\rho c)_{\mathrm{f}}\alpha),$$
$$\mathrm{Nt}=(\rho c)_{\mathrm{p}}D_{\mathrm{T}}(T_{\mathrm{H}})/[(\rho c)_{\mathrm{f}}\alpha T_{\mathrm{c}}].$$

The Nusselt number is defined as:

$$\mathrm{Nu}=\frac{-\ell k\left(\dfrac{\partial T}{\partial y}\right)_{y=h(t)}}{T_{\mathrm{H}}} \tag{3.97}$$

In terms of Eq. (3.91), we obtain

$$\mathrm{Nu}^*=\sqrt{1-\alpha t}\,\mathrm{Nu}=-\theta'(1) \tag{3.98}$$

3.3.1.2 Solution with the HPM

According to the HPM, we have

$$H(f,p)=(1-p)(f^{iv}-f_0^{iv})+p(f^{iv}-S(\eta f'''+3f''+f'f''-ff''')-\mathrm{Ha}^2 f''), \tag{3.99}$$

$$H(\theta,p)=(1-p)(\theta''-\theta_0'')$$
$$+p(\theta''+\mathrm{PrS}(f\theta'-\eta\theta')+\mathrm{PrEc}(f''^2)+\mathrm{Nb}\phi'\theta'+\mathrm{Nt}\theta'^2), \tag{3.100}$$

$$H(\phi,p)=(1-p)(\phi''-\phi_0'')+p\left(\phi''+S.\mathrm{Sc}(f\phi'-\eta\phi')+\frac{\mathrm{Nt}}{\mathrm{Nb}}\theta''=0,\right). \tag{3.101}$$

We consider f, θ, and ϕ as follows:

$$f(\eta)=f_0(\eta)+f_1(\eta)+\ldots=\sum_{i=0}^{n}f_i(\eta) \tag{3.102}$$

$$\theta(\eta)=\theta_0(\eta)+\theta_1(\eta)+\ldots=\sum_{i=0}^{n}\theta_i(\eta) \tag{3.103}$$

$$\phi(\eta)=\phi_0(\eta)+\phi_1(\eta)+\ldots=\sum_{i=0}^{n}\phi_i(\eta) \tag{3.104}$$

Substituting F from Eqs (3.102)–(3.104) and (3.99)–(3.101) into some simplification and rearranging based on powers of p terms, we have:

$$\begin{aligned}&p^0:\\&f_0^{iv}=0,\quad \theta_0''=0,\quad \phi_0''=0,\end{aligned} \tag{3.105}$$

And boundary conditions are

$$\begin{aligned}&f_0(0)=0,\quad f_0''(0)=0,\quad \theta_0'(0)=0,\\&\phi_0'(0)=0,\quad f_0(1)=1,\quad f_0'(1)=0,\\&\theta_0(1)=\phi_0(1)=1.\end{aligned} \tag{3.106}$$

$$\begin{aligned}&p^1:\\&f_1^{iv}+S\,f_0'f_0''-3S\,f_0''-S\eta\,f_0'''+Sf_0''-Sf_0f_0'''-\mathrm{Ha}^2f''=0,\\&\theta_1''+\Pr\mathrm{Ec}\,(f_0'')^2-\Pr S\,\eta\theta_0'+\Pr S\,f_0\theta_0'+\mathrm{Nb}\,\theta_0'\phi_0'+\mathrm{Nt}(\theta_0')^2=0,\\&\phi_1''+\mathrm{Sc}\,S\,f_0\phi_0'-\eta\phi_0'+(\mathrm{Nt}/\mathrm{Nb})\phi_0''=0.\end{aligned} \tag{3.107}$$

And boundary conditions are

$$\begin{aligned}&f_1(0)=0,\quad f_1''(0)=0,\quad \theta_1'(0)=0,\\&\phi_1'(0)=0,\quad f_1(1)=0,\quad f_1'(1)=0,\\&\theta_1(1)=\phi_1(1)=0.\end{aligned} \tag{3.108}$$

Solving Eqs (3.105) and (3.107) with boundary conditions, we have:

$$\begin{aligned}&f_0(\eta)=-0.5\eta^3+1.5\eta,\\&\theta_0(\eta)=1,\\&\phi_0(\eta)=1.\end{aligned} \tag{3.109}$$

$$\begin{aligned}f_1(\eta)=\;&S(0.003571428571\eta^7-(0.1+0.025\mathrm{Ha}^2)\eta^5\\&+(0.1892857143+0.05\mathrm{Ha}^2)\eta^3-0.09285714286\eta-0.025\mathrm{Ha}^2\eta),\\\theta_1(\eta)=\;&\Pr\mathrm{Ec}(-0.75\eta^4+0.75),\\\phi_1(\eta)=\;&0\end{aligned} \tag{3.110}$$

The terms $f_i(\eta)$, $\theta_i(\eta)$, and $\phi_i(\eta)$ when $i\geq 2$ are too large that they are mentioned graphically.

The solution of this equation when $p\to 1$ is as follows:

$$\begin{aligned}&f(\eta)=f_0(\eta)+f_1(\eta)+\ldots+f_{10}(\eta)\\&\theta(\eta)=\theta_0(\eta)+\theta_1(\eta)+\ldots+\theta_{10}(\eta)\\&\phi(\eta)=\phi_0(\eta)+\phi_1(\eta)+\ldots+\phi_{10}(\eta)\end{aligned} \tag{3.111}$$

3.3.1.3 Effects of active parameters

The effect of the squeeze number and Hartmann number on the velocity profiles and skin friction coefficient is shown in Fig. 3.2. It is important to note that the squeeze number (S) describes the movement of the plates ($S > 0$ corresponds to the plates moving apart, and $S < 0$ corresponds to the plates moving together [the so-called squeezing flow]). In this study, positive values of squeeze number are considered. As the squeeze number increases, the horizontal velocity decreases slightly. It is worthwhile mentioning that the effect of the magnetic field decreases the value of the velocity magnitude throughout the enclosure because the presence of the magnetic field introduces a force called the Lorentz force, which acts against the flow if the magnetic field is applied in the normal direction. This type of resisting force slows down the fluid velocity. Also, Fig. 3.2 shows that the absolute values of the skin friction coefficient have a reverse relationship with the squeeze number and Hartmann number.

Fig. 3.3 shows the effect of the squeeze number and Hartmann number on the temperature profile and Nusselt number. An increase in the squeeze number can be related to a decrease in the kinematic viscosity, an increase in the distance between

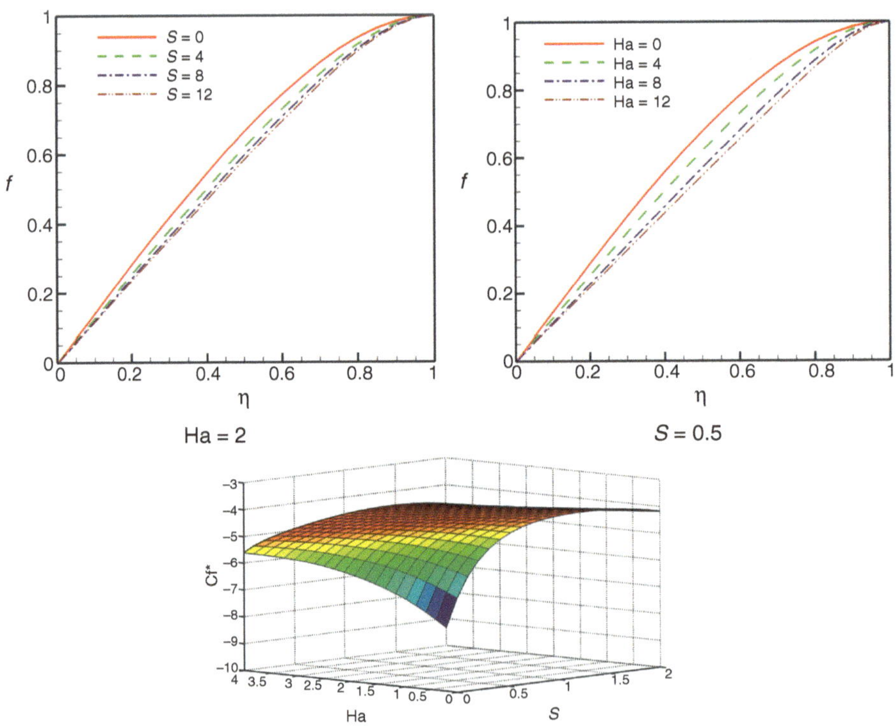

FIGURE 3.2 Effect of the Squeeze Number and Hartmann Number on the Velocity Profiles and Skin Friction Coefficient When Ha = 2

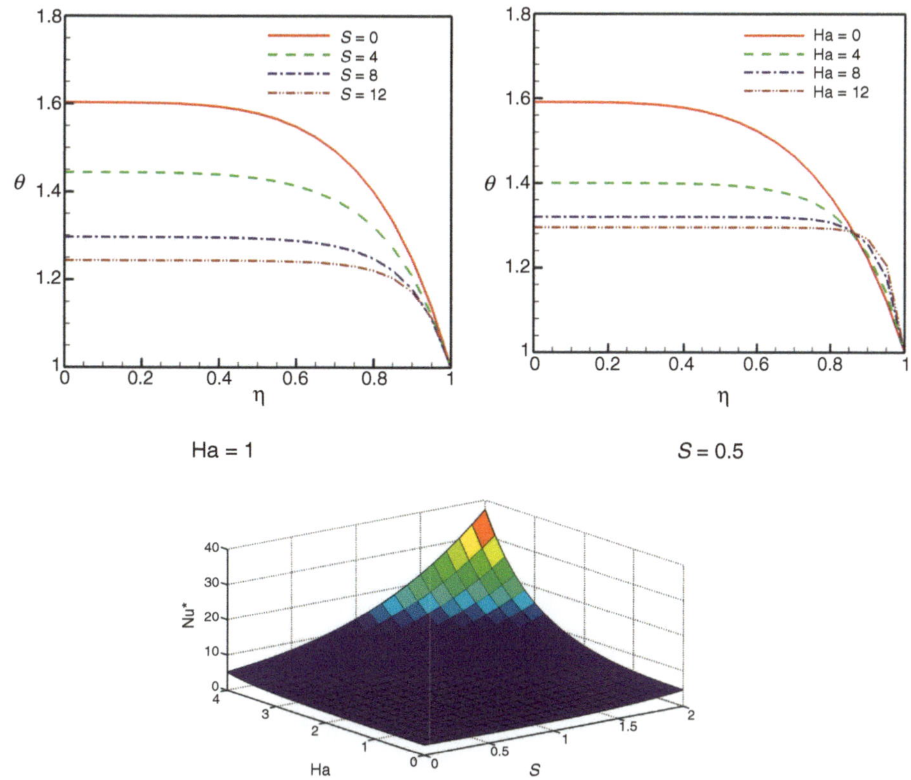

FIGURE 3.3 Effect of the Squeeze Number and Hartmann Number on the Temperature Profile and Nusselt Number When Ec = 0.1, Nt = 0.1, Nb = 0.1, Sc = 0.5, and Pr = 10

the plates, and an increase in the speed at which the plates move. Temperature increases as the squeeze number increases. A similar effect is observed for the Hartmann number. Thus, the Nusselt number increases with an increase of the squeeze number and Hartmann number. The effect of the Eckert number on the temperature profile and Nusselt number is depicted in Fig. 3.4. The presence of viscous dissipation effects significantly increases the temperature. The Nusselt number increases with an increase of the Eckert number because of reduction of thermal boundary layer thickness near the upper plate.

The effect of the squeeze number and Hartmann number on the concentration profile is shown in Fig. 3.5. The concentration profile increase with an augment of the squeeze number and Hartmann number when $\eta < 0.5$, but it decreases with an increase of these parameters when $\eta > 0.5$. Fig. 3.6 shows the effect of the Schmidt number and Eckert number on the concentration profile. As the Schmidt number increases, the concentration profile increases, but the opposite trend is observed for the Eckert number.

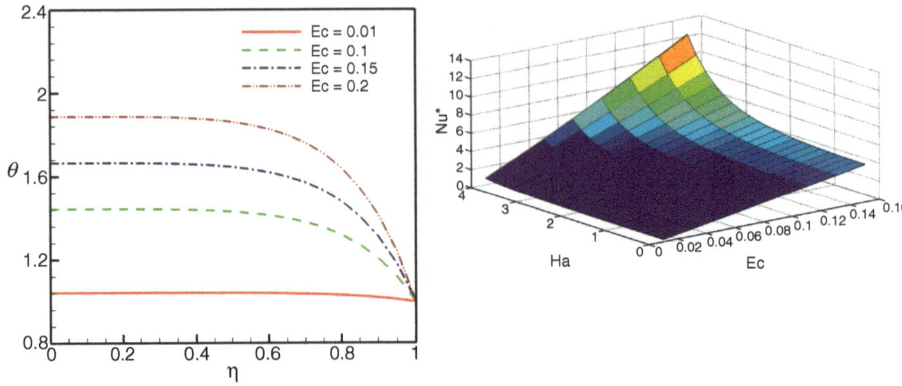

FIGURE 3.4 Effect of Eckert Number on the Temperature Profile and Nusselt Number When $Nt = Nb = 0.1, Ha = 1, S = 0.5, Sc = 0.5$, and $Pr = 10$

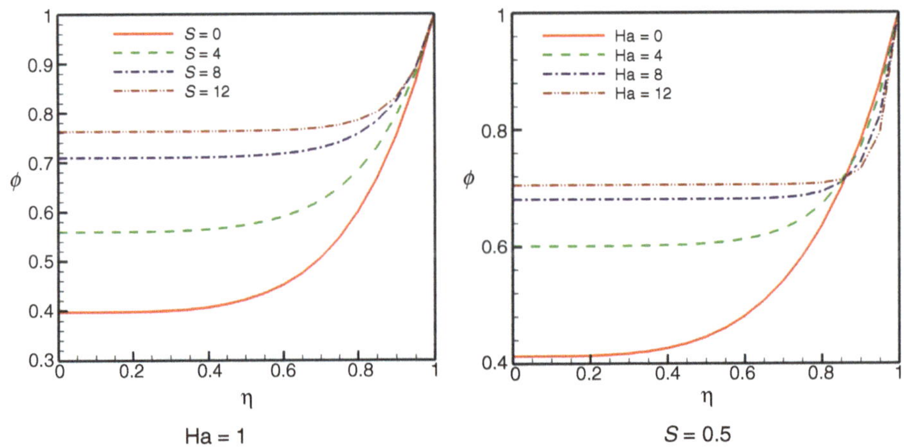

FIGURE 3.5 Effect of the Squeeze Number and Hartmann Number on the Concentration Profile When $Nt = Nb = 0.1, Ha = 1, S = 0.5, Sc = 0.5$, and $Pr = 10$

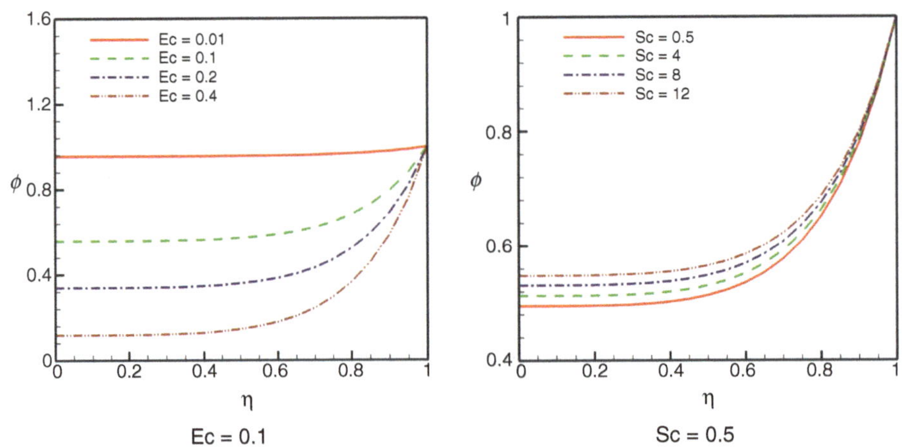

FIGURE 3.6 Effect of the Schmidt Number and Eckert Number on the Concentration Profile When $Nt = Nb = 0.1, Ha = 1, S = 0.5$, and $Pr = 10$

3.3.2 NANOFLUID FLOW IN A SEMIPOROUS CHANNEL IN THE PRESENCE OF A UNIFORM MAGNETIC FIELD

3.3.2.1 Definition of the problem

Consider the laminar two-dimensional stationary flow of an electrically conducting incompressible viscous fluid in a semiporous channel made by a long rectangular plate with length of L_x in uniform translation in x^* direction and an infinite porous plate. The distance between the two plates is h. We observe a normal velocity q on the porous wall. A uniform magnetic field B is assumed to be applied toward direction y^* (Fig. 3.7). In the case of a short circuit to neglect the electrical field and perturbations to the basic normal field and without any gravity forces, the governing equations are [6]

$$\frac{\partial u^*}{\partial x^*}+\frac{\partial v^*}{\partial y^*}=0, \tag{3.112}$$

$$u^*\frac{\partial u^*}{\partial x^*}+v^*\frac{\partial u^*}{\partial y^*}=-\frac{1}{\rho_{\mathrm{nf}}}\frac{\partial P^*}{\partial x^*}+\frac{\mu_{\mathrm{nf}}}{\rho_{\mathrm{nf}}}\left(\frac{\partial^2 u^*}{\partial x^{*2}}+\frac{\partial^2 u^*}{\partial y^{*2}}\right)-u^*\frac{\sigma_{\mathrm{nf}}B^2}{\rho_{\mathrm{nf}}}, \tag{3.113}$$

$$u^*\frac{\partial v^*}{\partial x^*}+v^*\frac{\partial v^*}{\partial y^*}=-\frac{1}{\rho_{\mathrm{nf}}}\frac{\partial P^*}{\partial y^*}+\frac{\mu_{\mathrm{nf}}}{\rho_{\mathrm{nf}}}\left(\frac{\partial^2 v^*}{\partial x^{*2}}+\frac{\partial^2 v^*}{\partial y^{*2}}\right), \tag{3.114}$$

The appropriate boundary conditions for the velocity are

$$y^*=0: u^*=u_0^*, v^*=0, \tag{3.115}$$

$$y^*=h: u^*=0,\ v^*=-q, \tag{3.116}$$

Calculating a mean velocity U by the relation:

$$y^*=0: u^*=u_0^*, v^*=0, \tag{3.117}$$

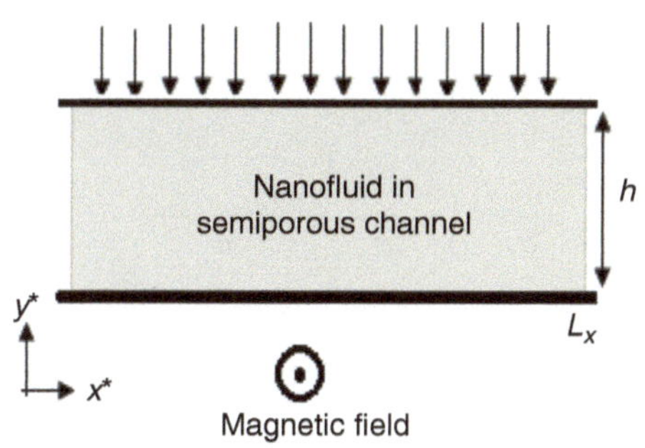

FIGURE 3.7 Schematic of the Problem (Nanofluid in a Porous Media Between Parallel Plates and the Magnetic Field)

We consider the following transformations:

$$x = \frac{x^*}{L_x};\; y = \frac{y^*}{h}, \tag{3.118}$$

$$u = \frac{u^*}{U};\; v = \frac{v^*}{q},\; P_y = \frac{P^*}{\rho_f . q^2} \tag{3.119}$$

Then we can consider two dimensionless numbers: the Hartman number, Ha, for the description of magnetic forces [6] and the Reynolds number, Re, for dynamic forces:

$$Ha = Bh\sqrt{\frac{\sigma_f}{\rho_f . \upsilon_f}}, \tag{3.120}$$

$$\mathrm{Re} = \frac{hq}{\mu_{nf}}\rho_{nf}. \tag{3.121}$$

where the effective density (ρ_{nf}) is defined as:

$$\rho_{nf} = \rho_f(1-\phi) + \rho_s\phi \tag{3.122}$$

where ϕ is the solid volume fraction of nanoparticles. The dynamic viscosity of the nanofluids given by Brinkman [6] is

$$\mu_{nf} = \frac{\mu_f}{(1-\phi)^{2.5}} \tag{3.123}$$

The effective thermal conductivity of the nanofluid can be approximated by the Maxwell–Garnetts (MG) model as [6]

$$\frac{k_{nf}}{k_f} = \frac{k_s + 2k_f - 2\phi(k_f - k_s)}{k_s + 2k_f + \phi(k_f - k_s)} \tag{3.124}$$

The effective electrical conductivity of nanofluid was presented by Maxwell [6] as follows:

$$\frac{\sigma_{nf}}{\sigma_f} = 1 + \frac{3\left(\frac{\sigma_s}{\sigma_f} - 1\right)\phi}{\left(\frac{\sigma_s}{\sigma_f} + 2\right) - \left(\frac{\sigma_s}{\sigma_f} - 1\right)\phi} \tag{3.125}$$

Introducing Eqs (3.118) and (3.119) into Eqs (3.112) and (3.114) leads to the dimensionless equations:

$$\frac{\partial u}{\partial x} + \frac{\partial v}{\partial y} = 0, \tag{3.126}$$

$$u\frac{\partial u}{\partial x}+v\frac{\partial u}{\partial y}=-\varepsilon^2\frac{\partial P_y}{\partial x}+\frac{\mu_{\rm nf}}{\rho_{\rm nf}}\frac{1}{hq}\left(\varepsilon^2\frac{\partial^2 u}{\partial x^2}+\frac{\partial^2 u}{\partial y^2}\right)-u\frac{{\rm Ha}^2}{\rm Re}\frac{B^*}{A^*}, \tag{3.127}$$

$$u\frac{\partial v}{\partial x}+v\frac{\partial v}{\partial y}=-\frac{\partial P_y}{\partial x}+\frac{\mu_{\rm nf}}{\rho_{\rm nf}}\frac{1}{hq}\left(\varepsilon^2\frac{\partial^2 v}{\partial x^2}+\frac{\partial^2 v}{\partial y^2}\right). \tag{3.128}$$

where A^* and B^* are constant parameters:

$$A^*=(1-\phi)+\frac{\rho_{\rm s}}{\rho_{\rm f}}\phi,\quad B^*=1+\frac{3\left(\frac{\sigma_{\rm s}}{\sigma_{\rm f}}-1\right)\phi}{\left(\frac{\sigma_{\rm s}}{\sigma_{\rm f}}+2\right)-\left(\frac{\sigma_{\rm s}}{\sigma_{\rm f}}-1\right)\phi} \tag{3.129}$$

The quantity of ε is defined as the aspect ratio between distance h and a characteristic length L_x of the slider. This ratio is normally small. Berman's similarity transformation is used to be free from the aspect ratio of ε:

$$v=-V(y);u=\frac{u^*}{U}=u_0U(y)+x\frac{dV}{dy}. \tag{3.130}$$

Introducing Eq. (3.130) in the second momentum equation, Eq. (3.17) shows that quantity $\partial P_y/\partial y$ does not depend on the longitudinal variable x. With the first momentum equation, we also observe that $\partial^2 P_y/\partial x^2$ is independent of x. We omit asterisks for simplicity. Then a separation of variables leads to [6]:

$$V'^2-VV''-\frac{1}{\rm Re}\frac{1}{A^*(1-\phi)^{2.5}}V'''+\frac{{\rm Ha}^2}{\rm Re}\frac{B^*}{A^*}V'=\varepsilon^2\frac{\partial^2 P_y}{\partial x^2}=\varepsilon^2\frac{1}{x}\frac{\partial P_y}{\partial x}, \tag{3.131}$$

$$UV'-VU'=\frac{1}{\rm Re}\frac{1}{A^*(1-\phi)^{2.5}}[U''-{\rm Ha}^2B^*(1-\phi)^{2.5}U]. \tag{3.132}$$

The right-hand side of Eq. (3.131) is constant. So, we derive this equation with respect to x. This gives:

$$V^{\rm IV}={\rm Ha}^2B^*(1-\phi)^{2.5}V''+{\rm Re}\,A^*(1-\phi)^{2.5}[V'V''-VV'''], \tag{3.133}$$

where primes denote differentiation with respect to y and asterisks have been omitted for simplicity. The dynamic boundary conditions are:

$$y=0:\quad U=1;V=0;V'=0, \tag{3.134}$$

$$y=1:\quad U=0;V=1;V'=0. \tag{3.135}$$

3.3.2.2 Solution with the HPM

According to HPM, we have:

$$H(V,p)=(1-p)(V^{\mathrm{IV}}-V_0^{\mathrm{IV}})+p(-V^{\mathrm{IV}}+\mathrm{Ha}^2B^*(1-\phi)^{2.5}V'' \\ +\mathrm{Re}\,A^*(1-\phi)^{2.5}[V'V''-VV'''])=0, \tag{3.136}$$

$$H(U,p)=(1-p)(U''-U_0'') \\ +p\left(-UV'+VU'+\frac{1}{\mathrm{Re}}\frac{1}{A^*(1-\phi)^{2.5}}[U''-\mathrm{Ha}^2B^*(1-\phi)^{2.5}U]\right)=0, \tag{3.137}$$

We consider V and U follows:

$$V(y)=V_0(y)+V_1(y)+\ldots=\sum_{i=0}^{n}V_i(y) \tag{3.138}$$

$$U(y)=U_0(y)+U_1(y)+\ldots=\sum_{i=0}^{n}U_i(y) \tag{3.139}$$

with substituting U, V from Eqs (3.138) and Eq. (3.139) into Eqs (3.136) and (3.237) and some simplification and rearranging based on powers of p terms, we have:

$$p^0: \\ V_0^{iv}=0, \\ U_0''=0, \tag{3.140}$$

And boundary conditions are:

$$y=0:\quad U_0=1; V_0=0; V_0'=0, \\ y=1:\quad U_0=0; V_0=1; V_0'=0. \tag{3.141}$$

$$p^1: \\ -\mathrm{Ha}^2(1-\phi)^{-2.5}B^*V_0''+V_1^{iv}+\mathrm{Re}\,A^*(1-\phi)^{-2.5}V_0'''V_0-\mathrm{Re}\,A^*(1-\phi)^{-2.5}V_0''V_0'=0, \\ -\mathrm{Ha}^2(1-\phi)^{-2.5}B^*U_0''+U_1''-\mathrm{Re}\,A^*(1-\phi)^{-2.5}V_0'U_0+\mathrm{Re}\,A^*(1-\phi)^{-2.5}V_0U_0'=0. \tag{3.142}$$

And boundary conditions are:

$$y=0:\quad U_1=0; V_1=0; V_1'=0, \\ y=1:\quad U_1=0; V_1=0; V_1'=0. \tag{3.143}$$

Solving Eqs (3.140) and (3.142) with boundary conditions, we have:

$$V_0(y)=-2y^3+3y^2, \\ U_0(y)=-y+1. \tag{3.144}$$

$$V_1(y)=0.0571428\,\mathrm{Re}\,A^*(1-\phi)^{-2.5}y^7-0.2\,\mathrm{Re}\,A^*(1-\phi)^{-2.5}y^6-0.1\,\mathrm{Ha}^2(1-\phi)^{-2.5}B^*y^5 \\ +0.3\,\mathrm{Re}\,A^*(1-\phi)^{-2.5}y^4-0.385714\,\mathrm{Re}\,A^*(1-\phi)^{-2.5}y^3-0.2\,\mathrm{Ha}^2(1-\phi)^{-2.5}B^*y^3 \\ +0.22857142\,\mathrm{Re}\,A^*(1-\phi)^{-2.5}y^2+0.5\,\mathrm{Ha}^2(1-\phi)^{-2.5}B^*y^2, \\ U_1(y)=-0.2\,\mathrm{Re}\,A^*(1-\phi)^{-2.5}y^5-0.7\,\mathrm{Re}\,A^*(1-\phi)^{-2.5}y^4-0.16667\,\mathrm{Ha}^2(1-\phi)^{-2.5}B^*y^3 \\ +0.1\,\mathrm{Re}\,A^*(1-\phi)^{-2.5}y^3+0.5\,\mathrm{Ha}^2(1-\phi)^{-2.5}B^*y^2-0.45\,\mathrm{Re}\,A^*(1-\phi)^{-2.5}y \\ -0.3333\,\mathrm{Ha}^2(1-\phi)^{-2.5}B^*y. \tag{3.145}$$

The terms $V_i(y)$, $U_i(y)$ when $i \geq 2$ are too large, it is mentioned graphically. The solution of this equation, when $p \rightarrow 1$, is as follows:

$$\begin{aligned} V(y) &= V_0(y) + V_1(y) + \ldots = \sum_{i=0}^{n} V_i(y) \\ U(y) &= U_0(y) + U_1(y) + \ldots = \sum_{i=0}^{n} U_i(y) \end{aligned} \tag{3.146}$$

3.3.2.3 Effects of active parameters

Effect of the nanoparticle volume fraction on $U(y)$ is shown in Fig. 3.8. For both cases (the presence and absence of a magnetic field), the velocity boundary layer thickness decreases with an increase of the nanoparticle volume fraction. Also, it can be seen that increasing the nanoparticle volume friction leads to a decrease in the values of $U(y)$, and this decrement is more sensible in the absence of a magnetic field.

The effect of various values of Hartmann numbers on $V(y)$ and $U(y)$ is shown in Fig. 3.9. Generally, when the magnetic field is imposed on the enclosure, the velocity field is suppressed owing to the retarding effect of the Lorenz force. For a low Reynolds number, as the Hartmann number increases, $V(y)$ decreases for $y > y_m$, but the opposite trend is observed for $y < y_m$; y_m is a meeting point that all curves joint together at this point. When Reynolds number increases this meeting point shifts to the solid wall, and it can be seen that $V(y)$ decreases with an increase of the Hartmann number. As the Hartmann number increases, $U(y)$ decreases for all values of Reynolds number. Besides, this figure shows that this change is more pronounced for a low Reynolds number.

Fig. 3.10 shows the effect of the Reynolds number on $V(y)$ and $U(y)$ profiles at a constant Hartmann number. Contour plots of V and U are shown in Figs 3.11 and 3.12. It is worth mentioning that the Reynolds number indicates the relative significance of the inertia effect compared with the viscous effect. Thus, the velocity profile decreases

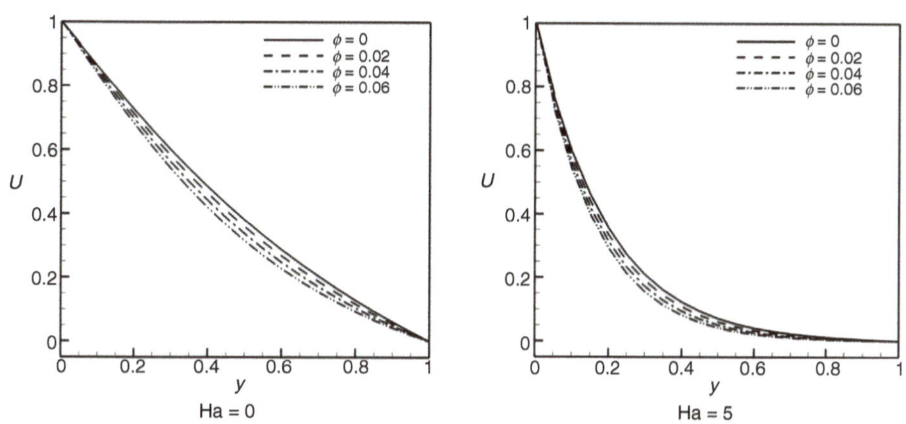

FIGURE 3.8 Effect of Nanoparticle Volume Fraction (ϕ) on $U(y)$ when Re = 1

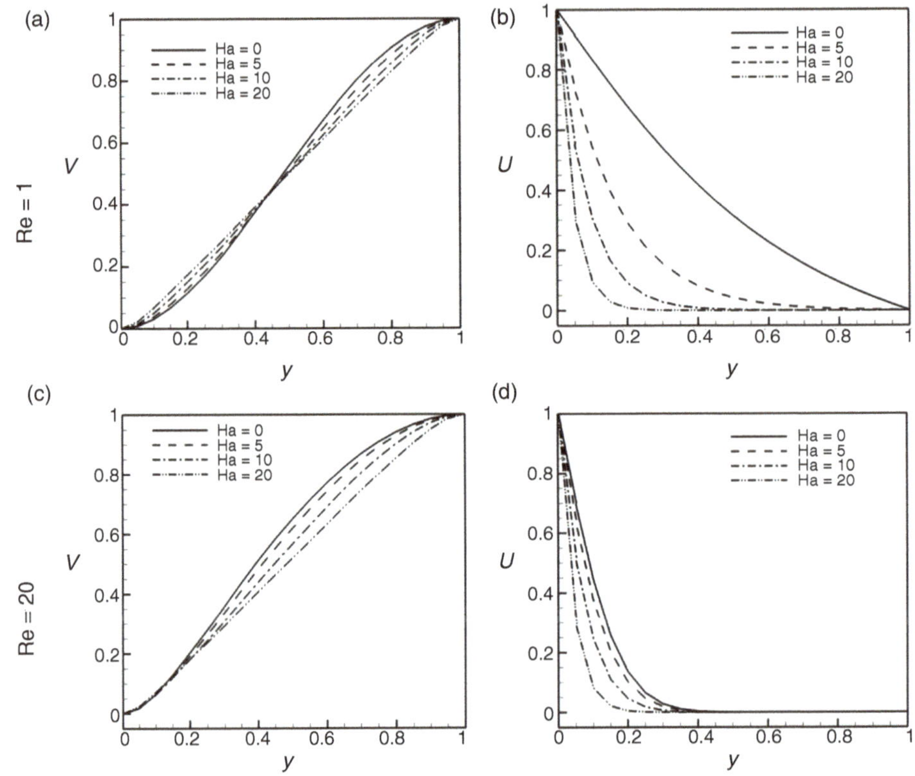

FIGURE 3.9 Effect of Various Values of Hartmann Numbers (Ha) on $V(y)$ and $U(y)$ When $\phi = 0.06$

as Re increases, and in turn increasing Re leads to increase in the magnitude of the skin friction coefficient. With an increasing Reynolds number, $V(y)$ and $U(y)$ increase. These effects become less at higher Hartmann numbers. Also, it shows that increasing Hartmann number leads to increasing the curve of the velocity profile.

3.3.3 NANOFLUID FLOW AND HEAT TRANSFER BETWEEN PARALLEL PLATES CONSIDERING BROWNIAN MOTION

3.3.3.1 Definition of the problem

We consider the flow and heat transfer analysis in the unsteady two-dimensional squeezing flow of an incompressible nanofluid between the infinite parallel plates. The two plates are placed at $z = \pm\ell(1-\alpha t)^{1/2} = \pm h(t)$ (ℓ is distant of plate at $t = 0$, and α is squeezed parameter). For $\alpha > 0$, the two plates are squeezed until they touch $t = 1/\alpha$, and for $\alpha < 0$, the two plates are separated. The viscous dissipation effect, the generation of heat caused by fraction caused by shear in the flow, is neglected. Furthermore, the symmetric nature of the flow is adopted (Fig. 3.13).

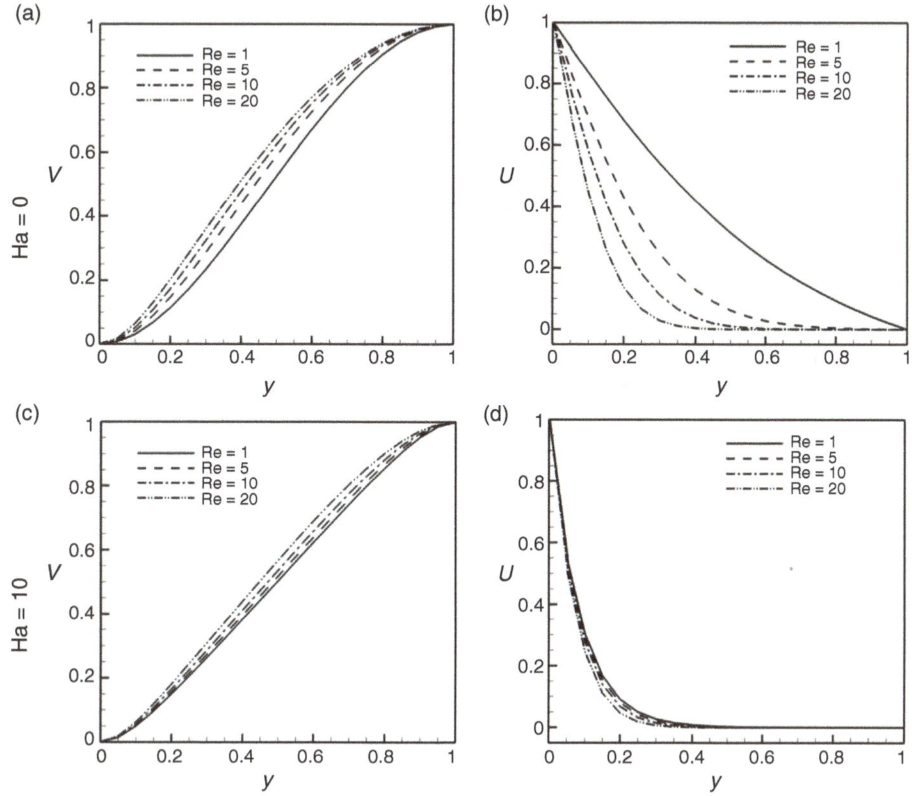

FIGURE 3.10 Effects of Various Values of Reynolds Numbers (Re) on $V(y)$ and $U(y)$ When $\phi = 0.06$

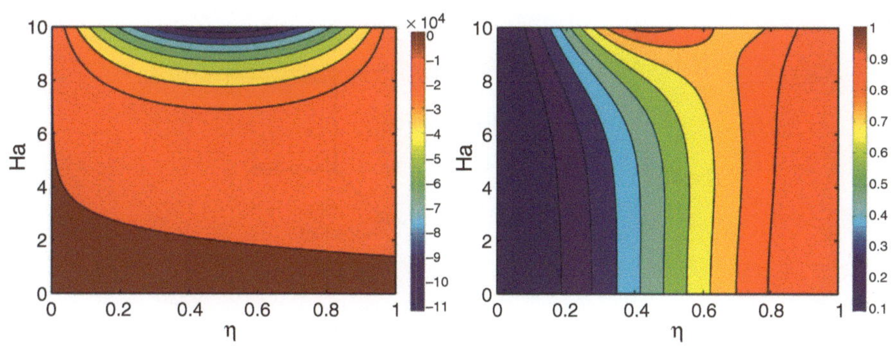

FIGURE 3.11 Contour Plots of V (Left) and U (Right) When $\phi = 0.06$, Re = 1 for the Cu–Water Case

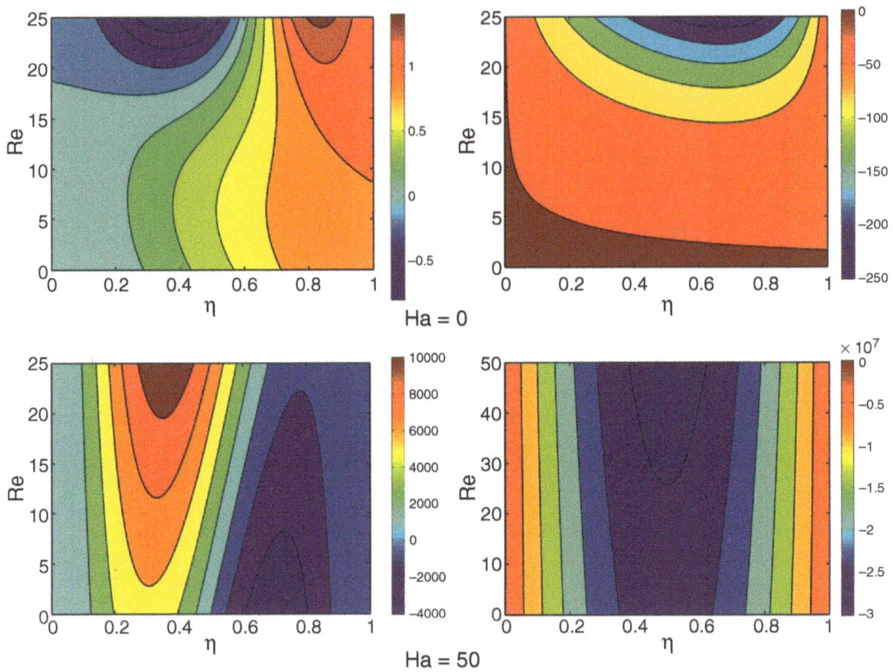

FIGURE 3.12 Contour Plots of V (Left) and U (Right) When $\phi = 0.06$ for the Cu–Water Case

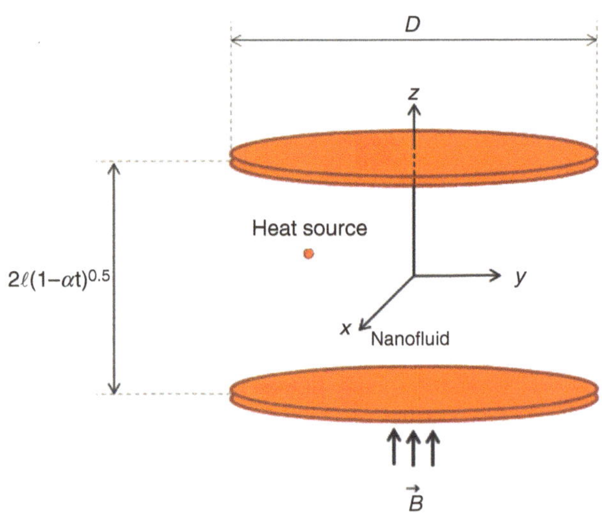

FIGURE 3.13 Geometry of Problem

Table 3.2 Thermophysical Properties of Water and Nanoparticles at Room Temperature [33]

	ρ (kg/m^3)	C_p (j/kgk)	k (W/m.k)	d_p (nm)
Pure water	997.1	4179	0.613	–
Al_2O_3	3970	765	25	47
CuO	6500	540	18	29

It is also assumed that the time-variable magnetic field ($\vec{B} = B\vec{e}_y$, $B = B_0(1-\alpha t)^{-0.5}$) is applied, where $\vec{e}_y$ is unit vectors in the Cartesian coordinate system. The electric current J and the electromagnetic force F are defined by $J = \sigma\left(\vec{V}\times\vec{B}\right)$ and $F = \sigma\left(\vec{V}\times\vec{B}\right)\times\vec{B}$, respectively. Also, a heat source $\left(Q = Q_0/(1-\alpha t)\right)$ is applied between two plates.

The nanofluid is a two-component mixture with the following assumptions: incompressible, no chemical reaction, negligible radiative heat transfer, and anosolid particles and the base fluid are in thermal equilibrium and no slip occurs between them. The thermophysical properties of the nanofluid are given in Table 3.2 [33]. The governing equations for conservative momentum and energy in unsteady two-dimensional flow of a nanofluid fluid are [33]:

$$\frac{\partial u}{\partial x}+\frac{\partial v}{\partial y}=0, \tag{3.147}$$

$$\rho_{\mathrm{nf}}\left(\frac{\partial u}{\partial t}+u\frac{\partial u}{\partial x}+v\frac{\partial u}{\partial y}\right)=-\frac{\partial p}{\partial x}+\mu_{\mathrm{nf}}\left(\frac{\partial^2 u}{\partial x^2}+\frac{\partial^2 u}{\partial y^2}\right)-\sigma_{\mathrm{nf}}B^2u, \tag{3.148}$$

$$\rho_{\mathrm{nf}}\left(\frac{\partial v}{\partial t}+u\frac{\partial v}{\partial x}+v\frac{\partial v}{\partial y}\right)=-\frac{\partial p}{\partial y}+\mu_{\mathrm{nf}}\left(\frac{\partial^2 v}{\partial x^2}+\frac{\partial^2 v}{\partial y^2}\right), \tag{3.149}$$

$$\frac{\partial T}{\partial t}+u\frac{\partial T}{\partial x}+v\frac{\partial T}{\partial y}=\frac{k_{\mathrm{nf}}}{(\rho C_{\mathrm{p}})_{\mathrm{nf}}}\left(\frac{\partial^2 T}{\partial x^2}+\frac{\partial^2 T}{\partial y^2}\right)+\frac{Q}{(\rho C_{\mathrm{p}})_{\mathrm{nf}}}T \tag{3.150}$$

where u and v are the velocities in the x and y directions, respectively, and T is the temperature, P is the pressure. The effective density (ρ_{nf}), effective heat capacity $(\rho C_{\mathrm{p}})_{\mathrm{nf}}$, and electrical conductivity (σ_{nf}) of the nanofluid are defined as [33]:

$$\begin{aligned}
&\rho_{\mathrm{nf}}=(1-\phi)\rho_{\mathrm{f}}+\phi\rho_{\mathrm{p}},\\
&(\rho C_{\mathrm{p}})_{\mathrm{nf}}=(1-\phi)(\rho C_{\mathrm{p}})_{\mathrm{f}}+\phi(\rho C_{\mathrm{p}})_{\mathrm{p}}\\
&\frac{\sigma_{\mathrm{nf}}}{\sigma_{\mathrm{f}}}=1+\frac{3\left(\dfrac{\sigma_{\mathrm{p}}}{\sigma_{\mathrm{f}}}-1\right)\phi}{\left(\dfrac{\sigma_{\mathrm{p}}}{\sigma_{\mathrm{f}}}+2\right)-\left(\dfrac{\sigma_{\mathrm{p}}}{\sigma_{\mathrm{f}}}-1\right)\phi}
\end{aligned} \tag{3.151}$$

The Brownian motion has a significant impact on the effective thermal conductivity. The thermal conductivity model takes into account the effects of particle size, particle volume fraction, and temperature dependence as well as types of particle and base fluid combinations.

$$k_{\mathrm{eff}} = k_{\mathrm{static}} + k_{\mathrm{Brownian}} \tag{3.152}$$

$$\frac{k_{\mathrm{static}}}{k_{\mathrm{f}}} = 1 + \frac{3\left(\dfrac{k_{\mathrm{p}}}{k_{\mathrm{f}}} - 1\right)\phi}{\left(\dfrac{k_{\mathrm{p}}}{k_{\mathrm{f}}} + 2\right) - \left(\dfrac{k_{\mathrm{p}}}{k_{\mathrm{f}}} - 1\right)\phi} \tag{3.153}$$

where k_{static} is the static thermal conductivity based on Maxwell classical correlation. The enhanced thermal conductivity component generated by microscale convective heat transfer of a particle's Brownian motion and affected by ambient fluid motion is obtained via simulating Stokes' flow around a sphere (nanoparticle). By introducing two empirical functions (β and f), we combined the interaction between nanoparticles with to the temperature effect in the model, leading to:

$$k_{\mathrm{Brownian}} = 5\times10^{4}\,\beta\phi\rho_{\mathrm{f}}c_{\mathrm{p,f}}\sqrt{\frac{\kappa_{\mathrm{b}}T}{\rho_{\mathrm{p}}d_{\mathrm{p}}}}\,f(T,\phi) \tag{3.154}$$

The empirical g′-function depends on the type of nanofluid [33]. Also, by introducing a thermal interfacial resistance $R_{\mathrm{f}} = 4\times10^{-8}\,km^2/W$, the original k_{p} in Eq. (3.153) was replaced by a new $k_{\mathrm{p,eff}}$ in the form:

$$R_{\mathrm{f}} + \frac{d_{\mathrm{p}}}{k_{\mathrm{p}}} = \frac{d_{\mathrm{p}}}{k_{\mathrm{p,eff}}} \tag{3.155}$$

For different base fluids and different nanoparticles, the function should be different. Only water-based nanofluids are considered in the current study. For Al_2O_3–water nanofluids and CuO–water nanofluids, this function follows the format:

$$\begin{aligned} g'(T,\phi,d_{\mathrm{p}}) = {} & (a_1 + a_2\ln(d_{\mathrm{p}}) + a_3\ln(\phi) + a_4\ln(\phi)\ln(d_{\mathrm{p}}) + a_5\ln(d_{\mathrm{p}})^2)\ln(T) \\ & + (a_6 + a_7\ln(d_{\mathrm{p}}) + a_8\ln(\phi) + a_9\ln(\phi)\ln(d_{\mathrm{p}}) + a_{10}\ln(d_{\mathrm{p}})^2) \end{aligned} \tag{3.156}$$

The coefficients a_i (i = 0..10) are based on the type of nanoparticles, and with these coefficients, Al_2O_3–water nanofluids and CuO–water nanofluids have an R^2 of 96% and 98%, respectively [33] (Table 3.3). Finally, the KKL (Koo–Kleinstreuer–Li) correlation is written as:

$$k_{\mathrm{Brownian}} = 5\times10^{4}\,\phi\rho_{\mathrm{f}}c_{\mathrm{p,f}}\sqrt{\frac{\kappa_{\mathrm{b}}T}{\rho_{\mathrm{p}}d_{\mathrm{p}}}}\,g'(T,\phi,d_{\mathrm{p}}) \tag{3.157}$$

Table 3.3 The Coefficient values of Al_2O_3–Water Nanofluids and CuO–Water Nanofluids [33]

Coefficient Values	Al_2O_3–Water	CuO–Water
a_1	52.813488759	−26.593310846
a_2	6.115637295	−0.403818333
a_3	0.6955745084	−33.3516805
a_4	4.17455552786E-02	−1.915825591
a_5	0.176919300241	6.42185846658E-02
a_6	−298.19819084	48.40336955
a_7	−34.532716906	−9.787756683
a_8	−3.9225289283	190.245610009
a_9	−0.2354329626	10.9285386565
a_{10}	−0.999063481	−0.72009983664

For the effective viscosity caused by micromixing in suspensions, we have:

$$\mu_{\text{eff}} = \mu_{\text{static}} + \mu_{\text{Brownian}} = \mu_{\text{static}} + \frac{k_{\text{Brownian}}}{k_{\text{f}}} \times \frac{\mu_{\text{f}}}{\text{Pr}_{\text{f}}} \tag{3.158}$$

where $\mu_{\text{static}} = \dfrac{\mu_{\text{f}}}{(1-\phi)^{2.5}}$ is the viscosity of the nanofluid, as given originally by Brinkman.

The relevant boundary conditions are:

$$\begin{aligned} &v = v_{\text{w}} = dh/dt, \quad T = T_{\text{H}} && \text{at } y = h(t), \\ &v = \partial u/\partial y = \partial T/\partial y = 0 && \text{at } y = 0. \end{aligned} \tag{3.159}$$

We introduce these parameters:

$$\begin{aligned} &\eta = \frac{y}{[l(1-\alpha t)^{1/2}]}, \qquad u = \frac{\alpha x}{[2(1-\alpha t)]} f'(\eta), \\ &v = -\frac{\alpha l}{[2(1-\alpha t)^{1/2}]} f(\eta), \quad \theta = \frac{T}{T_{\text{H}}}. \end{aligned} \tag{3.160}$$

Substituting the above variables into Eqs (3.148) and (3.149) and then eliminating the pressure gradient from the resulting equations gives

$$f^{iv} - S\,(A_1/A_4)(\eta f''' + 3f'' + f'f'' - ff''') - \text{Ha}^2 (A_5/A_4) f'' = 0, \tag{3.161}$$

Using Eq. (3.160), Eq. (3.150) reduces to the following differential equations:

$$\theta'' + \Pr S\left(\frac{A_2}{A_3}\right)(f\theta' - \eta\theta') + \frac{Hs}{A_3}\theta = 0 \tag{3.162}$$

where A_1, A_2, A_3, A_4, and A_5 are dimensionless constants given by:

$$A_1 = \frac{\rho_{\mathrm{nf}}}{\rho_{\mathrm{f}}},\quad A_2 = \frac{(\rho C_{\mathrm{p}})_{\mathrm{nf}}}{(\rho C_{\mathrm{p}})_{\mathrm{f}}},\quad A_3 = \frac{k_{\mathrm{eff}}}{k_{\mathrm{f}}},\quad A_4 = \frac{\mu_{\mathrm{eff}}}{\mu_{\mathrm{f}}},\quad A_5 = \frac{\sigma_{\mathrm{nf}}}{\sigma_{\mathrm{f}}} \tag{3.163}$$

With these boundary conditions:

$$\begin{aligned} &f(0)=0,\quad f''(0)=0,\\ &f(1)=1,\quad f'(1)=0,\\ &\theta'(0)=0,\quad \theta(1)=1. \end{aligned} \tag{3.164}$$

where S is the squeeze number, Pr is the Prandtl number, Ha is the Hartmann number, and Hs is the heat source parameter, which are defined as:

$$S = \frac{\alpha \ell^2}{2\upsilon_{\mathrm{f}}},\quad \mathrm{Pr} = \frac{\mu_{\mathrm{f}}(\rho C_{\mathrm{p}})_{\mathrm{f}}}{\rho_{\mathrm{f}} k_{\mathrm{f}}},\quad \mathrm{Ha} = \ell B_0\sqrt{\frac{\sigma_{\mathrm{f}}}{\mu_{\mathrm{f}}}},\quad \mathrm{Hs} = \frac{Q_0 \ell^2}{k_{\mathrm{f}}} \tag{3.165}$$

Physical quantities of interest are the skin fraction coefficient and Nusselt number, which are defined as:

$$Cf^* = \frac{\mu_{\mathrm{nf}}\left(\dfrac{\partial u}{\partial y}\right)_{y=h(t)}}{\rho_{\mathrm{nf}} v_{\mathrm{w}}^2},\quad Nu^* = \frac{-lk_{\mathrm{nf}}\left(\dfrac{\partial T}{\partial y}\right)_{y=h(t)}}{k_{\mathrm{f}} T_{\mathrm{H}}}. \tag{3.166}$$

In terms of Eq. (3.160), we obtain

$$\begin{aligned} C_f &= \left|(A_1 / A_4) f''(1)\right|,\\ Nu &= \left|A_3 \theta'(1)\right|. \end{aligned} \tag{3.167}$$

3.3.3.2 Solution with the DTM

Now the DTM has been applied into governing equations 3.15 and 3.16. Taking the differential transforms of Eqs (3.15) and (3.16) with respect to χ and considering $H=1$ gives:

$$\begin{aligned} &(k+1)(k+2)(k+3)(k+4)F[k+4]\\ &+S(A_1/A_4)\sum_{m=0}^{k}\left(\Delta[k-m-1](m+1)(m+2)(m+3)F[m+3]\right)\\ &-3S(A_1/A_4)(k+1)(k+2)F[k+2]\\ &-S(A_1/A_4)\sum_{m=0}^{k}\left((k-m+1)F[k-m+1](m+1)(m+2)F[m+2]\right)\\ &+S(A_1/A_4)\sum_{m=0}^{k}\left(F[k-m](m+1)(m+2)(m+3)F[m+3]\right)\\ &-Ha^2(A_5/A_4)(k+1)(k+2)F[k+2]=0,\\ &\Delta[m]=\begin{cases}1 & m=1\\ 0 & m\neq 1\end{cases} \end{aligned} \tag{3.168}$$

$$F[0]=0,\ F[1]=a_1,\ F[2]=0,\ F[3]=a_2 \tag{3.169}$$

$$
\begin{aligned}
&(k+1)(k+2)\Theta[k+2]+\Pr.S.\left(\frac{A_2}{A_3}\right)\sum_{m=0}^{k}(F[k-m](m+1)\Theta[m+1])\\
&-\Pr.S.\left(\frac{A_2}{A_3}\right)\sum_{m=0}^{k}(\Delta[k-m](m+1)\Theta[m+1])\\
&+\frac{\mathrm{Hs}}{A_3}\Theta[k]=0,\\
&\Delta[m]=\begin{cases}1 & m=1\\ 0 & m\neq 1\end{cases}
\end{aligned}
\tag{3.170}
$$

$$
\Theta[0]=a_3,\ \Theta[1]=0 \tag{3.171}
$$

where F [k] and Θ [k] are the differential transforms of $f(\eta)$, $\theta(\eta)$ and a_1, a_2, a_3 are constants that can be obtained through boundary condition. This problem can be solved as followed:

$$
\begin{aligned}
&F[0]=0,\ F[1]=a_1,\ F[2]=0,\ F[3]=a_2, F[4]=0\\
&F[5]=\frac{3}{20}S\,a_2\,(A_1/A_2)+\frac{1}{20}Sa_1a_2\ (A_1/A_2)+\frac{1}{20}a_1a_2+\frac{1}{20}\mathrm{Ha}^2\,a_2,\ldots
\end{aligned}
\tag{3.172}
$$

$$
\begin{aligned}
&\Theta[0]=a_3,\ \Theta[1]=0, \Theta[2]=-\frac{1}{2}\frac{\mathrm{Hs}}{A_3}\mathrm{a}_3, \Theta[3]=0.0,\\
&\Theta[4]=\frac{1}{12}\Pr S\left(\frac{A_2}{A_3}\right)\frac{\mathrm{Hs}}{A_3}a_3a_1, \Theta[5]=0,\ldots
\end{aligned}
\tag{3.173}
$$

The above process is continuous. By substituting Eqs (3.172) and (3.173) into the main equation based on DTM, it can be obtained that the closed form of the solutions is:

$$
\begin{aligned}
F(\eta)=\,&a_1\eta+a_2\eta^3\\
&+\left(\frac{3}{20}S\,a_2\,(A_1/A_2)+\frac{1}{20}Sa_1a_2\ (A_1/A_2)+\frac{1}{20}a_1a_2+\frac{1}{20}\mathrm{Ha}^2a_2\right)\eta^4+\ldots
\end{aligned}
\tag{3.174}
$$

$$
\theta(\eta)=a_3+\left(-\frac{1}{2}\frac{\mathrm{Hs}}{A_3}\mathrm{a}_3\right)\eta^2+\left(\frac{1}{12}\Pr S\left(\frac{A_2}{A_3}\right)\frac{\mathrm{Hs}}{A_3}a_3a_1\right)\eta^4+\ldots \tag{3.175}
$$

By substituting the boundary condition from Eq. (3.164) into Eqs (3.174) and (3.175) in point η = 1, the values of a_1, a_2, a_3 can be obtained. By substituting the obtained a_1, a_2, a_3 into Eqs (3.174) and (3.175), it the expression of F (η) and Θ (η) can be obtained.

3.3.3.3 Effects of active parameters

The kind of nanoparticle is a key factor for heat transfer enhancement. Tables 3.4 and 3.5 show a comparison among two different types of nanoparticles to find selecting which of them leads to higher enhancement for this problem. It is also found that choosing CuO

Table 3.4 Comparison of the Skin Friction Coefficient Between Al_2O_3–Water and CuO–Water at $\phi = 0.04$, Pr = 6.2

		C_f	
S	**Ha**	**CuO**	**Al_2O_3**
1	0	3.779263	3.629287
1	8	8.809621	8.676915
10	0	7.23826	6.897281
10	8	10.61304	10.33316

Table 3.5 Comparison of the Nusselt Number Between Al_2O_3–Water and CuO–Water at $\phi = 0.04$, Pr = 6.2

			Nu	
Hs	**S**	**Ha**	**CuO**	**Al_2O_3**
−1	1	0	0.717776	0.632233
−1	1	8	0.78598	0.700244
−1	10	0	0.291036	0.243075
−1	10	8	0.348638	0.293618
−10	10	8	3.532066	2.130418

as the nanoparticle leads to a greater amount of the skin friction coefficient and Nusselt number. So we select CuO–water as the nanofluid. The influences of the squeeze number, Hartmann number, heat source, and nanofluid volume fraction on flow and heat transfer characteristics are studied. Fig. 3.14 shows the effect of nanofluid volume fraction on the skin friction coefficient and Nusselt number. As the nanofluid volume fraction increases, the Nusselt number increases, and the skin friction coefficient decreases.

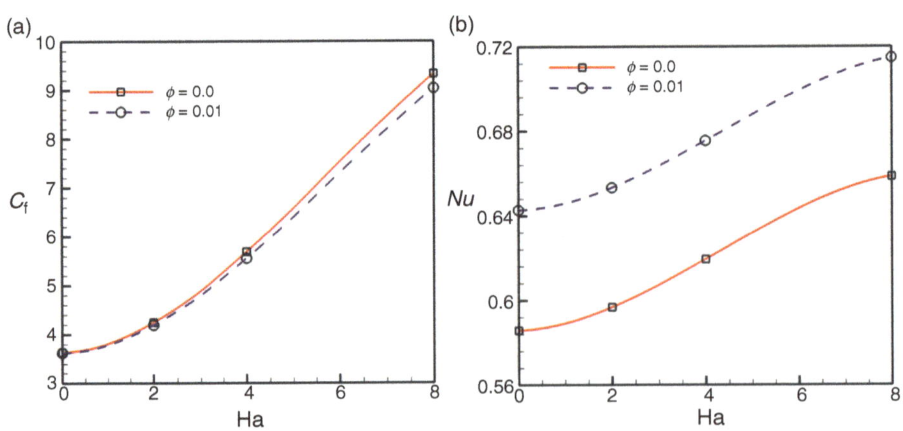

FIGURE 3.14 Effect of Volume Fraction of Nanofluid on the Skin Friction Coefficient and Nusselt Number When S = 1, Hs = −1, Pr = 6.2 (CuO–Water)

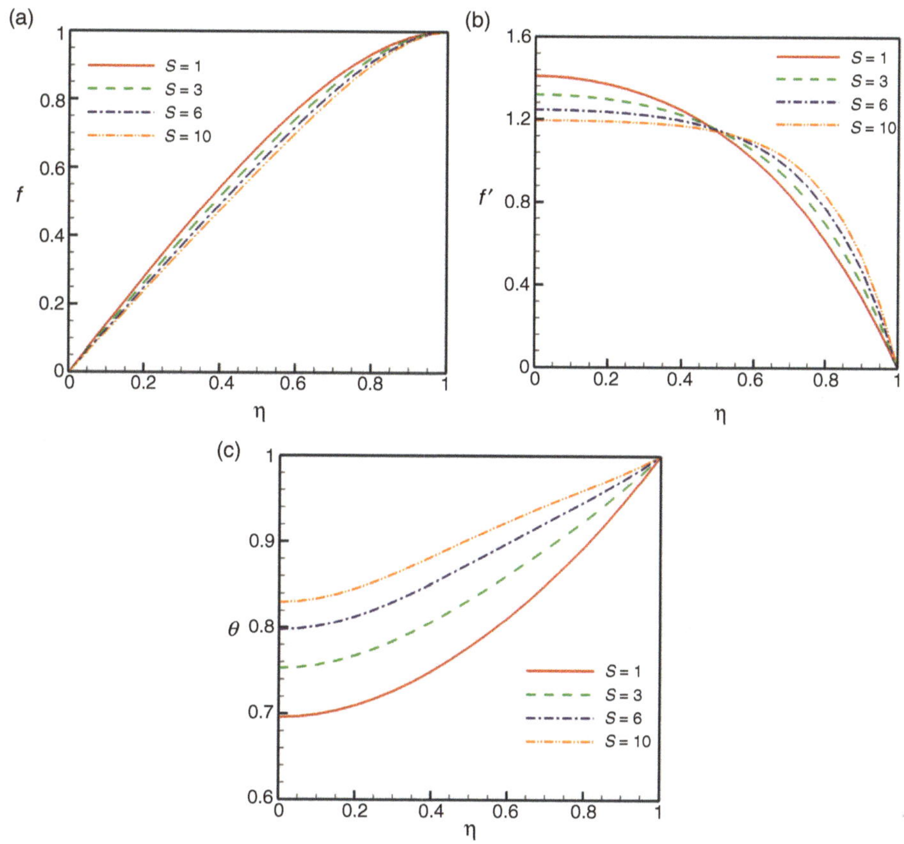

FIGURE 3.15 Effect of the Squeeze Number on Velocity and Temperature Profiles When Ha = 1, Hs = −1, Pr = 6.2 (CuO–Water)

Fig. 3.15 shows the effect of the squeeze number on the velocity and temperature profiles. It is important to note that the squeeze number (S) describes the movement of the plates ($S > 0$ corresponds to the plates moving apart, while $S < 0$ corresponds to the plates moving together the so-called squeezing flow). Vertical velocity decreases with increase of squeeze number while horizontal velocity has different behavior. It means that horizontal velocity decrease with increase of S when $\eta <$ 0.5 while opposite trend is observed for $\eta > 0.5$. Thermal boundary layer thickness increases with increase of squeeze number. Effect of the Hartmann number on velocity and temperature profiles is shown in Fig. 3.16. Effects of Hartmann number on velocity profiles are similar to that of squeeze number. While increasing in Hartmann number leads to decrease in thermal boundary layer thickness.

Fig. 3.17 depicts the effects of Hartmann number and squeeze number on the skin friction coefficient and Nusselt number. This figure shows that the Hartmann number has direct relationship with both the skin friction coefficient and Nusselt number. However, the squeeze number has a direct relationship with the skin

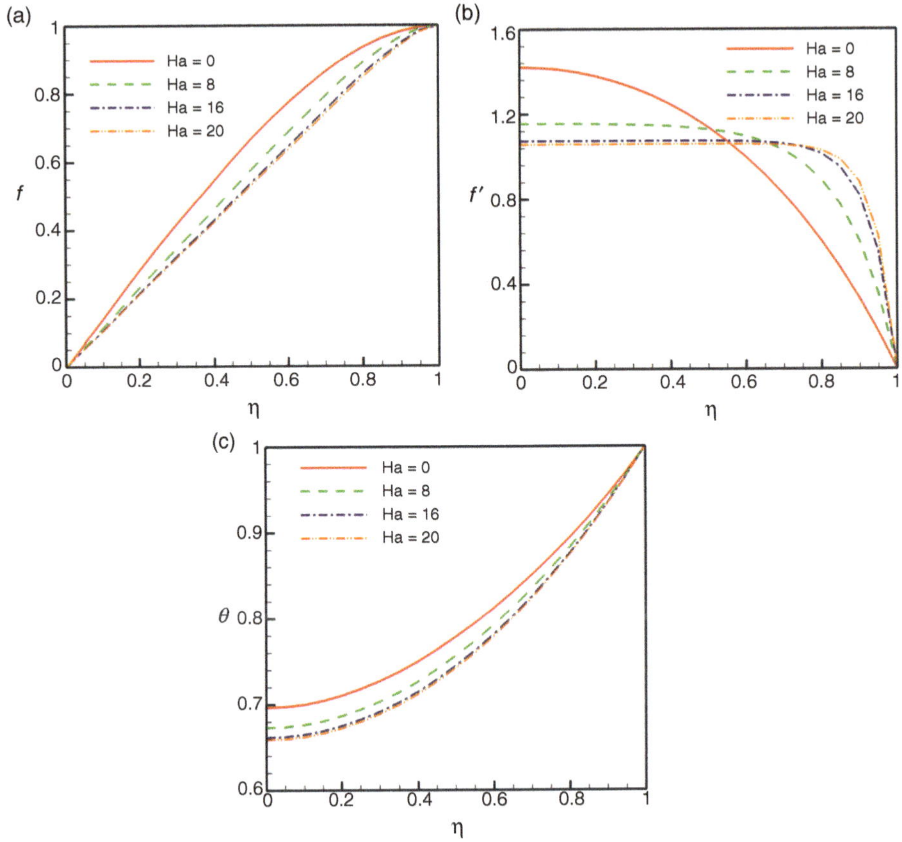

FIGURE 3.16 Effect of the Hartmann Number on Velocity and Temperature Profiles When $S = 1, Hs = -1, Pr = 6.2$ (CuO–Water)

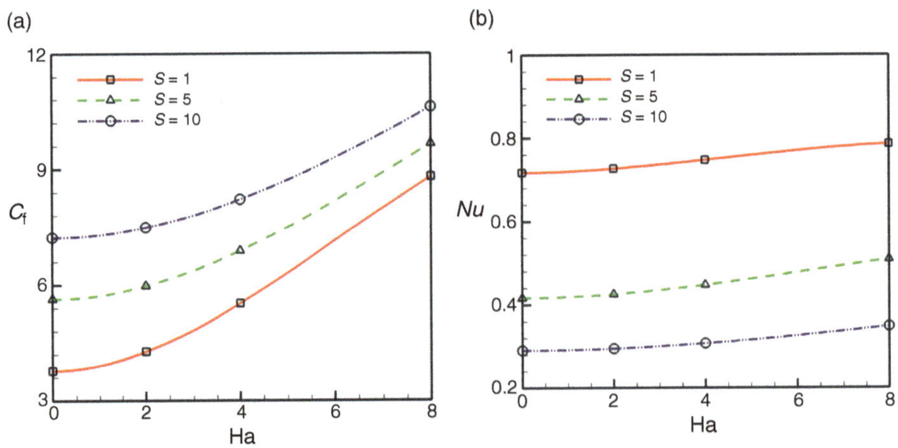

FIGURE 3.17 Effects of Hartmann Number and Squeeze Number on Skin Friction Coefficient and Nusselt Number When $\phi = 0.04, Hs = -1, Pr = 6.2$ (CuO–Water)

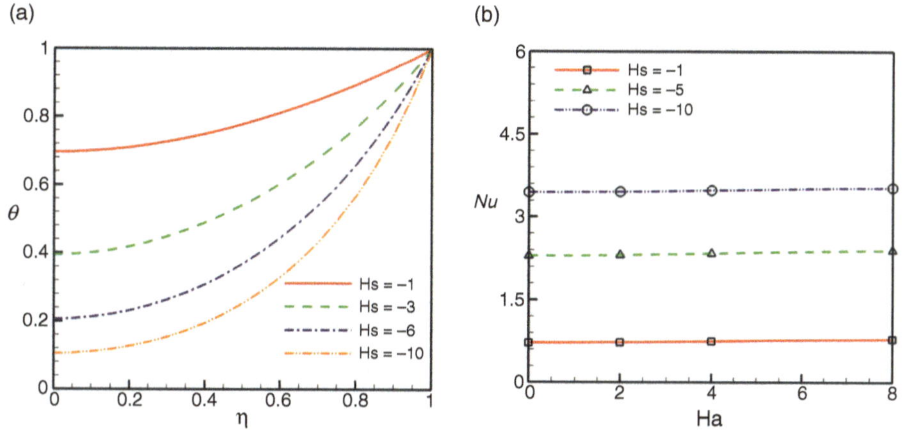

FIGURE 3.18 Effects of Heat Source Parameter on Temperature Profile and Nusselt Number When $\phi = 0.04$, Hs = −1, Pr = 6.2 (CuO–Water)

friction coefficient and a reverse relationship with the Nusselt number. The effects of heat source parameter on the temperature profile and Nusselt number are shown in Fig. 3.18. As the heat source parameter increases, the temperature boundary layer thickness decrease, and in turn, the Nusselt number increases.

3.3.4 NANOFLUID FLOW AND HEAT TRANSFER IN THE PRESENCE OF A MAGNETIC FIELD IN A ROTATING SYSTEM

3.3.4.1 Definition of the problem

Consider the steady flow of nanofluid between two horizontal parallel plates when the fluid and the plates rotate together around the axis which is normal to the plates with a constant angular velocity Ω. A Cartesian coordinate system (x, y, z) is considered as follows: the x-axis is along the plate, the y-axis is perpendicular to it, and the z-axis is normal to the x–y plane (see Fig. 3.19). The origin is located

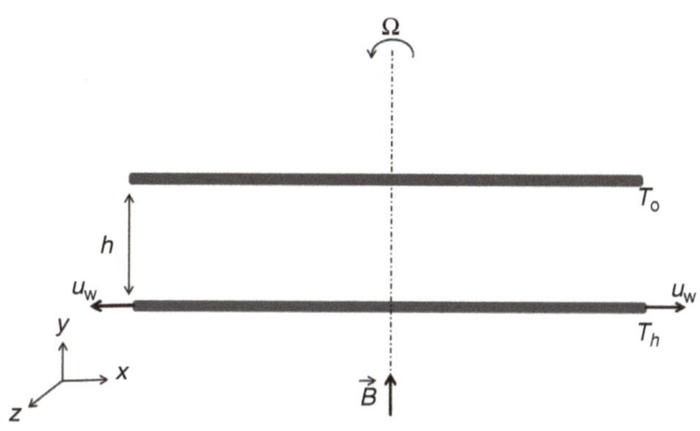

FIGURE 3.19 Schematic Theme of the Problem Geometry

at the lower plate, and the plates are located at $y = 0$ and $y = h$. The lower plate is being stretched by two equal opposite forces so that the position of the point (0,0,0) remains unchanged. A uniform magnetic flux with density of B_0 is acting along the y-axis. Under these assumptions, the Navier–Stokes and energy equations are [57]:

$$\frac{\partial u}{\partial x}+\frac{\partial v}{\partial y}+\frac{\partial w}{\partial z}=0 \tag{3.176}$$

$$\rho_{\mathrm{nf}}\left(u\frac{\partial u}{\partial x}+v\frac{\partial u}{\partial y}+2\Omega w\right)=-\frac{\partial p^*}{\partial x}+\mu_{\mathrm{nf}}\left(\frac{\partial^2 u}{\partial x^2}+\frac{\partial^2 u}{\partial y^2}\right)-\sigma_{\mathrm{nf}}B_0^2 u, \tag{3.177}$$

$$\rho_{\mathrm{nf}}\left(u\frac{\partial v}{\partial y}\right)=-\frac{\partial p^*}{\partial y}+\mu_{\mathrm{nf}}\left(\frac{\partial^2 v}{\partial x^2}+\frac{\partial^2 v}{\partial y^2}\right) \tag{3.178}$$

$$\rho_{\mathrm{nf}}\left(u\frac{\partial w}{\partial x}+v\frac{\partial w}{\partial y}-2\Omega w\right)=\mu_{\mathrm{nf}}\left(\frac{\partial^2 w}{\partial x^2}+\frac{\partial^2 w}{\partial y^2}\right)-\sigma_{\mathrm{nf}}B_0^2 w, \tag{3.179}$$

$$\begin{aligned}(\rho C_{\mathrm{p}})_{\mathrm{nf}}\left(u\frac{\partial T}{\partial x}+v\frac{\partial T}{\partial y}+w\frac{\partial T}{\partial z}\right)&=k_{\mathrm{nf}}\left(\frac{\partial^2 T}{\partial x^2}+\frac{\partial^2 T}{\partial y^2}+\frac{\partial^2 T}{\partial z^2}\right)\\&+\mu_{\mathrm{nf}}\left(2\left[\left(\frac{\partial u}{\partial x}\right)^2+\left(\frac{\partial v}{\partial y}\right)^2+\left(\frac{\partial w}{\partial z}\right)^2\right]+\left(\frac{\partial v}{\partial x}+\frac{\partial u}{\partial y}\right)^2+\left(\frac{\partial w}{\partial y}+\frac{\partial v}{\partial z}\right)^2+\left(\frac{\partial w}{\partial x}+\frac{\partial u}{\partial z}\right)^2\right)\end{aligned} \tag{3.180}$$

where u, v, and w denote the fluid velocity components along the x, y, and z directions; p^* is the modified fluid pressure; and T is temperature. The physical meanings of the other quantities are mentioned in the nomenclature. The absence of $\partial p^*/\partial z$ in Eq. (3.179) implies that there is a net cross-flow along the z-axis. The corresponding boundary conditions are:

$$\begin{aligned}&u=ax,\quad v=0,\quad w=0,\quad T=T_{\mathrm{h}}\quad \text{at}\quad y=0\\&u=0,\quad v=0,\quad w=0,\quad T=T_o\quad \text{at}\quad y=h\end{aligned} \tag{3.181}$$

The effective density, effective heat capacity, and electrical conductivity of the nanofluid are defined as [58]:

$$\begin{aligned}&\rho_{\mathrm{nf}}=(1-\phi)\rho_{\mathrm{f}}+\phi\rho_{\mathrm{p}},\\&(\rho C_{\mathrm{p}})_{\mathrm{nf}}=(1-\phi)(\rho C_{\mathrm{p}})_{\mathrm{f}}+\phi(\rho C_{\mathrm{p}})_{\mathrm{p}}\\&\frac{\sigma_{\mathrm{nf}}}{\sigma_{\mathrm{f}}}=1+\frac{3\left(\frac{\sigma_{\mathrm{s}}}{\sigma_{\mathrm{f}}}-1\right)\phi}{\left(\frac{\sigma_{\mathrm{s}}}{\sigma_{\mathrm{f}}}+2\right)-\left(\frac{\sigma_{\mathrm{s}}}{\sigma_{\mathrm{f}}}-1\right)\phi}\end{aligned} \tag{3.182}$$

$$k_{\mathrm{eff}}=k_{\mathrm{static}}+k_{\mathrm{Brownian}} \tag{3.183}$$

$$\frac{k_{\text{static}}}{k_{\text{f}}}=1+\frac{3\left(\frac{k_{\text{p}}}{k_{\text{f}}}-1\right)\phi}{\left(\frac{k_{\text{p}}}{k_{\text{f}}}+2\right)-\left(\frac{k_{\text{p}}}{k_{\text{f}}}-1\right)\phi} \tag{3.184}$$

$$k_{\text{Brownian}}=5\times10^{4}\,\beta\phi\rho_{\text{f}}c_{\text{p,f}}\sqrt{\frac{\kappa_{\text{b}}T}{\rho_{\text{p}}d_{\text{p}}}}f(T,\phi) \tag{3.185}$$

$$R_{\text{f}}+\frac{d_{\text{p}}}{k_{\text{p}}}=\frac{d_{\text{p}}}{k_{\text{p,eff}}} \tag{3.186}$$

$$\begin{aligned}g'(T,\phi,d_{\text{p}})&=(a_1+a_2\ln(d_{\text{p}})+a_3\ln(\phi)+a_4\ln(\phi)\ln(d_{\text{p}})+a_5\ln(d_{\text{p}})^2)\ln(T)\\&\quad+(a_6+a_7\ln(d_{\text{p}})+a_8\ln(\phi)+a_9\ln(\phi)\ln(d_{\text{p}})+a_{10}\ln(d_{\text{p}})^2)\end{aligned} \tag{3.187}$$

The coefficients a_i $(i=0\ldots10)$ are based on the type of nanoparticles Finally, the KKL correlation is written as:

$$k_{\text{Brownian}}=5\times10^{4}\,\phi\rho_{\text{f}}c_{\text{p,f}}\sqrt{\frac{\kappa_{\text{b}}T}{\rho_{\text{p}}d_{\text{p}}}}g'(T,\phi,d_{\text{p}}) \tag{3.188}$$

For the effective viscosity caused by micromixing in suspensions, we have:

$$\mu_{\text{eff}}=\mu_{\text{static}}+\mu_{\text{Brownian}}=\mu_{\text{static}}+\frac{k_{\text{Brownian}}}{k_{\text{f}}}\times\frac{\mu_{\text{f}}}{\text{Pr}_{\text{f}}} \tag{3.189}$$

where $\mu_{\text{static}}=\frac{\mu_{\text{f}}}{(1-\phi)^{2.5}}$ is the viscosity of the nanofluid, as given originally by Brinkman. The following nondimensional variables are introduced now:

$$\begin{aligned}&\eta=\frac{y}{h},\qquad u=axf'(\eta),\qquad v=-ah\,f(\eta),\qquad w=ax\,g(\eta)\\&\theta(\eta)=\frac{T-T_{\text{h}}}{T_0-T_{\text{h}}}\end{aligned} \tag{3.190}$$

where the prime denotes differentiation with respect to η.

Therefore, the governing momentum and energy equations for this problem are given in dimensionless form by:

$$f^{\text{iv}}-R\frac{A_1}{A_2}(f'f''-f\,f''')-2\,\text{Kr}\frac{A_1}{A_2}g'-A_5\frac{M}{A_2}f''=0 \tag{3.191}$$

$$g''-R\frac{A_1}{A_2}(f'g-fg')+2\text{Kr}\frac{A_1}{A_2}f'-A_5\frac{M}{A_2}g=0 \tag{3.192}$$

$$\theta''+\text{Pr}\frac{A_2A_3}{A_1A_4}\left[R\frac{A_1}{A_2}f\theta'+\text{Ec}\frac{A_1}{A_3}(4f'^2+g^2)+\text{Ec}_x\frac{A_1}{A_3}(f''^2+g'^2)\right]=0 \tag{3.193}$$

The dimensionless quantities in these equations are:

$$R=\frac{ah^2}{\upsilon_f},\ M=\frac{\sigma_f B_0^2 h^2}{\rho_f \upsilon_f},\ \mathrm{Kr}=\frac{\Omega h^2}{\upsilon_f},\mathrm{Pr}=\frac{\mu_f (\rho C_p)_f}{\rho_f k_f},$$
$$\mathrm{Ec}=\frac{\rho_f a^2 h^2}{(\rho C_p)_f(\theta_0-\theta_h)},\mathrm{Ec}_x=\frac{\rho_f a^2 x^2}{(\rho C_p)_f(\theta_0-\theta_h)} \tag{3.194}$$
$$A_1=\frac{\rho_{nf}}{\rho_f},A_2=\frac{\mu_{nf}}{\mu_f},A_3=\frac{(\rho C_p)_{nf}}{(\rho C_p)_f},A_4=\frac{k_{nf}}{k_f},A_5=\frac{\sigma_{nf}}{\sigma_f}$$

Subject to the boundary conditions:

$$\begin{aligned} &f=0, f'=1, g=0, \theta=1 \quad \text{at} \quad \eta=0 \\ &f=0, f'=0, g=0, \theta=0 \quad \text{at} \quad \eta=1 \end{aligned} \tag{3.195}$$

The physical quantity of interest in this problem is the skin friction coefficient C_f along the stretching wall, which is defined as

$$C_f=\frac{A_2}{A_1}f''(0) \tag{3.196}$$

The Nusselt number at the lower plate is defined as

$$\mathrm{Nu}=\left(\frac{k_{nf}}{k_f}\right)\theta'(0) \tag{3.197}$$

3.3.4.2 Solution with the GOHAM

In this section, according to GOHAM, we have

$$\begin{aligned}
&H(f,p)=(1-p)\left(f^{iv}-f_0^{iv}\right) \\
&+H_1(p)\left(f^{iv}-R\frac{A_1}{A_2}\left(f'f''-ff'''\right)-2Kr\frac{A_1}{A_2}g'-A_5\frac{M}{A_2}f''\right)=0, \\
&H(g,p)=(1-p)\left(g''-g_0''\right) \\
&+H_2(p)\left(g''-R\frac{A_1}{A_2}\left(f'g-fg'\right)+2Kr\frac{A_1}{A_2}f'-A_5\frac{M}{A_2}g\right)=0, \\
&H(\theta,p)=(1-p)\left(\theta''-\theta_0''\right) \\
&+H_3(p)\left(\begin{aligned}&\theta''+\mathrm{Pr}\frac{A_2A_3}{A_1A_4}\left[R\frac{A_1}{A_2}f\theta'+Ec\frac{A_1}{A_3}\left(4f'^2+g^2\right)\right.\\ &\left.+Ec_x\frac{A_1}{A_3}\left(f''^2+g'^2\right)\right]\end{aligned}\right)=0.
\end{aligned} \tag{3.198}$$

where primes denote differentiation with respect to η.

We consider f, θ, ϕ, and $H_1(p)$, $H_2(p)$, $H_3(p)$ as follows:

$$\begin{aligned} &f=f_0+pf_1+p^2f_2, g=g_0+pg_1+p^2g_2, \theta=\theta_0+p\theta_1+p^2\theta_2 \\ &H_1(p)=pC_1+p^2C_2, H_2(p)=pC_3+p^2C_4, H_3(p)=pC_5+p^2C_6. \end{aligned} \tag{3.199}$$

Substituting Eq. (3.199) into Eq. (3.198) and with some simplification and rearranging based on powers of p terms, we have:

$$
\begin{aligned}
&p^0:\\
&\quad f^{iv}=0,\\
&\quad g_0''=0,\\
&\quad \theta_0''=0.
\end{aligned}
\tag{3.200}
$$

$$
\begin{aligned}
&f(0)=0, \quad f'(0)=1, \quad f(1)=0, \qquad f'(1)=0\\
&g(0)=1, \quad g'(1)=0\\
&\theta(0)=1, \quad \theta(1)=0
\end{aligned}
$$

$$
\begin{aligned}
&p^1:\\
&C_1 f_0^{iv}-f_0^{iv}+f_1^{iv}-C_1\,R\,A_1/A_2 f_0' f_0''+C_1\,R\,A_1/A_2 f_0 f_0'''\\
&-2C_1\mathrm{Kr}\,A_1/A_2\,g_0'-C_1 M A_5/A_2 f_0''=0\\
&g_1''-g_0''+2C_3\mathrm{Kr}\,A_1/A_2 f_0'+C_3 g_0''-C_3 R\,A_1/A_2 g_0 f_0'\\
&+C_3 R\,A_1/A_2 g_0' f_0-C_3 M\,A_5/A_2\,g_0=0\\
&\theta_1''+\theta_0''(C_5-1)+C_5\,\mathrm{Pr}\frac{A_2}{A_4}\mathrm{Ec}_x\,(g_0')^2+C_5\,\mathrm{Pr}\frac{A_3}{A_4}Rf_0\theta_0'\\
&+C_5\,\mathrm{Pr}\frac{A_2}{A_4}(\mathrm{Ec}_x+4\mathrm{Ec})(f_0'')^2+C_5\,\mathrm{Pr}\frac{A_2}{A_4}Ec\,(g_0')^2=0\\
&f(0)=0, \quad f'(0)=0, \quad f(1)=0, \quad f'(1)=0\\
&g(0)=0, \quad g'(1)=0\\
&\theta(0)=0, \quad \theta(1)=0
\end{aligned}
\tag{3.201}
$$

Solving Eqs (3.200) and (3.201) with boundary conditions leads to:

$$
\begin{aligned}
&f_0(\eta)=\eta^3-2\eta^2+\eta\\
&g_0(\eta)=0\\
&\theta_0(\eta)=-\eta+1
\end{aligned}
\tag{3.202}
$$

$$
\begin{aligned}
f_1(\eta)=\;&2\frac{C_1}{A_2}\left(\frac{1}{140}RA_1\eta^7-\frac{1}{30}RA_1\eta^6+\frac{1}{15}RA_1\eta^5+\frac{1}{40}MA_5\eta^5\right.\\
&\left.-\frac{1}{12}MA_5\eta^4-\frac{1}{12}RA_1\eta^4\right)+\frac{1}{420A_2}C_1(54RA_1+77M\,A_5)\eta^3\\
&-\frac{1}{120A_2}C_1(9RA_1+14M\,A_5)\eta^2\\
g_1(\eta)=\;&-2\frac{C_3\mathrm{Kr}}{A_2}\left(\frac{1}{2}\eta^2+\frac{1}{4}\eta^4-\frac{2}{3}\eta^3\right)+\frac{1}{6}\frac{C_3\mathrm{Kr}A_1}{A_2}\eta\\
\theta_1(\eta)=\;&-C_3\frac{\mathrm{Pr}}{A_4}\left(-\frac{1}{20}A_3R\eta^5+\frac{1}{6}A_3R\eta^4-\frac{1}{6}A_3R\eta^3+3A_2Ec_x\eta^4\right.\\
&\left.-8A_2\mathrm{Ec}_x\eta^3+8A_2\mathrm{Ec}_x\eta^2+\frac{6}{5}A_2\mathrm{Ec}\,\eta^5+\frac{22}{3}A_2\mathrm{Ec}\,\eta^4-\frac{16}{3}A_2\mathrm{Ec}\,\eta^3+2A_2\mathrm{Ec}\,\eta^2\right)\\
&+\frac{1}{20}C_5\frac{\mathrm{Pr}}{A_4}\eta(-RA_3+60A_2\,\mathrm{Ec}_x+8A_2\,\mathrm{Ec})
\end{aligned}
\tag{3.203}
$$

The terms $f_2(\eta)$, $\theta_2(\eta)$, and $g_2(\eta)$ are too large to be mentioned. Therefore, the final expression for $f(\eta)$, $g(\eta)$, $\theta(\eta)$ is:

$$\begin{aligned} f(\eta) &= f_0(\eta)+f_1(\eta)+f_2(\eta) \\ g(\eta) &= g_0(\eta)+g_1(\eta)+g_2(\eta) \\ \theta(\eta) &= \theta_0(\eta)+\theta_1(\eta)+\theta_2(\eta) \end{aligned} \tag{3.204}$$

From Eq. (3.204), by substituting $f(\eta)$, $g(\eta)$, $\theta(\eta)$ into main equations, $R_1(\eta, C_1, C_2)$, $R_2(\eta, C_3, C_4)$ and $R_3(\eta, C_5, C_6)$ are obtained, and J is obtained in the following manner:

$$\begin{aligned} J_1(C_1,C_2) &= \int_a^b w_i \,.R_1(\tau,C_1,C_2)d\tau = 0, \\ J_2(C_3,C_4) &= \int_a^b w_i \,.R_2(\tau,C_3,C_4)d\tau = 0, \\ J_3(C_5,C_6) &= \int_a^b w_i \,.R_3(\tau,C_5,C_6)d\tau = 0. \end{aligned} \tag{3.205}$$

The constant C_i,(i = 1..6) are obtained from Eq. (3.205). By subsisting constants C_i,(i = 1..6) into Eq. (3.204), an expression for $f(\eta)$, $g(\eta)$, $\theta(\eta)$ is obtained.

3.3.4.3 Effects of active parameters

The effect of the magnetic parameter on velocity and temperature profiles is shown in Fig. 3.20. Rotational velocity increases with augmentation of the magnetic parameter. Thus, the presence of the magnetic field decreases the momentum boundary layer thickness because of the Lorenz force. The thermal boundary layer thickness increases with an increase of the magnetic parameter.

Fig. 3.21 shows the effects of the magnetic parameter and nanoparticle volume fraction on the skin friction coefficient and Nusselt number. Both the thermal and velocity boundary layer thicknesses increase with an increase of the nanoparticle volume fraction. Therefore, the skin friction coefficient has a reverse relationship with the nanoparticle volume fraction. The sensitivity of the thermal boundary layer thickness to the volume fraction of nanoparticles is related to the increased thermal conductivity of the nanofluid. In fact, higher values of thermal conductivity are accompanied by higher values of thermal diffusivity. The high value of thermal diffusivity causes a drop in the temperature gradients and accordingly increases the boundary thickness. This increase in the thermal boundary layer thickness reduces the Nusselt number; however, the Nusselt number is a multiplication of the temperature gradient, and the thermal conductivity ratio (conductivity of the nanofluid to the conductivity of the base fluid). Because the reduction in temperature gradient caused by the presence of nanoparticles is much smaller than the thermal conductivity ratio, an enhancement in the Nusselt number takes place by increasing the volume fraction of nanoparticles. The effect of the magnetic parameter on the skin friction coefficient and Nusselt number is the opposite of the volume fraction of nanoparticles.

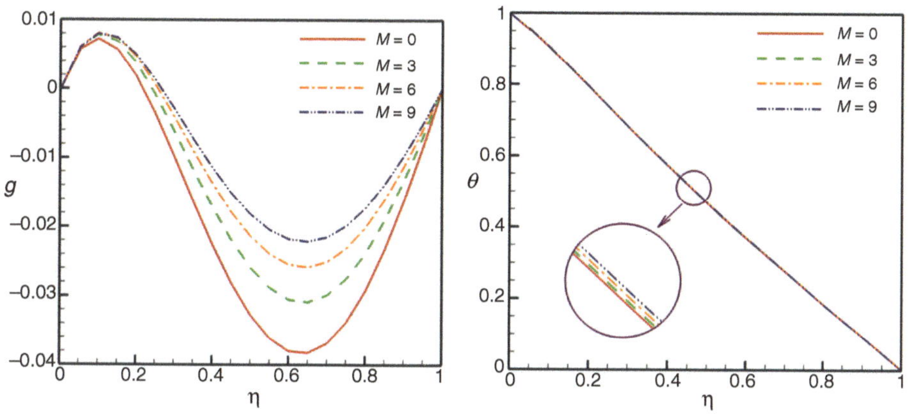

FIGURE 3.20 Effect of the Magnetic Parameter on Velocity and Temperature Profiles When Kr = 1, $R = 0.1$, Ec = Ec_x = 0.01 and $\phi = 0.04$ (Al_2O_3–Water)

Fig. 3.22 shows the effect of the Reynolds number on velocity, temperature profiles, skin friction coefficient, and Nusselt number. It is worth mentioning that the Reynolds number indicates the relative significance of the inertia effect compared with the viscous effect. Thus, both the velocity and temperature profiles decrease as Re increases, and in turn, increasing Reynolds number leads to an increase in the magnitude of the skin friction coefficient and Nusselt number.

The effect of the rotation parameter on the velocity, temperature profiles, skin friction coefficient, and Nusselt number is depicted in Fig. 3.23. As the rotation parameter increases, the Coriolis force increases, which causes both the rotational velocity and temperature profiles to increase. Also, increasing the rotation parameter leads to a decrease in the Nusselt number. The skin friction coefficient increases as the rotation parameter increases. The effect of the Eckert number on the temperature profile is shown in Fig. 3.24. The temperature increases with an increase of the viscous dissipation.

The effect of different kind of nanoparticles on velocity and temperature profiles is shown in Fig. 3.25. The skin friction coefficient and Nusselt number for different kind of nanoparticles are shown in Tables 3.6 and 3.7. It can be said that the skin friction coefficient and Nusselt number change by using different types of nanofluid. This means that the type of nanofluids has a great effect on the cooling and heating processes. Rotation velocity for CuO–water is lower than Al_2O_3–water; the opposite trend is observed for the temperature profile. By selecting CuO as the nanoparticle, a higher Nusselt number will be obtained; reach higher values of skin friction coefficient, Al_2O_3 should be selected as the nanoparticle.

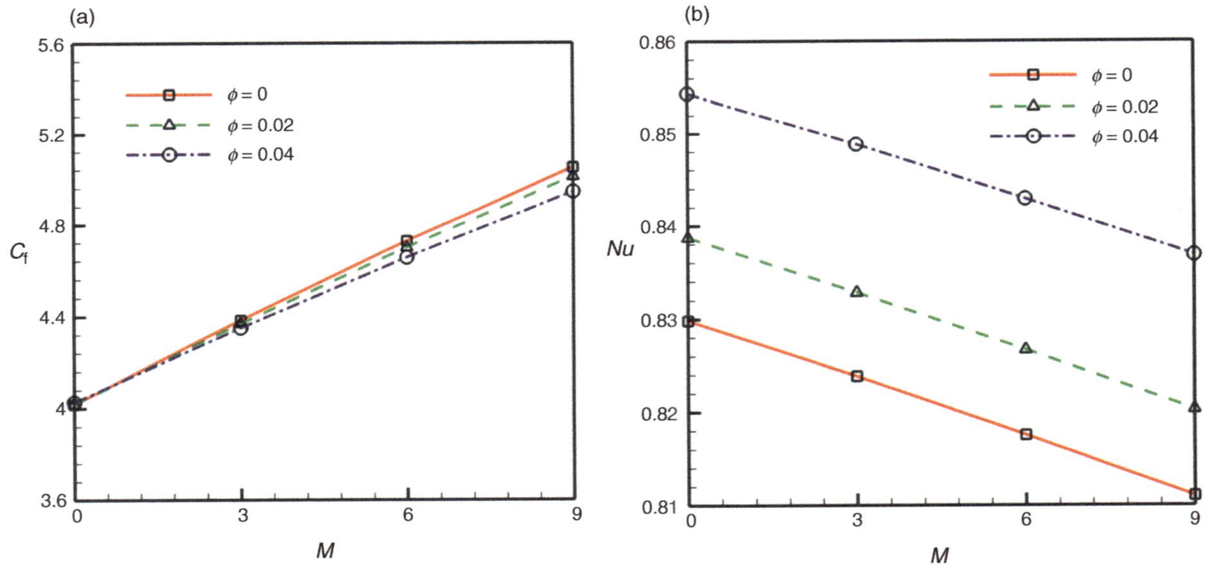

FIGURE 3.21 Effects of the Magnetic Parameter and Nanoparticle Volume Fraction on the Skin Friction Coefficient and Nusselt Number When Kr = 1, $R = 0.1$, Ec = Ec_x = 0.01 (Al_2O_3–Water)

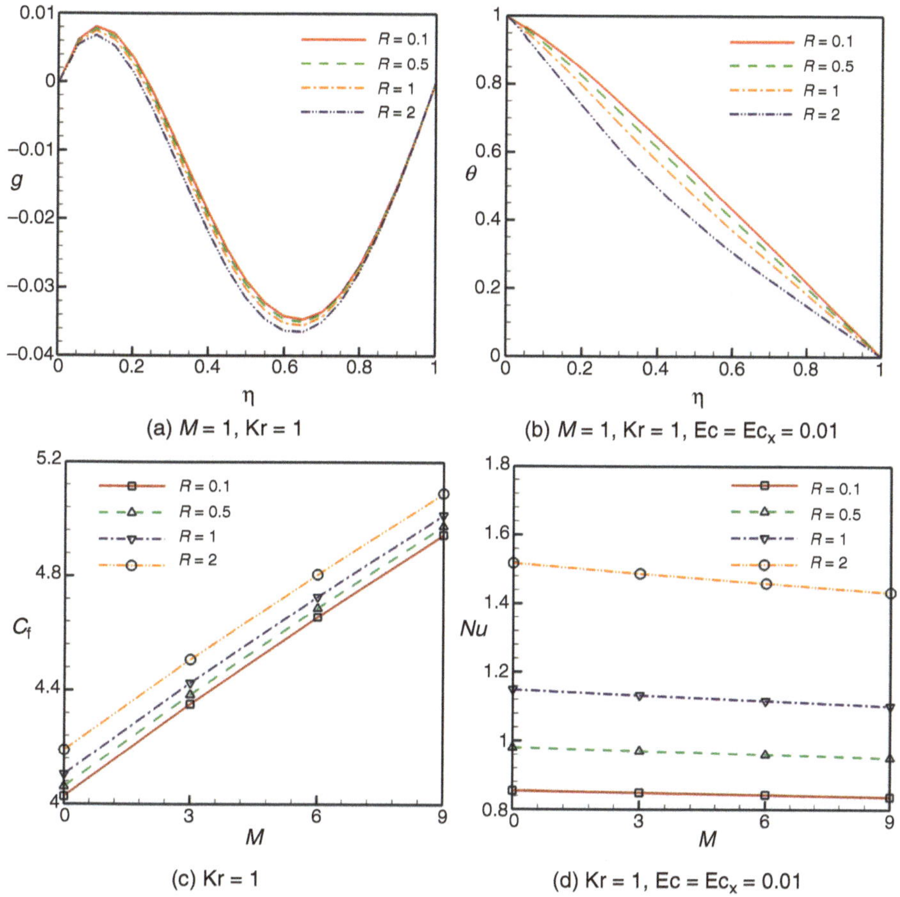

FIGURE 3.22 Effect of the Reynolds Number on Velocity (a), Temperature Profiles (b), Skin Friction Coefficient (c), and Nusselt Number (d) When $\phi = 0.04$ (Al_2O_3–Water)

3.3.5 NANOFLUID JEFFERY–HAMEL FLOW WITH A HIGH MAGNETIC FIELD DIVERGENT CHANNEL

3.3.5.1 Definition of the problem

Consider a system of cylindrical polar coordinates (r,θ,z) in which steady two-dimensional flow of an incompressible conducting viscous fluid from a source or sink at channel walls lie in planes and intersect in the axis of z. Assuming purely radial motion, which means that there is no change in the flow parameter along the z direction, the flow depends on r and θ, and we further assume that there is no magnetic field in the z direction.

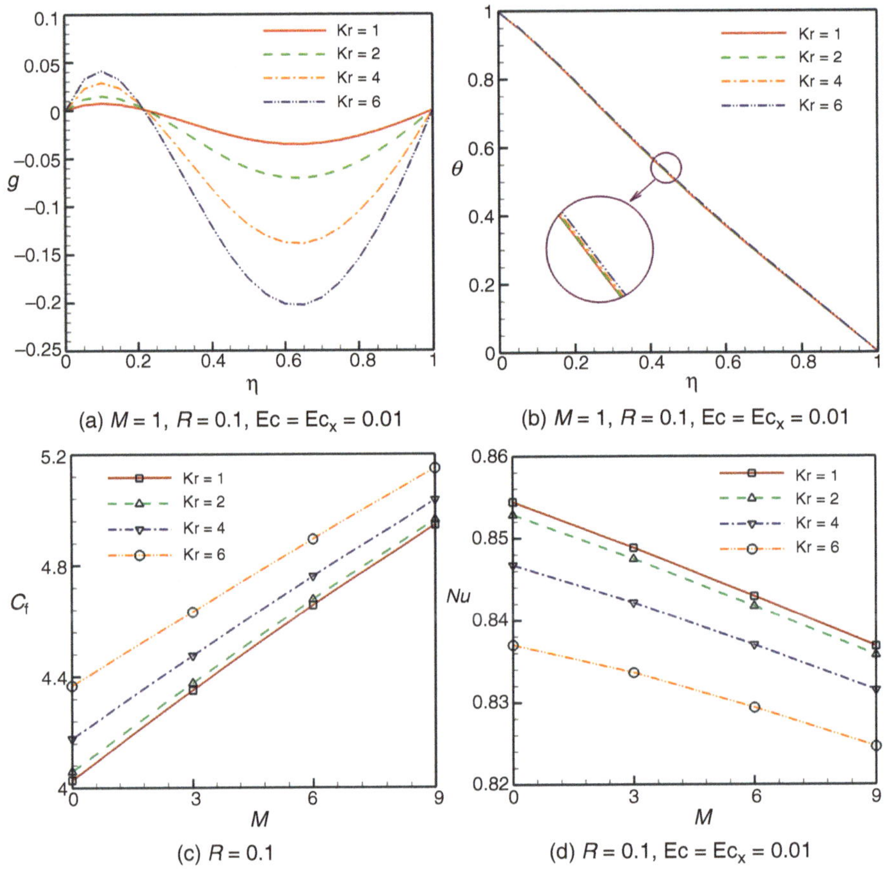

FIGURE 3.23 Effect of the Rotation Parameter on (Velocity (a), Temperature Profiles (b), Skin Friction Coefficient (c), and Nusselt Number (d) When $\phi = 0.04$ (Al_2O_3–Water)

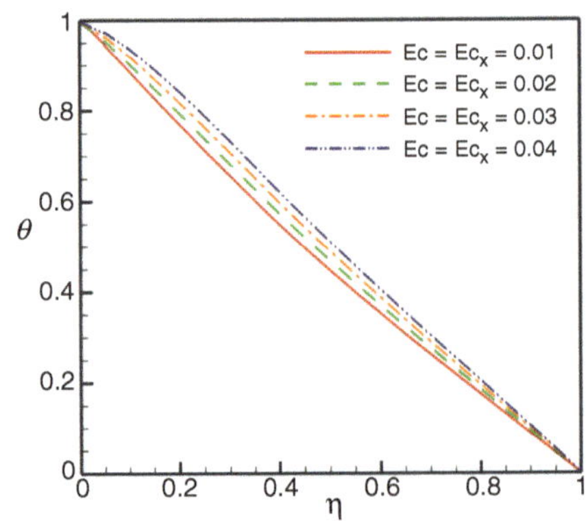

FIGURE 3.24 Effect of the Eckert Number on Temperature Profile When $M = 1$, $\mathrm{Kr} = 1$, $R = 0.1$ and $\phi = 0.04$ (Al_2O_3–Water)

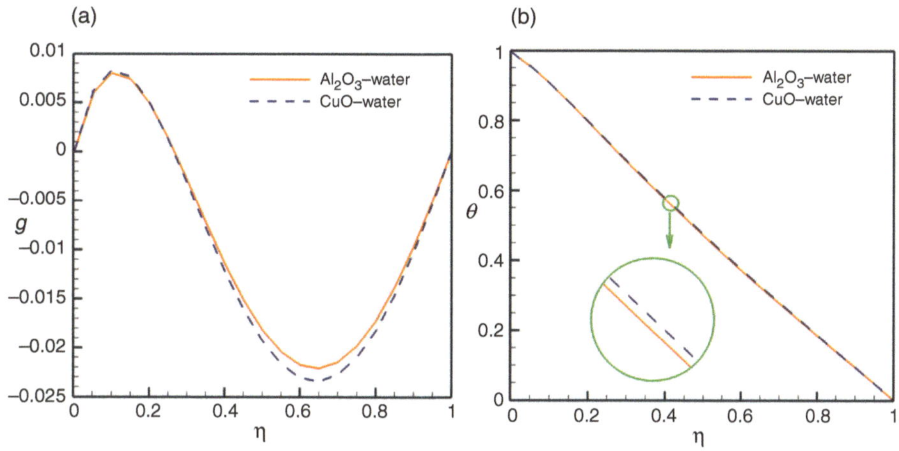

FIGURE 3.25 Velocity and Temperature Profiles for Different Kinds of Nanoparticles When $M = 1, \mathrm{Kr} = 1, R = 0.1, \mathrm{Ec} = \mathrm{Ec}_x = 0.01$ and $\phi = 0.04$

Table 3.6 Skin Friction Coefficient for Different Kinds of Nanoparticle When $R = 0.1, \mathrm{Kr} = 1, \mathrm{Ec} = \mathrm{Ec}_x = 0.01$

	$\phi = 0.01$		$\phi = 0.04$	
M	**Al_2O_3**	**CuO**	**Al_2O_3**	**CuO**
0	4.018423	4.009853	4.027509	3.890326
3	4.385039	4.384991	4.351153	4.187378
6	4.728087	4.723649	4.656276	4.46823
9	5.050931	5.039964	4.945242	4.734851

Table 3.7 Nusselt Number for Different Kinds of Nanoparticle When $R = 0.1, \mathrm{Kr} = 1, \mathrm{Ec} = \mathrm{Ec}_x = 0.01$

	$\phi = 0.01$		$\phi = 0.04$	
M	**Al_2O_3**	**CuO**	**Al_2O_3**	**CuO**
0	0.829851	0.884405	0.854409	0.958332
3	0.823884	0.878473	0.84884	0.952823
6	0.817558	0.872184	0.842939	0.946988
9	0.811036	0.865695	0.836839	0.940949

The reduced form of the continuity, Navier–Stokes, and Maxwell's equations is [44]:

$$\frac{\rho_{\text{nf}}}{r}\frac{\partial(ru(r,\theta))}{\partial r}(ru(r,\theta))=0 \tag{3.206}$$

$$u(r,\theta)\frac{\partial u(r,\theta)}{\partial r}=-\frac{1}{\rho_{\text{nf}}}\frac{\partial P}{\partial r}+\upsilon_{\text{nf}}\left[\frac{\partial^2 u(r,\theta)}{\partial r^2}+\frac{1}{r}\frac{\partial u(r,\theta)}{\partial r}+\frac{1}{r^2}\frac{\partial^2 u(r,\theta)}{\partial \theta^2}-\frac{u(r,\theta)}{r^2}\right]-\frac{\sigma B_0^2}{\rho_{\text{nf}} r^2}u(r,\theta) \tag{3.207}$$

$$\frac{1}{\rho_{\text{nf}} r}\frac{\partial P}{\partial \theta}-\frac{2\upsilon_{\text{nf}}}{r^2}\frac{\partial u(r,\theta)}{\partial \theta}=0 \tag{3.208}$$

where B_0 is the electromagnetic induction, σ is the conductivity of the fluid, $u(r)$ is the velocity along a radial direction, P is the fluid pressure, υ_{nf} is the coefficient of kinematic viscosity, and ρ_{nf} is the fluid density. The effective density ρ_{nf}, the effective dynamic viscosity μ_{nf}, and kinematic viscosity υ_{nf} of the nanofluid are given as [44]:

$$\rho_{\text{nf}}=\rho_{\text{f}}(1-\phi)+\rho_{\text{s}}\phi,\ \mu_{\text{nf}}=\frac{\mu_{\text{f}}}{(1-\phi)^{2.5}},\ \upsilon_{\text{nf}}=\frac{\mu_{\text{f}}}{\rho_{\text{nf}}} \tag{3.209}$$

where ϕ is the solid volume fraction.

Considering $u_\theta=0$ for a purely radial flow, one can define the velocity parameter as:

$$f(\theta)=ru(r) \tag{3.210}$$

Introducing $\eta=\frac{\theta}{\alpha}$ as the dimensionless degree, the dimensionless form of the velocity parameter can be obtained by dividing it to its maximum value as:

$$f(\eta)=\frac{f(\theta)}{f_{\max}} \tag{3.211}$$

Substituting Eq. (3.211) into Eqs (3.208) and (3.207) and eliminating P, one can obtain the ordinary differential equation for the normalized function profile as [44]:

$$f'''(\eta)+2\alpha\,\text{Re}.A^*(1-\phi)^{2.5}f(\eta)f'(\eta)+(4-(1-\phi)^{1.25}Ha)\alpha^2 f'(\eta)=0 \tag{3.212}$$

where A^* is a parameter Reynolds number, and the Hartmann number based on the electromagnetic parameter is introduced as the following form:

$$A^*=(1-\phi)+\frac{\rho_{\text{s}}}{\rho_{\text{f}}}\phi \tag{3.213}$$

$$\text{Re}=\frac{f_{\max}\alpha}{\upsilon_{\text{f}}}=\frac{U_{\max}r\alpha}{\upsilon_{\text{f}}}\begin{pmatrix}\text{divergent-channel}:\alpha>0,f_{\max}>0\\ \text{convergent-channel}:\alpha<0,f_{\max}<0\end{pmatrix} \tag{3.214}$$

$$\text{Ha}=\sqrt{\frac{\sigma B_0^2}{\rho_{\text{f}}\upsilon_{\text{f}}}} \tag{3.215}$$

With the following reduced form of boundary conditions:

$$f(0)=1, f'(0)=0, f(1)=0 \tag{3.216}$$

Physically, these boundary conditions mean that maximum values of velocity are observed at centerline (η = 0) as shown in Fig. 3.26, and we consider a fully developed velocity profile; thus, the rate of velocity is zero at (η = 0). Also, in fluid dynamics, the no-slip condition for fluid states that at a solid boundary, the fluid will have zero velocity relative to the boundary. The fluid velocity at all fluid–solid boundaries is equal to that of the solid boundary, so we can see that the value of the velocity is zero at (η = 1).

3.3.5.2 Solution with the ADM

According to Eq. (3.44), Eq. (3.212) must be written as following:

$$Lf = -2\,\mathrm{Re}.\alpha.A^{*}(1-\phi)^{2.5} f f' - (4-(1-\phi)^{1.25}\,\mathrm{Ha})\alpha^2 f' \tag{3.217}$$

where the differential operator L is given by $L = d^3/d\eta^3$. Assume the inverse of the operator L exists and that it can be integrated from 0 to η, that is, $L^{-1} = \int_0^{\eta}\int_0^{\eta}\int_0^{\eta}(\bullet)\,d\eta\,d\eta\,d\eta$.

Operating with L^{-1} on Eq. (3.217) and after exerting the boundary condition on it, we have:

$$f(\eta) = f(0) + f'(0)\eta + f''(0)\frac{\eta^2}{2} + L^{-1}(Nu) \tag{3.218}$$

where

$$\mathrm{Nu} = -2\,\mathrm{Re}\,\alpha.A^{*}(1-\phi)^{2.5} f(\eta) f'(\eta) - (4-(1-\phi)^{1.25}\,\mathrm{Ha}) f'(\eta) \tag{3.219}$$

ADM introduces the following expression:

$$f(\eta) = \sum_{m=0}^{\infty} f_m(\eta) \tag{3.220}$$

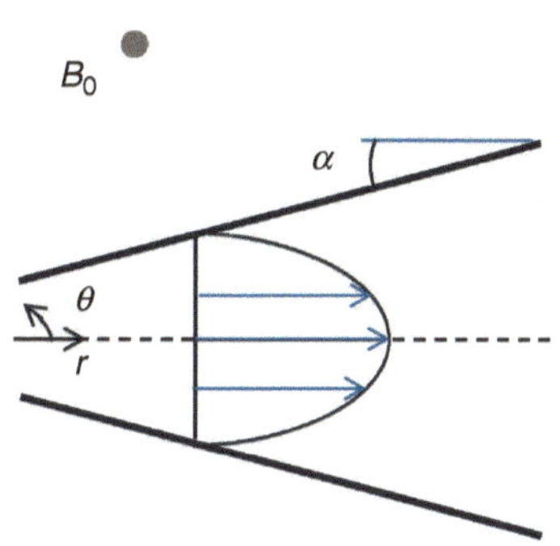

FIGURE 3.26 Geometry of the Magnetohydrodynamic Jeffery–Hamel Flow

$$f(\eta)=\sum_{m=0}^{\infty} f_m = f_0 + L^{-1}(\mathrm{Nu}) \tag{3.221}$$

To determine the components of $f_m(\eta)$, the $f_0(\eta)$ is defined by applying the boundary condition of Eq. (3.216) and by assuming $f''(0)=\beta$.

$$f_0(\eta)=1+\beta\frac{\eta^2}{2} \tag{3.222}$$

$$\begin{aligned} f_1(\eta)=&\left(\frac{-1}{120}\right)\alpha\,\mathrm{Re}.A^*(1-\phi)^{2.5}\beta^2\eta^6-\frac{1}{4}\left(\frac{1}{3}\alpha\,\mathrm{Re}.A^*(1-\phi)^{2.5}\beta\right.\\ &\left.+\frac{1}{6}(4-(1-\phi)^{1.25}\,\mathrm{Ha})\alpha^2\beta\right)\eta^4-\frac{1}{120}(4-(1-\phi)^{1.25}\,\mathrm{Ha})\alpha^2\beta\eta^5\\ &-\frac{1}{6}(4-\mathrm{Ha})\alpha^2\eta^3 \end{aligned} \tag{3.223}$$

$$\begin{aligned} f_2(\eta)=&\left(\frac{1}{10800}\right)\alpha^2(\mathrm{Re}.A^*(1-\phi)^{2.5})^2\beta^3\eta^{10}+\frac{1}{7560}\alpha^3\,\mathrm{Re}.A^*(1-\phi)^{2.5}(4-(1-\phi)^{1.25}\,\mathrm{Ha})\beta^2\eta\\ &-\frac{1}{8}\left(\frac{1}{21}\alpha\,\mathrm{Re}.A^*(1-\phi)^{2.5}\left(-\frac{1}{12}\alpha\,\mathrm{Re}.A^*(1-\phi)^{2.5}\beta-\frac{1}{24}(4-(1-\phi)^{1.25}\,\mathrm{Ha}\right)\alpha^2\beta\right)\beta\\ &-\frac{1}{420}\alpha^2(\mathrm{Re}.A^*(1-\phi)^{2.5})^2+\frac{1}{42}\alpha^2\,\mathrm{Re}.A^*(1-\phi)^{2.5}\beta\left(-\frac{1}{3}\alpha\,\mathrm{Re}.A^*(1-\phi)^{2.5}\beta\right.\\ &\left.-\frac{1}{6}(4-(1-\phi)^{1.25}\,\mathrm{Ha})\alpha^2\beta\right)-\frac{1}{840}\alpha^3\,\mathrm{Re}.A^*(1-\phi)^{2.5}((4-(1-\phi)^{1.25}\,\mathrm{Ha})\beta^2)\eta^8\\ &-\frac{1}{7}\left(-\frac{11}{360}\alpha^3\,\mathrm{Re}.A^*(1-\phi)^{2.5}(4-(1-\phi)^{1.25}\,\mathrm{Ha})\right)\beta-\frac{1}{720}(4-(1-\phi)^{1.25}\,\mathrm{Ha})^2\alpha^4\beta)\eta^7\\ &-\frac{1}{6}\left(\frac{1}{10}\alpha\,\mathrm{Re}.A^*(1-\phi)^{2.5}\left(-\frac{1}{3}\alpha\,\mathrm{Re}.A^*(1-\phi)^{2.5}\beta-\frac{1}{6}(4-(1-\phi)^{1.25}\,\mathrm{Ha})\alpha^2\beta\right)\right.\\ &\left.+\frac{1}{20}(4-(1-\phi)^{1.25}\,\mathrm{Ha})\alpha^2\left(-\frac{1}{3}\alpha\,\mathrm{Re}.A^*(1-\phi)^{2.5}\beta-\frac{1}{6}(4-(1-\phi)^{1.25}\,\mathrm{Ha})\alpha^2\beta\right)\right)\eta^6\\ &-\frac{1}{5}\left(-\frac{1}{12}\alpha^3\,\mathrm{Re}.A^*(1-\phi)^{2.5}(4-(1-\phi)^{1.25}\,Ha)-\frac{1}{720}(4-(1-\phi)^{1.25}\,\mathrm{Ha})^2\alpha^4\right)\eta^5\\ &+\frac{1}{40320}(4-(1-\phi)^{1.25}\,\mathrm{Ha})^2\alpha^4\beta\eta^8-\frac{1}{120}(4-(1-\phi)^{1.25}\,\mathrm{Ha})\alpha^2\left(-\frac{1}{12}\alpha\,\mathrm{Re}\beta\right.\\ &\left.-\frac{1}{24}(4-(1-\phi)^{1.25}\,\mathrm{Ha})\alpha^2\beta\right)\eta^7+\frac{1}{720}(4-(1-\phi)^{1.25}\,\mathrm{Ha})^2\alpha^4\eta^6 \end{aligned} \tag{3.224}$$

$f_3(\eta), f_4(\eta), \ldots$ can be determined in similar way.

Using $f(\eta)=\sum_{m=0}^{\infty} f_m(\eta)=f_0(\eta)+f_1(\eta)+f_2(\eta)+f_3(\eta)+\ldots.$, thus

$$\begin{aligned} f(\eta)=&1+\beta\frac{\eta^2}{2}+\left(\frac{-1}{120}\right)\alpha\,\mathrm{Re}.A^*(1-\phi)^{2.5}\beta^2\eta^6-\frac{1}{4}\left(\frac{1}{3}\alpha\,\mathrm{Re}.A^*(1-\phi)^{2.5}\beta\right.\\ &\left.+\frac{1}{6}(4-(1-\phi)^{1.25}\,\mathrm{Ha})\alpha^2\beta\right)\eta^4-\frac{1}{120}(4-(1-\phi)^{1.25}\,\mathrm{Ha})\alpha^2\beta\eta^5\\ &-\frac{1}{6}(4-\mathrm{Ha})\alpha^2\eta^3+\ldots \end{aligned} \tag{3.225}$$

For the complete solution of Eq. (3.225), β should be determined, with boundary condition of $f(1) = 0$.

3.3.5.3 Effects of active parameters

Fig. 3.27 shows the effect of nanoparticle volume fraction on velocity profiles. This figure indicates that increasing nanoparticle volume fraction leads to a decrease in the velocity boundary layer thickness. Also, it can be seen that for higher values of Reynolds number, this decrement is more sensible. The effect of the Hartmann number, Reynolds number, and angle of the channel on velocity profiles is shown in Fig. 3.28. It is worth mentioning that the Reynolds number indicates the relative significance of the inertia effect compared with the viscous effect. Thus, the velocity profile decreases as Re increases, and in turn, increasing Re leads to increase in the magnitude of the skin friction coefficient. Generally, when the magnetic field is imposed on the channel, the velocity field is suppressed owing to the retarding effect of the Lorenz force. Thus, the presence of a magnetic field increases the momentum boundary layer thickness. The results show moderate increases in the velocity with increasing Hartmann numbers at a small angle ($\alpha = 2.5$ degrees), and the difference between velocity profiles is more noticeable at greater angles. Backflow is excluded in converging channels, but it may occur for large Reynolds numbers in diverging channels. For a specified opening angle, after a critical Reynolds number, we observe that separation and backflow start. By increasing the Hartmann number, this phenomenon is eliminated, and higher values of Reynolds number need a greater magnetic field to eliminate the back flow. The effects of the angle of the channel, Reynolds number, and Hartmann number on the skin friction coefficient ($f''(0)$) are shown in Fig. 3.29. The skin friction coefficient is an increasing function of the Reynolds number, opening angle, and nanoparticle volume friction but decease the function of the Hartmann number.

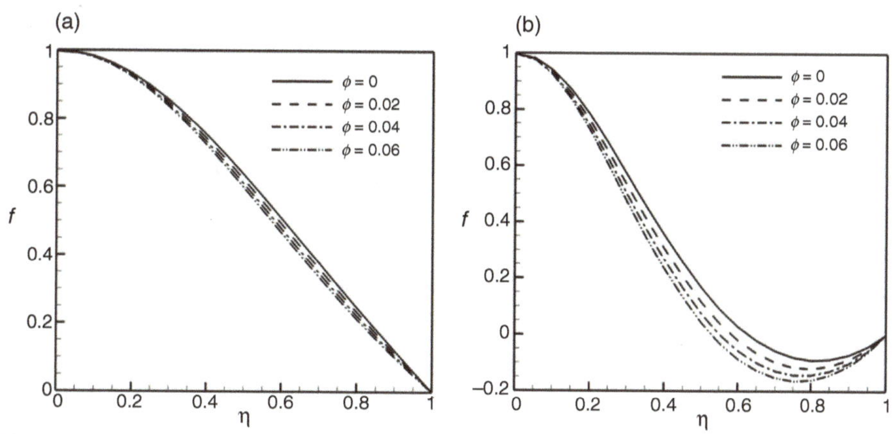

FIGURE 3.27 Effect of Nanoparticle Volume Fraction on Velocity Profiles for Cu–Water When (a) Ha = 50, Re = 50; (b) Ha = 50, Re = 100

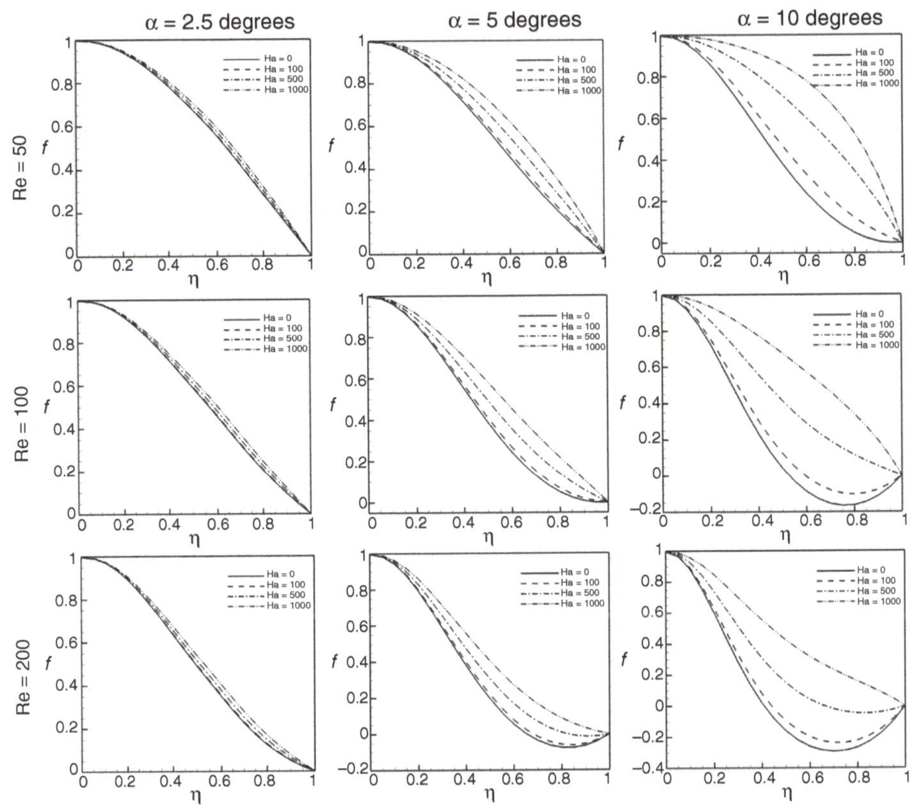

FIGURE 3.28 Effect of the Hartmann Number, Reynolds Number, and Angle of the Channel on Velocity Profiles for Cu–Water When $\phi = 0.06$

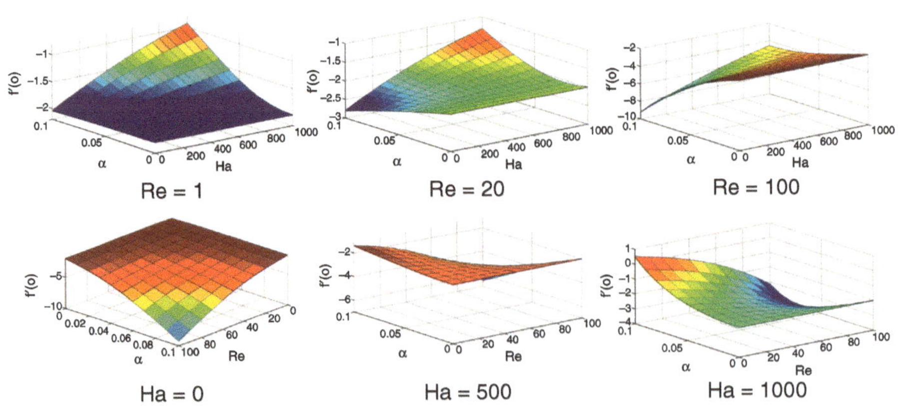

FIGURE 3.29 Effects of the Angle of the Channel, Reynolds Number, and Hartmann Number on the Skin Friction Coefficient $f''(0)$ When $\phi = 0.06$

3.3.6 EFFECT OF THERMAL RADIATION ON MAGNETOHYDRODYNAMIC NANOFLUID FLOW AND HEAT TRANSFER BY MEANS OF THE TWO-PHASE MODEL

3.3.6.1 Definition of the problem

Consider the steady nanofluid flow between two horizontal parallel plates when the fluid and the plates rotate together around the y-axis, which is normal for the plates with an angular velocity. A Cartesian coordinate system is considered as followed: the x-axis is along the plate, the y-axis is perpendicular to it, and the z-axis is normal to the x–y plane (see Fig. 3.30). The plates are located at $y = 0$ and $y = h$. The lower plate is being stretched by two equal and opposite forces so that the position of the point (0,0,0) remains unchanged. A uniform magnetic flux with density B_0 is acting along the y-axis about which the system is rotating. The governing equations in a rotating frame of reference are [59]:

$$\frac{\partial u}{\partial x}+\frac{\partial v}{\partial y}+\frac{\partial w}{\partial z}=0 \tag{3.226}$$

$$\rho_{\mathrm{f}}\left(u\frac{\partial u}{\partial x}+v\frac{\partial u}{\partial y}+2\Omega w\right)=-\frac{\partial p^{*}}{\partial x}+\mu\left(\frac{\partial^2 u}{\partial x^2}+\frac{\partial^2 u}{\partial y^2}\right)-\sigma B_0^2 u, \tag{3.227}$$

$$\rho_{\mathrm{f}}\left(u\frac{\partial v}{\partial y}\right)=-\frac{\partial p^{*}}{\partial y}+\mu\left(\frac{\partial^2 v}{\partial x^2}+\frac{\partial^2 v}{\partial y^2}\right), \tag{3.228}$$

$$\rho_{\mathrm{f}}\left(u\frac{\partial w}{\partial x}+v\frac{\partial w}{\partial y}-2\Omega u\right)=\mu\left(\frac{\partial^2 w}{\partial x^2}+\frac{\partial^2 w}{\partial y^2}\right)-\sigma B_0^2 w, \tag{3.229}$$

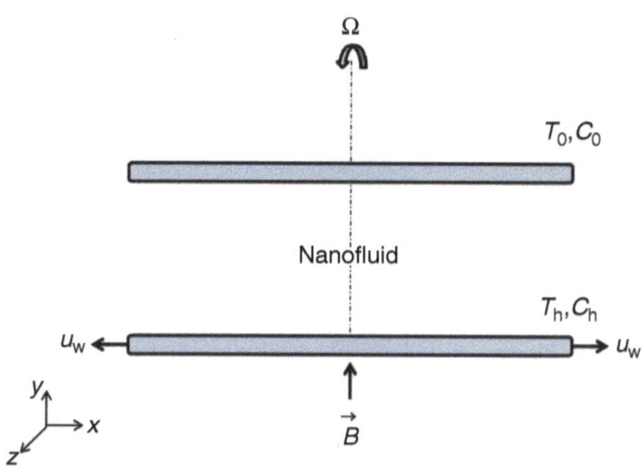

FIGURE 3.30 Geometry of Problem

$$u\frac{\partial T}{\partial x}+v\frac{\partial T}{\partial y}+w\frac{\partial T}{\partial z}=\alpha\left(\frac{\partial^2 T}{\partial x^2}+\frac{\partial^2 T}{\partial y^2}+\frac{\partial^2 T}{\partial z^2}\right)+ \\ +\frac{(\rho c_P)_p}{(\rho c_P)_f}\left[\begin{array}{l} D_B\left\{\frac{\partial C}{\partial x}.\frac{\partial T}{\partial x}+\frac{\partial C}{\partial y}.\frac{\partial T}{\partial y}+\frac{\partial C}{\partial z}.\frac{\partial T}{\partial z}\right\} \\ +(D_T/T_c)\left\{\left(\frac{\partial T}{\partial x}\right)^2+\left(\frac{\partial T}{\partial y}\right)^2+\left(\frac{\partial T}{\partial z}\right)^2\right\}\end{array}\right]-\frac{1}{(\rho C_p)_f}\frac{\partial q_r}{\partial y}, \tag{3.230}$$

$$u\frac{\partial C}{\partial x}+v\frac{\partial C}{\partial y}+w\frac{\partial C}{\partial z}=D_B\left(\frac{\partial^2 C}{\partial x^2}+\frac{\partial^2 C}{\partial y^2}+\frac{\partial^2 C}{\partial z^2}\right)+\left(\frac{D_T}{T_0}\right)\left\{\frac{\partial^2 T}{\partial x^2}+\frac{\partial^2 T}{\partial y^2}+\frac{\partial^2 T}{\partial z^2}\right\} \tag{3.231}$$

where the radiation heat flux q_r is considered according to the Rosseland approximation such that $q_r=-\frac{4\sigma_e}{3\beta_R}\frac{\partial T^4}{\partial y}$, where σ_e and β_R are the Stefan–Boltzmann constant and the mean absorption coefficient, respectively. According to Sheikholeslami and Ganji [60], the fluid-phase temperature differences within the flow are assumed to be sufficiently small so that T^4 may be expressed as a linear function of temperature. This is done by expanding T^4 in a Taylor series about the temperature T_c and neglecting higher order terms to yield $T^4 \cong 4T_c^3T-3T_c^4$.
where u, v, and z are the velocities in the x, y, and z directions, respectively; T is the temperature; C is the concentration; ρ_f is the base fluid's density; μ is the dynamic viscosity; k is the thermal conductivity; c_P is the specific heat of nanofluid; and D_B is the diffusion coefficient of the diffusing species. Also, p^* is the modified fluid pressure. The absence of $\frac{\partial p^*}{\partial z}$ in Eq. (3.229) implies that there is a net cross-flow along the z-axis. The relevant boundary conditions are:

$$\begin{array}{ll} u=ax,\ v=0, w=0, T=T_h, C=C_h & \text{at}\quad y=0 \\ u=0,\ v=0, w=0, T=T_0, C=C_0 & \text{at}\quad y=+h \end{array} \tag{3.232}$$

The following nondimensional variables are introduced:

$$\begin{array}{llll} \eta=\frac{y}{h}, & u=axf'(\eta), & v=-ah\,f(\eta), & w=ax\,g(\eta) \\ \theta(\eta)=\frac{T-T_h}{T_0-T_h}, & \phi(\eta)=\frac{C-C_h}{C_0-C_h} & & \end{array} \tag{3.233}$$

where a prime denotes differentiation with respect to η.

Substituting (3.233) in Eqs (3.226)–(3.231), we have [61]:

$$-\frac{1}{\rho h}\frac{\partial p^*}{\partial \eta}=a^2x\left[f'-ff''-\frac{f'''}{R}+\frac{M}{R}+\frac{2K_r}{R}g\right], \tag{3.234}$$

$$-\frac{1}{\rho h}\frac{\partial p^*}{\partial \eta}=a^2h\left[ff'+\frac{1}{R}f''\right], \tag{3.235}$$

$$g'' - R(f'g - fg') + 2K_r f' - Mg = 0 \tag{3.236}$$

and the nondimensional quantities are defined through in which R is the viscosity parameter, M is the magnetic parameter, and K_r is the rotation parameter.

$$R = \frac{ah^2}{\upsilon},\ M = \frac{\sigma B_0^2 h^2}{\rho\upsilon},\ K_r = \frac{\Omega h^2}{\upsilon} \tag{3.237}$$

Eq. (3.234) with the help of (3.235) can be written as:

$$f''' - R\left[f'^2 - ff''\right] - 2K_r^2 g - M^2 f' = A \tag{3.238}$$

Differentiation of Eq. (3.238) with respect to η gives:

$$f^{iv} - R(2f'f'' - f'f'' - ff''') - 2K_r g' - Mf'' = 0 \tag{3.239}$$

Therefore, the governing equations and boundary conditions for this case in nondimensional form are given by:

$$f^{iv} - R(f'f'' - ff''') - 2K_r g' - Mf'' = 0 \tag{3.240}$$

$$g'' - R(f'g - fg') + 2K_r f' - Mg = 0 \tag{3.241}$$

$$\left(1 + \frac{4}{3}\mathrm{Rd}\right)\theta'' + \mathrm{Pr}\,Rf\theta' + \mathrm{Nb}\phi'\theta' + \mathrm{Nt}\theta'^2 = 0, \tag{3.242}$$

$$\phi'' + R.Scf\phi' + \frac{\mathrm{Nt}}{\mathrm{Nb}}\theta'' = 0, \tag{3.243}$$

With these boundary conditions:

$$\begin{aligned} &f = 0, f' = 1, g = 0, \theta = 1, \phi = 1 && \text{at } \eta = 0 \\ &f = 0, f' = 0, g = 0, \theta = 0, \phi = 0 && \text{at } \eta = 1 \end{aligned} \tag{3.244}$$

where Pr is the Prandtl number, Sc is the Schmidt number, Rd is Radiation parameter, Nb is the Brownian motion parameter, and Nt is the thermophoretic parameter, which are defined as:

$$\begin{aligned} &\mathrm{Pr} = \frac{\mu}{\rho_f\alpha},\quad Sc = \frac{\mu}{\rho_f D},\quad \mathrm{Rd} = 4\sigma_e T_c^3/(\beta_R k), \\ &\mathrm{Nb} = (\rho c)_p D_B (C_h)/((\rho c)_f \alpha), \\ &\mathrm{Nt} = (\rho c)_p D_T (T_H)/[(\rho c)_f \alpha T_c]. \end{aligned} \tag{3.245}$$

The skin friction coefficient C_f along the stretching wall and Nusselt number Nu along the stretching wall are defined as

$$\tilde{C}_f = (Rx/h)C_f = f''(0),\quad \mathrm{Nu} = -\left(1 + \frac{4}{3}\mathrm{Rd}\right)\theta'(0) \tag{3.246}$$

3.3.6.2 Solution with the HAM

For HAM solutions of the governing equations, we choose the initial approximations of $f(\eta)$, $g(\eta)$, and $\theta(\eta)$ as follows:

$$f_0(\eta)=\eta^3-2\eta^2+\eta \tag{3.247}$$

$$g_0(\eta)=0 \tag{3.248}$$

$$\begin{aligned}\theta_0(\eta)&=1-\eta\\ \phi_0(\eta)&=1-\eta\end{aligned} \tag{3.249}$$

And the auxiliary linear operators are

$$L_1(f)=f^{iv} \tag{3.250}$$

$$L_2(\theta)=g'' \tag{3.251}$$

$$L_3(\theta)=\theta'' \tag{3.252}$$

$$L_4(\theta)=\phi'' \tag{3.253}$$

These auxiliary linear operators satisfy

$$L_1(C_0+C_1\eta+C_2\eta^2+C_3\eta^3) \tag{3.254}$$

$$L_2(C_4+C_5\eta) \tag{3.255}$$

$$L_3(C_6+C_7\eta) \tag{3.256}$$

$$L_4(C_8+C_9\eta) \tag{3.257}$$

where $C_i\,(i=0..9)$ are constants. Introducing nonzero auxiliary parameters $\hbar_1,\hbar_2,\hbar_3$, and $\hbar_4$, we develop the zeroth-order deformation problems as follows:

$$(1-p)L\left[f(\eta;p)-f_0(\eta)\right]=p\hbar_1 N_1\left[f(\eta;p),g(\eta;p),\theta(\eta;p),\phi(\eta;p)\right] \tag{3.258}$$

$$f(0;p)=0,f(1;p)=0,f'(0;p)=0,f'(1;p)=1 \tag{3.259}$$

$$(1-p)L[g(\eta;p)-g_0(\eta)]=p\hbar_2 N_2[f(\eta;p),g(\eta;p),\theta(\eta;p),\phi(\eta;p)] \tag{3.260}$$

$$g(0;p)=0,g(1;p)=0 \tag{3.261}$$

$$(1-p)L[\theta(\eta;p)-\theta_0(\eta)]=p\hbar_3 N_3[f(\eta;p),g(\eta;p),\theta(\eta;p),\phi(\eta;p)] \tag{3.262}$$

$$\theta(0;p)=1,\theta(1;p)=0 \tag{3.263}$$

$$(1-p)L\left[\phi(\eta;p)-\phi_0(\eta)\right]=p\hbar_4 N_4\left[f(\eta;p),g(\eta;p),\theta(\eta;p),\phi(\eta;p)\right] \tag{3.264}$$

$$\phi(0;p)=1,\phi(1;p)=0 \tag{3.265}$$

where nonlinear operators, N_1, N_2, N_3, and N_4 are defined as

$$N_1[f(\eta;p),g(\eta;p),\theta(\eta;p)]=\frac{\partial^4 f(\eta;p)}{\partial\eta^4}-R\left(\left(\frac{\partial f(\eta;p)}{\partial\eta}\right)\left(\frac{\partial^2 f(\eta;p)}{\partial\eta^2}\right)\right.$$
$$\left.f(\eta;p)\frac{\partial^3 f(\eta;p)}{\partial\eta^3}\right)-2Kr\frac{\partial g(\eta;p)}{\partial\eta}-M\frac{\partial^2 f(\eta;p)}{\partial\eta^2} \tag{3.266}$$

$$N_2[f(\eta;p),g(\eta;p),\theta(\eta;p)]=\frac{\partial^2 g(\eta;p)}{\partial\eta^2}-R\left(\left(\frac{\partial f(\eta;p)}{\partial\eta}\right)g(\eta;p)\right.$$
$$\left.-\left(\frac{\partial g(\eta;p)}{\partial\eta}\right)f(\eta;p)\right)+2Kr\frac{\partial f(\eta;p)}{\partial\eta}-Mg(\eta;p) \tag{3.267}$$

$$N_3[f(\eta;p),g(\eta;p),\theta(\eta;p)]=\left(1+\frac{4}{3}Rd\right)\frac{\partial^2\theta(\eta;p)}{\partial\eta^2}+R\,\mathrm{Pr}\left(\frac{\partial\theta(\eta;p)}{\partial\eta}\right)f(\eta;p)$$
$$+Nb\frac{\partial\theta(\eta;p)}{\partial\eta}\frac{\partial\phi(\eta;p)}{\partial\eta}+Nt\frac{\partial\theta(\eta;p)}{\partial\eta}\frac{\partial\theta(\eta;p)}{\partial\eta} \tag{3.268}$$

$$N_4[f(\eta;p),g(\eta;p),\theta(\eta;p)]=\frac{\partial^2\phi(\eta;p)}{\partial\eta^2}+RSc\left(\frac{\partial\phi(\eta;p)}{\partial\eta}\right)f(\eta;p)+\frac{Nt}{Nb}\frac{\partial^2\theta(\eta;p)}{\partial\eta^2} \tag{3.269}$$

For $p = 0$ and $p = 1$, respectively, we have:

$$f(\eta;p)=f_0(\eta) \qquad f(\eta;1)=f(\eta) \tag{3.270}$$

$$g(\eta;p)=g_0(\eta) \qquad g(\eta;1)=g(\eta) \tag{3.271}$$

$$\theta(\eta;p)=\theta_0(\eta) \qquad \theta(\eta;1)=\theta(\eta) \tag{3.272}$$

$$\phi(\eta;p)=\phi_0(\eta) \qquad \phi(\eta;1)=\phi(\eta) \tag{3.273}$$

As p increases from 0 to 1, $f(\eta;p), g(\eta;p), \theta(\eta;p)$, and $\phi(\eta;p)$ vary, respectively, from $f_0(\eta), g_0(\eta), \theta_0(\eta)$ and $\phi_0(\eta)$ to $f(\eta), g(\eta), \theta(\eta)$, and $\phi(\eta)$. By Taylor's theorem and using Eqs (3.266)–(3.269), $f(\eta), g(\eta), \theta(\eta)$, and $\phi(\eta)$ can be expanded in a power series of p as follows:

$$f(\eta;p)=f_0(\eta)+\sum_{m=1}^{\infty}(f_m(\eta)p^m)$$
$$f_m(\tau)=\frac{1}{m!}\frac{\partial^m f(\eta;p)}{\partial p^m} \tag{3.274}$$

$$g(\eta;p)=g_0(\eta)+\sum_{m=1}^{\infty}(g_m(\eta)p^m)$$
$$g_m(\tau)=\frac{1}{m!}\frac{\partial^m g(\eta;p)}{\partial p^m} \tag{3.275}$$

$$\theta(\eta;p)=\theta_0(\eta)+\sum_{m=1}^{\infty}(\theta_m(\eta)p^m)$$
$$\theta_m(\tau)=\frac{1}{m!}\frac{\partial^m\theta(\eta;p)}{\partial p^m} \tag{3.276}$$

$$\phi(\eta;p)=\phi_0(\eta)+\sum_{m=1}^{\infty}(\phi_m(\eta)p^m)$$
$$\phi_m(\tau)=\frac{1}{m!}\frac{\partial^m\phi(\eta;p)}{\partial p^m} \tag{3.277}$$

In which $\hbar_1,\hbar_2,\hbar_3$, and $\hbar_4$ are chosen in such a way that these series are convergent at $p = 1$. Convergence of the solution depends on the auxiliary parameters $\hbar_1,\hbar_2,\hbar_3$, and $\hbar_4$.

$$f(\eta)=f_0(\eta)+\sum_{m=1}^{\infty}(f_m(\eta)) \tag{3.278}$$

$$g(\eta)=g_0(\eta)+\sum_{m=1}^{\infty}(g_m(\eta)) \tag{3.279}$$

$$\theta(\eta)=\theta_0(\eta)+\sum_{m=1}^{\infty}(\theta_m(\eta)) \tag{3.280}$$

$$\phi(\eta)=\phi_0(\eta)+\sum_{m=1}^{\infty}(\phi_m(\eta)) \tag{3.281}$$

We have the following mth-order deformation problem:

$$L_1[f_m(\eta)-\chi_m f_{m-1}(\eta)]=\hbar_1 R_m^f(\eta)$$
$$f(0;p)=0, f(1;p)=\lambda, f'(0;p)=0, f'(1;p)=1$$
$$R_m^f(\eta)=f_{m-1}^{IV}-R\left(\left(\sum_{n=0}^{m-1}f_{m-1-n}f_n''\right)-\left(\sum_{n=0}^{m-1}f_{m-1-n}f_n'''\right)\right)-2\,\mathrm{Kr}\,g_{m-1}'-M\,f_{m-1}'' \tag{3.282}$$

$$L_1\left[g_m(\eta)-\chi_m g_{m-1}(\eta)\right]=\hbar_2 R_m^g(\eta)$$
$$g(0;p)=0, g(1;p)=0$$
$$R_m^g(\eta)=g_{m-1}''-R\left(\left(\sum_{n=0}^{m-1}g_{m-1-n}f_n'\right)-\left(\sum_{n=0}^{m-1}f_{m-1-n}g_n'\right)\right)+2Kr\,f_{m-1}'-Mg_{m-1} \tag{3.283}$$

$$L_3[\theta_m(\eta)-\chi_m\theta_{m-1}(\eta)]=\hbar_3 R_m^\theta(\eta)$$
$$\theta(0;p)=1, \theta(1;p)=0$$
$$R_m^\theta(\eta)=\left(1+\frac{4}{3}Rd\right)\theta_{m-1}''+R\,\mathrm{Pr}\left(\sum_{n=0}^{m-1}f_{m-1-n}\theta_n'\right)+Nb\left(\sum_{n=0}^{m-1}\phi_{m-1-n}'\theta_n'\right)+Nt\left(\sum_{n=0}^{m-1}\theta_{m-1-n}'\theta_n'\right) \tag{3.284}$$

$$\begin{aligned}&L_4[\phi_m(\eta)-\chi_m\phi_{m-1}(\eta)]=\hbar_4R_m^{\phi}(\eta)\\&\phi(0;p)=1,\phi(1;p)=0\\&R_m^{\phi}(\eta)=\phi''_{m-1}+RSc\left(\sum_{n=0}^{m-1}f_{m-1-n}\phi'_n\right)+\frac{Nt}{Nb}\theta''_{m-1}\end{aligned} \quad (3.285)$$

We use Maple software to obtain the solution of these equations. We assume $\hbar_1=\hbar_2=\hbar_3=\hbar_4=\hbar$. The solutions $f_1(\eta), g_1(\eta), \theta_1(\eta)$, and $\phi_1(\eta)$ are too long to be mentioned here; therefore, they are shown graphically.

3.3.6.3 Effects of active parameters

The effects of Reynolds number, magnetic parameter, rotation parameter, Schmidt number, thermophoretic parameter, Brownian parameter, and radiation parameter on heat and mass characteristics are examined.

Fig. 3.31 shows the effects of Reynolds number, magnetic parameter, and rotation parameter on velocity profiles. The Reynolds number shows the relative significance of the inertia effect compared with the viscous effect. Thus, rotation velocity boundary layer thicknesses decrease with an increase of the Reynolds number. By increasing the magnetic parameter velocity in y direction decreases while the

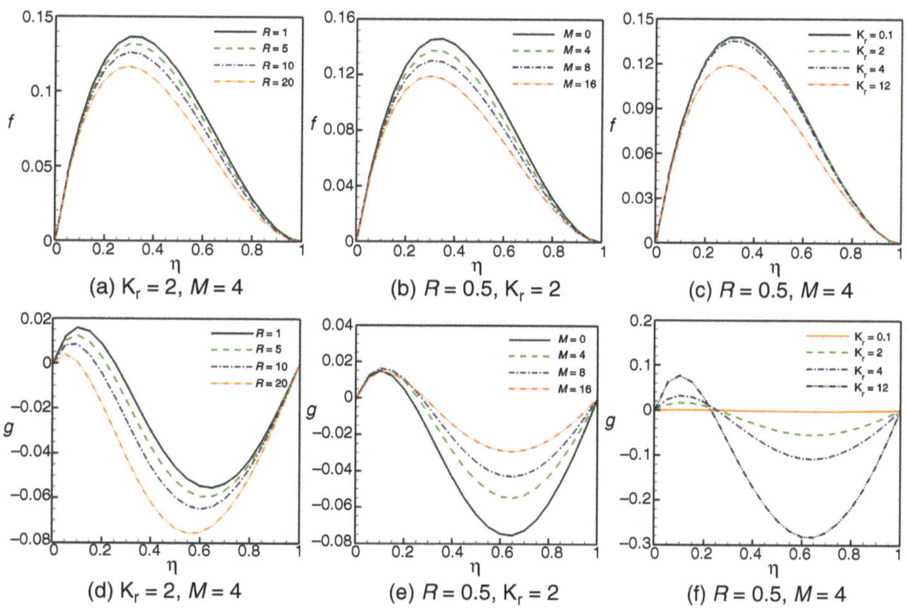

FIGURE 3.31 Effects of the Reynolds Number, Magnetic Parameter, and Rotation Parameter on Velocity Profiles (*f,g*)

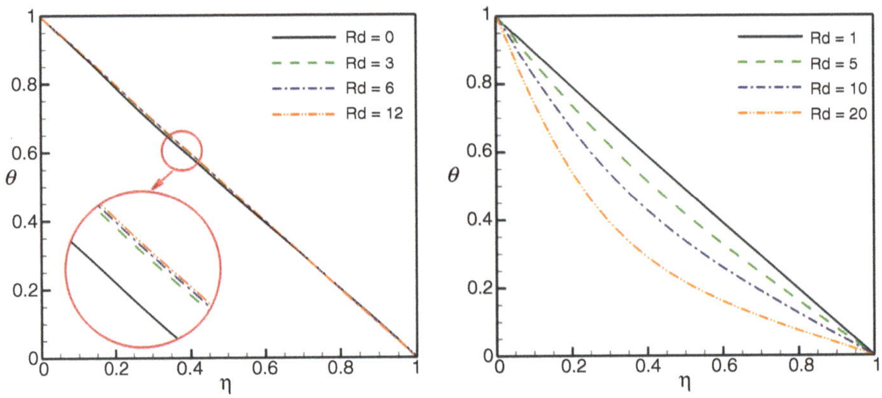

(a) $R = 0.5$ (b) Rd = 4

FIGURE 3.32 Effects of the Reynolds Number and Radiation Parameter on the Temperature Profile When $K_r = 2$, Sc = 0.5, Nt = 0.2, Nb = 0.2, $M = 4$ and Pr = 10

velocity in z direction increases. This phenomenon is due to the point that variation of magnetic parameter leads to the difference of the Lorentz force because of the magnetic field and the Lorentz force produces more resistance to transport phenomena. The effect of the rotation parameter on $f(\eta)$ is similar to that of the magnetic parameter. The Carioles force has an inverse effect on $g(\eta)$ in comparison with the Lorenz force, which means that with increasing rotation parameter, the transverse velocity increases.

The effects of the Reynolds number and radiation parameter on the temperature profile is shown in Fig. 3.32. As the Reynolds number increase, the temperature profile decreases between two plates; the opposite trend is observed for the radiation parameter. Fig. 3.33 depicts the effects of the radiation parameter, Reynolds number, thermophoretic parameter, Brownian parameter, and Schmidt number on the concentration profile. The concentration profile at the middle point increases as the Reynolds number and thermophoretic parameter increase, and it decreases with an increase of the radiation parameter, Brownian parameter, and Schmidt number. The effects of active parameters on the Nusselt number are shown in Fig. 3.34 and Tables 3.8–3.10. Thee thermal boundary layer thickness decreases with an increase of the radiation parameter and Reynolds number. Therefore, the Nusselt number increases with increases of these parameters. Also, it can be concluded that the Nusselt number is a decreasing function of the Schmidt number, rotation, Brownian, thermophoretic, and magnetic parameters.

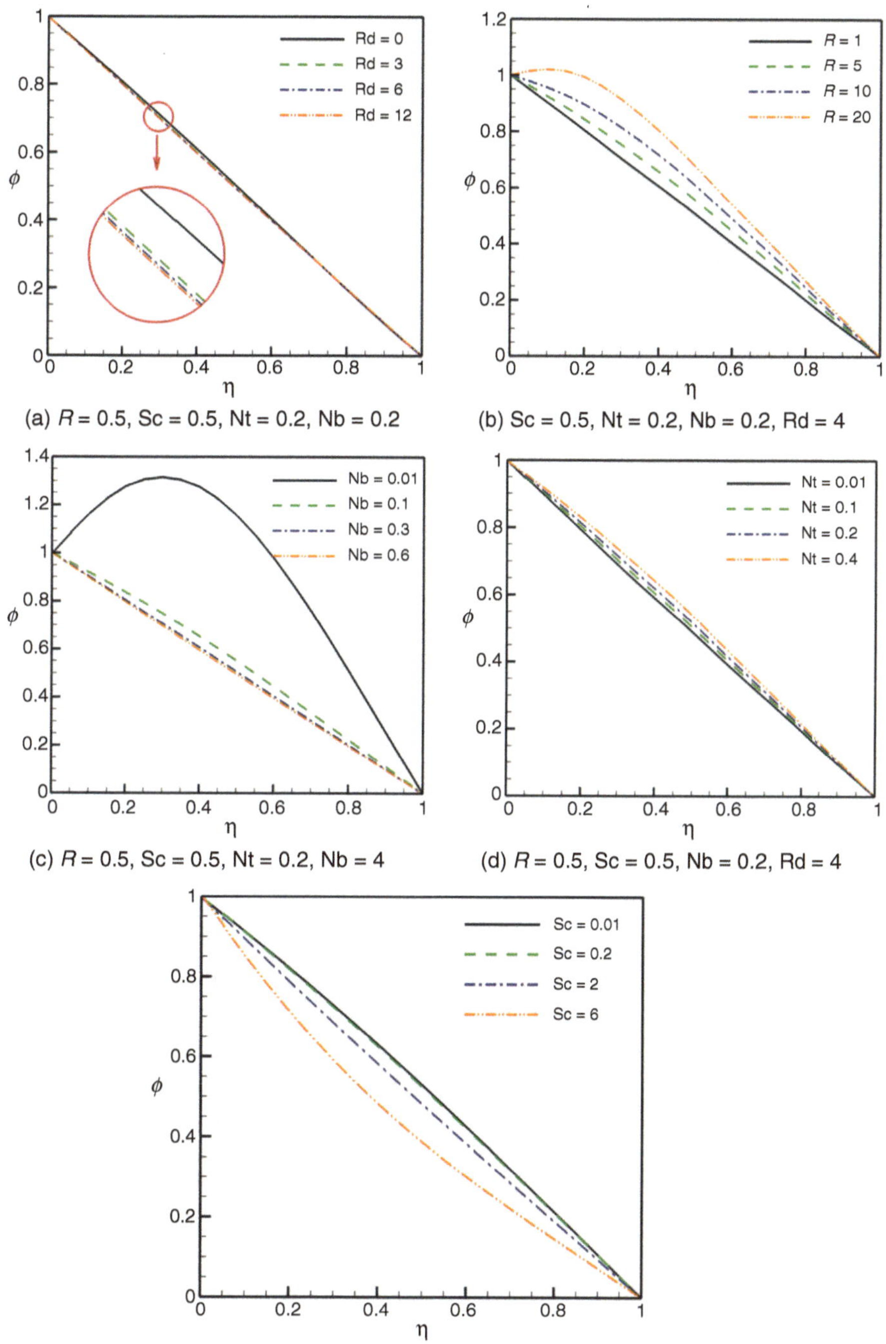

(a) $R = 0.5$, Sc = 0.5, Nt = 0.2, Nb = 0.2

(b) Sc = 0.5, Nt = 0.2, Nb = 0.2, Rd = 4

(c) $R = 0.5$, Sc = 0.5, Nt = 0.2, Nb = 4

(d) $R = 0.5$, Sc = 0.5, Nb = 0.2, Rd = 4

(e) $R = 0.5$, Nt = 0.2, Nb = 0.2, Rd = 4

FIGURE 3.33 Effects of the Radiation Parameter, Reynolds Number, Thermophoretic Parameter, Brownian Parameter, and Schmidt Number on Concentration Profile When $K_r = 2, M = 4$, and Pr = 10

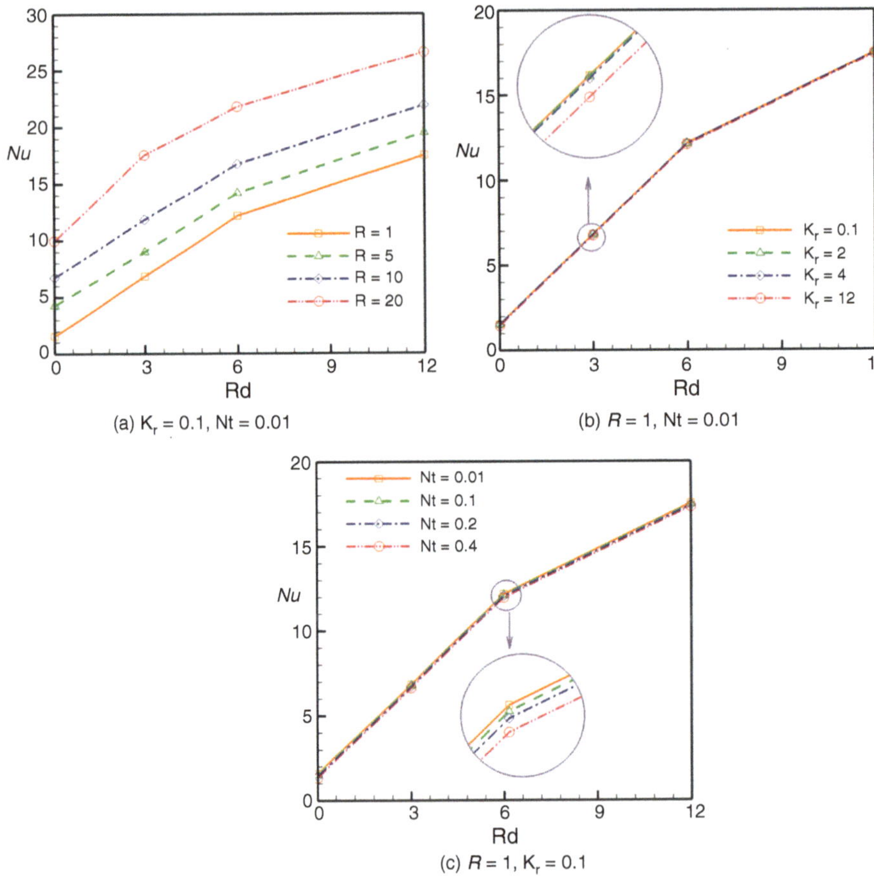

FIGURE 3.34 Effects of the Radiation Parameter, Reynolds Number, Rotation Parameter, and Thermophoretic Parameter on Nusselt Number When Sc = 0.5, Nb = 0.01, $M = 0$, and Pr = 10

Table 3.8 Effects of the Radiation Parameter and Magnetic Parameter on the Nusselt number When $Sc = 0.5, Nb = 0.01, R = 1, K_r = 0.1, Nt = 0.01$, and Pr = 10

	M			
Rd	**0**	**4**	**8**	**16**
0	1.557044	1.523667	1.4954	1.449936
3	6.830242	6.802148	6.778303	6.739817
6	12.1581	12.13053	12.10713	12.06931
12	17.48938	17.46202	17.43877	17.40121

Table 3.9 Effects of the Radiation Parameter and Schmidt Number on the Nusselt number When $Nb = 0.01, M = 0, R = 1, K_r = 0.1, Nt = 0.01$, and $Pr = 10$

Rd	Sc			
	0.01	0.2	2	6
0	1.557044	1.557021	1.556798	1.556285
3	6.830242	6.830225	6.830065	6.829702
6	12.1581	12.15808	12.15792	12.15757
12	17.48938	17.48936	17.48921	17.48886

Table 3.10 Effects of the Radiation Parameter and Brownian Parameter on the Nusselt number When $Sc = 0.5, M = 0, R = 1, K_r = 0.1, Nt = 0.01$, and $Pr = 10$

Rd	Nb			
	0.01	0.1	0.3	0.6
0	1.557044	1.49911	1.37463	1.199351
3	6.830242	6.783053	6.678947	6.524752
6	12.1581	12.1119	12.00966	11.85736
12	17.48938	17.44356	17.34201	17.19043

REFERENCES

[1] He JH. Comparison of homotopy perturbation method and homotopy analysis method. Appl Math Comput 2004;156:527–39.

[2] Jalaal M, Ganji DD, Ahmadi G. Analytical investigation on acceleration motion of a vertically falling spherical particle in incompressible Newtonian media. Adv Powder Technol 2010;21:298–304.

[3] Sheikholeslami M, Ashorynejad HR, Ganji DD, Kolahdooz A. Investigation of rotating MHD viscous flow and heat transfer between stretching and porous surfaces using analytical method. Mathl Prob Eng 2011;1024–123.

[4] Sheikholeslami M, Ashorynejad HR, Ganji DD, Yıldırım A. Homotopy perturbation method for three-dimensional problem of condensation film on inclined rotating disk. Sci Iran B 2012;19(3):437–42.

[5] Sheikholeslami M, Ganji DD. Heat transfer of Cu-water nanofluid flow between parallel plates. Powder Technol 2013;235:873–9.

[6] Sheikholeslami M, Ganji DD, Rokni HB. Nanofluid flow in a semi-porous channel in the presence of uniform magnetic field. IJE Trans C 2013;26(6):653–62.

[7] Ashorynejad HR, Sheikholeslami M, Pop I, Ganji DD. Nanofluid flow and heat transfer due to a stretching cylinder in the presence of magnetic field. Heat Mass Transfer 2013;49:427–36.

[8] Sheikholeslami M, Hatami M, Domairry G. Numerical simulation of two phase unsteady nanofluid flow and heat transfer between parallel plates in presence of time dependent magnetic field. J Taiwan Inst Chem Eng 2015;46:43–50.

[9] Sheikholeslami M, Hatami M, Ganji DD. Micropolar fluid flow and heat transfer in a permeable channel using analytical method. J Mol Liq 2014;194:30–6.

[10] Haq S, Ishaq M. Solution of strongly nonlinear ordinary differential equations arising in heat transfer with optimal homotopy asymptotic method. Int J Heat Mass Transfer 2012;55(21):5737–43.

[11] Hashmi MS, Khan N, Iqbal S. Numerical solutions of weakly singular Volterra integral equations using the optimal homotopy asymptotic method. Comput Math Appl 2012;64(6):1567–74.

[12] Sheikholeslami M, Ashorynejad HR, Domairry D, Hashim I. Investigation of the laminar viscous flow in a semi-porous channel in the presence of uniform magnetic field using Optimal Homotopy Asymptotic Method. Sains Malaysiana 2012;41(10): 1177–229.

[13] Sheikholeslami M, Ganji DD. Magnetohydrodynamic flow in a permeable channel filled with nanofluid. Sci Iran B 2014;21(1):203–12.

[14] Sheikholeslami M, Ashorynejad HR, Domairry D, Hashim I. Investigation of the laminar viscous flow in a semi-porous channel in the presence of uniform magnetic field using optimal homotopy asymptotic method. Sains Malays 2012;41(10):1177–229.

[15] Sheikholeslami M, Sheykholeslami FB, Khoshhal S, Mola-Abasia H, Ganji DD, Rokni HB. Effect of magnetic field on Cu–water nanofluid heat transfer using GMDH-type neural network. Neural Comput Applic 2014;25(1):171–8.

[16] Sheikholeslami M, Hatami M, Ganji DD. Nanofluid flow and heat transfer in a rotating system in the presence of a magnetic field. J Mol Liq 2014;190:112–20.

[17] Zhou K. Differential transformation and its applications for electrical circuits. Wuhan, China: Huazhong University Press; 1986.

[18] Chen CK, Ho SH. Solving partial differential equations by two dimensional differential transform method. Appl Math Comput 1999;106:171–9.

[19] Ayaz F. Solutions of the systems of differential equations by differential transform method. Appl Math Comput 2004;147:547–67.

[20] Hassan Abdel-Halim IH. Comparison differential transformation technique with Adomian decomposition method for linear and nonlinear initial value problems. Chaos Soliton Fract 2008;36:53–65.

[21] Jang MJ, Chen CL, Liu YC. Two-dimensional differential transform for partial differential equations. Appl Math Comput 2001;121:261–70.

[22] Hassan Abdel-Halim IH. On solving some eigenvalue-problems by using a differential transformation. Appl Math Comput 2002;127:1–22.

[23] Momania S, Suat Ertürk V. Solutions of non-linear oscillators by the modified differential transform method. Comput Math Appl 2008;55:833–42.

[24] Abazari R, Borhanifar A. Numerical study of the solution of the Burgers and coupled Burgers equations by a differential transformation method. Comput Math Appl 2010;59:2711–22.

[25] Joneidi AA, Ganji DD, Babaelahi M. Differential Transformation Method to determine fin efficiency of convective straight fins with temperature dependent thermal conductivity. Int Commun Heat Mass Transfer 2009;36:757–62.

[26] Rashidi MM. The modified differential transform method for solving MHD boundary-layer equations. Comput Phys Commun 2009;180:2210–7.

[27] Rashidi MM, Erfani E. The modified differential transform method for investigating nano boundary-layers over stretching surfaces. Int J Numer Methods Heat Fluid Flow 2011;21(7):864–83.

[28] Basiri Parsa A, Rashidi MM, Anwar Bég O, Sadri SM. Semi-computational simulation of magneto-hemodynamic flow in a semi-porous channel using optimal homotopy and differential transform methods. Comput Biol Med 2013;43(9):1142–53.

[29] Erfani E, Rashidi MM, Basiri Parsa A. The modified differential transform method for solving off-centered stagnation flow towards a rotating disc. Int J Comput Methods 2010;7(4):655–70.

[30] Rashidi MM, Laraqi N, Basiri Parsa A. Analytical modeling of heat convection in magnetized micropolar fluid by using modified differential transform method. Heat Transfer Asian Res 2011;40(3):187–204.

[31] Rashidi MM, Anwar Bég O, Rahimzadeh N. A generalized differential transform method for combined free and forced convection flow about inclined surfaces in porous media. Chem Eng Commun 2012;199(2):257–82.

[32] Domairry D, Sheikholeslami M, Ashorynejad HR, Subba R, Gorla R, Khani M. Natural convection flow of a non-Newtonian nanofluid between two vertical flat plates. Proc IMechE N J Nanoeng Nanosyst 2012;225(3):115–22.

[33] Sheikholeslami M, Ganji, Domiri D. Nanofluid flow and heat transfer between parallel plates considering Brownian motion using DTM. Comput Methods Appl Mech Eng 2015;283:651–63.

[34] Hatami M, Sheikholeslami M, Domairry G. High accuracy analysis for motion of a spherical particle in plane Couette fluid flow by multi-step differential transformation method. Powder Technol 2014;260:59–67.

[35] Chen C,-K, Ju S-P. Application of differential transformation to transient advective–dispersive transport equation. Appl Math Comput 2004;155:25–38.

[36] Peng H-S, Chen C-L. Hybrid differential transformation and finite difference method to annular fin with temperature-dependent thermal conductivity. Int J Heat Mass Transfer 2011;54:2427–33.

[37] Mosayebidorcheh S, Sheikholeslami M, Hatami M, Ganji DD. Analysis of turbulent MHD Couette nanofluid flow and heat transfer using hybrid DTM–FDM. Particuology 2014; doi: 10.1016/j.partic.2014.07.004. Available from: http://www.sciencedirect.com/science/article/pii/S1674200114001679.

[38] Adomian G. A review of the decomposition method in applied mathematics. J Math Anal Appl 1988;135(2):501–44.

[39] Jafari H, Daftardar-Gejji V. Revised Adomian decomposition method for solving a system of non-linear equations. Appl Math Comput 2006;175:1–7.

[40] Allan FM, Syam MI. On the analytic solutions of the nonhomogeneous Blasius problem. J Comput Appl Math 2005;182:362–71.

[41] Hashim I. Adomian decomposition method for solving BVPs for fourth-order integro-differential equations. J Comput Appl Math 2006;193:658–64.

[42] Hashim I. Comments on a new algorithm for solving classical Blasius equation. J Comput Appl Math 2005;182:362–71.

[43] Kechil SA, Hashim I. Non-perturbative solution of free-convective boundary-layer equation by Adomian decomposition method. Phys Lett A 2007;363:110–4.

[44] Sheikholeslami M, Ganji DD, Ashorynejad HR, Rokni Houman B. Analytical investigation of Jeffery-Hamel flow with high magnetic field and nano particle by Adomian decomposition method. Appl Math Mech Eng 2012;33(1):1553–64.

[45] Sheikholeslami M, Ganji DD, Ashorynejad HR. Investigation of squeezing unsteady nanofluid flow using ADM. Powder Technol 2013;239:259–65.
[46] Liao, SJ. The proposed homotopy analysis technique for the solution of nonlinear problems. PhD thesis, Shanghai Jiao Tong University; 1992.
[47] Liao SJ. Beyond perturbation: introduction to the homotopy analysis method. Boca Raton: Chapman & Hall/CRC Press; 2003.
[48] Sheikholeslami M, Ellahi R, Ashorynejad HR, Domairry G, Hayat T. Effects of heat transfer in flow of nanofluids over a permeable stretching wall in a porous medium. Int Comput Theor Nanosci 2014;11:1–11.
[49] Sheikholeslami M, Ashorynejad HR, Domairry G, Hashim I. Flow and heat transfer of Cu–Water nanofluid between a stretching sheet and a porous surface in a rotating system. J Appl Math 2012;2012:18. Article ID 421320.
[50] Rashidi MM, Anwar Bég O, Rastegari MT. A study of non-Newtonian Flow and heat transfer over a non-isothermal wedge using the homotopy analysis method. Chem Eng Commun 2012;199(2):231–56.
[51] Rashidi MM, Freidoonimehr N, Hosseini A, Anwar Bég O, Hung T-K. Homotopy simulation of nanofluid dynamics from a non-linearly stretching isothermal permeable sheet with transpiration. Meccanica 2014;49(2):469–82.
[52] Ziabakhsh Z, Domairry G. Solution of the laminar viscous flow in a semi-porous channel in the presence of a uniform magnetic field by using the homotopy analysis method. Commun Nonlinear Sci Numer Simul 2009;14(4):1284–94.
[53] Ziabakhsh Z, Domairry G. Analytic solution of natural convection flow of a non-Newtonian fluid between two vertical flat plates using homotopy analysis method. Commun Nonlinear Sci Numer Simul 2009;14(5):1868–80.
[54] Ziabakhsh Z, Domairry G, Bararnia H. Analytical solution of non-Newtonian micropolar fluid flow with uniform suction/blowing and heat generation. J Taiwan Inst Chem Eng 2009;40(4):443–51.
[55] Ziabakhsh Z, Domairry G, Ghazizadeh HR. Analytical solution of the stagnation-point flow in a porous medium by using the homotopy analysis method. J Taiwan Inst Chem Eng January 2009;40(1):91–7.
[56] Joneidi AA, Domairry G, Babaelahi M. Analytical treatment of MHD free convective flow and mass transfer over a stretching sheet with chemical reaction. J Taiwan Inst Chem Eng 2010;41(1):35–43.
[57] Sheikholeslami M, Abelman S, Domiri G, Domairry D. Numerical simulation of MHD nanofluid flow and heat transfer considering viscous dissipation. Int J Heat Mass Transfer 2014;79:212–22.
[58] Mohsen SK. KKL correlation for simulation of nanofluid flow and heat transfer in a permeable channel. Phys Lett A 2014;378(45):3331–9.
[59] Sheikholeslami M, Ganji DD, Javed MY, Ellahi R. Effect of thermal radiation on magnetohydrodynamics nanofluid flow and heat transfer by means of two phase model. J Magn Magn Mater 2015;374:36–43.
[60] Sheikholeslami M, Ganji DD. Unsteady nanofluid flow and heat transfer in presence of magnetic field considering thermal radiation. J Braz Soc Mech Sci Eng 2015;37(3):895–902.
[61] Sheikholeslami M, Ganji DD. Three dimensional heat and mass transfer in a rotating system using nanofluid. Powder Technol 2014;253:789–96.

CHAPTER

Lattice Boltzmann method: application for MHD nanofluid hydrothermal behavior

4

CHAPTER OUTLINE

External Magnetic Field Effects on Hydrothermal Treatment of Nanofluid

4.1 INTRODUCTION

Transport equations (for heat, mass, and momentum) can be simulated on different scales. On a macroscopic scale, partial differential equations (PDEs) such as the Navier–Stokes equation are used. These kinds of equations are difficult to solve analytically because of nonlinearity, complicated geometry, and boundary conditions. With the help of numerical schemes such as the finite difference method (FDM), finite volume method (FVM), or finite element method (FEM), PDE is converted into a system of algebraic equations. We usually solve these equations iteratively until satisfying results are obtained.

The second approach is to simulate small particles on a microscopic scale. This is molecular dynamics. The governing equation is Hamilton's equation. With molecular dynamics, we have to identify the location and velocity of each particle. But there would be simply too much data to handle to simulate a problem that is interesting on a macroscopic scale (i.e., flow past an object). For example, 1 L of air contains about 10^{22} molecules. We do not have to know the position of every particle; the important thing is the resultant effect [1] (e.g., wind).

The lattice Boltzmann method (LBM) closes the gap between macroscale and microscale. LBM considers the behavior of a collection of particles as a unit [1]. LBM is based on microscopic models and mesoscopic kinetic equations. The fundamental idea of the LBM is to construct simplified kinetic models that incorporate the essential physics of microscopic processes so that the macroscopic averaged properties obey the desired macroscopic equations [2]. The reason why simplified kinetic models can be used is that the macroscopic dynamics of a fluid are the result of the collective behavior of many microscopic particles in the system [2]. The property of the collection of particles is represented by a distribution function [1]. The reason why LBM is becoming increasingly more popular in the field of computational fluid dynamics (CFD) (exponential growth of number of articles with LBM topic) is because LBM is solved locally. It has a high degree of parallelization; hence, it is ideal for parallel machines (computational clusters) (Fig. 4.1).

In traditional CFD, the Navier–Stokes equations solve mass, momentum, and energy conservation equations on discrete nodes, elements, or volumes. In other words, the nonlinear PDEs convert into a set of nonlinear algebraic equations, which are solved iteratively. In LBM, the fluid is replaced by fractious particles. These particles stream along given directions (i.e., lattice links) and collide at the lattice sites. LBM can be considered an explicit method. The collision and streaming processes are local. Hence, it can be programmed naturally for parallel processing machines. Another advantage of LBM is in handling complex phenomena such as moving boundaries (multiphase, solidification, and melting problems) naturally without a need for a face-tracing method as it is in the traditional CFD. Recently, this method was applied for simulating flow, heat, and mass transfer.

Mohamad and Kuzmin [3] used LBM to present a detailed analysis of a natural convection problem. They show the high efficiency of the LBM in simulating natural convection. Mohammad et al. [4] studied natural convection in an open-ended cavity

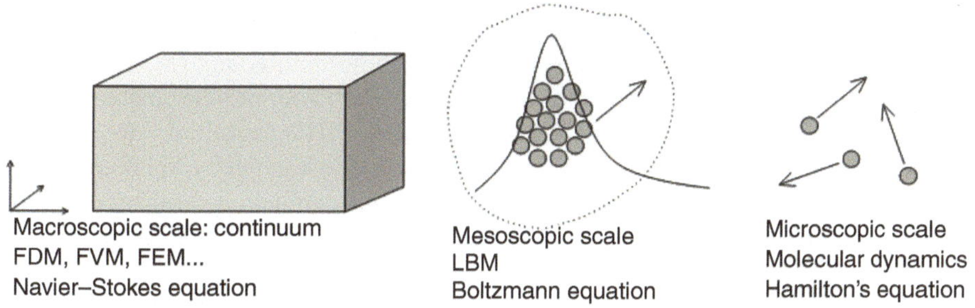

FIGURE 4.1 Techniques of Simulations on Different Scales [1]

using LBM. Their work was intended to address the physics of flow and heat transfer in open-end cavities and close-end slots. LBM was used to investigate magneto-hydrodynamic (MHD) flow using Cu–water nanofluid in a concentric annulus by Sheikholeslami et al. [5]. Their results proved that the enhancement ratio increases with a decrease of the Rayleigh number, and it increases with an augment of the Hartmann number.

Sheikholeslami et al. [6] studied the problem of MHD free convection in an eccentric semi-annulus filled with nanofluid. They showed that the Nusselt number decreases with an increase of the position of the inner cylinder at a high Rayleigh number. Sheikholeslami et al. [7] performed a numerical study to investigate natural convection in a square cavity with curved boundaries filled with Cu–water nanofluid. Their results proved that the change of inclination angle has a significant impact on the thermal and hydrodynamic flow fields. Natural convection in a nanofluid filled concentric annulus between an outer square cylinder and an inner elliptic cylinder was investigated by Sheikholeslami et al. [8]. Their results showed that the minimum value of enhancement of heat transfer occurs at $\varepsilon = 0.95$ for $Ra = 10^5$, but for other values of Rayleigh number, it occurs at $\varepsilon = 0.65$.

Sheikholeslami et al. [9] used the LBM to examine free convection of nanofluids. They found that choosing copper as the nanoparticle leads to the highest enhancement for this problem. Also, their results indicate that the maximum value of enhancement occurs in $\lambda = 2.5$ at $Ra = 10^6$, but for other Rayleigh numbers, it is obtained at $\lambda = 1.5$. Free convection of ferrofluid in a cavity heated from below in the presence of an external magnetic field was studied by Sheikholeslami and Gorji [10]. They found that particles with a smaller size have a better ability to dissipate heat, and a larger volume fraction would provide a stronger driving force, which leads to an increase in the temperature profile.

Sheikholeslami and Ellahi [11] studied three-dimensional mesoscopic simulation of magnetic field effect on natural convection of nanofluid. They found that thermal boundary layer thickness increases with an increase of Lorentz forces. The effect of a static radial magnetic field on natural convection heat transfer in a horizontal cylindrical annulus enclosure filled with nanofluid was investigated numerically using LBM by Ashorynejad et al. [12]. They found that the average Nusselt number increases as the nanoparticle volume fraction and Rayleigh number increase, and it decreases as the Hartmann number increases. Free convection heat transfer in a concentric annulus between a cold square and heated elliptic cylinders in the presence of a magnetic field was investigated by Sheikholeslami et al. [13]. They found that the enhancement in heat transfer increases as Hartmann number increases, but it decreases with an increase of the Rayleigh number. MHD nanofluid flow and heat transfer in an enclosure with curved boundaries was investigated by Sheikholeslami et al. [14]. They found that an increasing Lorentz force can increase enhancement in heat transfer. Sheikholeslami et al. [15] studied the magnetic field effect on CuO–water nanofluid flow and heat transfer in an enclosure that is heated from below. They found that the effect of Hartmann number and heat source length is more pronounced at a high Rayleigh number. Sheikholeslami and Ellahi [16] used LBM to simulate ferrofluid flow for magnetic drug targeting. They found that the skin friction coefficient is a decreasing function of the Reynolds number and magnetic number. Entropy generation of nanofluid in presence of magnetic field was investigated by Sheikholeslami and Ganji [17]. They found that the heat transfer rate and dimensionless entropy generation number increase with an increase of the Rayleigh number and nanoparticle volume fraction, but it decreases with an increase of the Hartmann number.

4.2 LATTICE GAS

The LBM has its roots in the lattice gas automata (LGA) kinetic model with discrete lattice and discrete time. Starting from the LGA on a hexagonal lattice, Frish et al. [18] obtained correct Navier–Stokes equations for the first time. This model is known as the Frisch, Hasslacher and Pomeau (FHP) model. As already discussed, the lattice gas automaton is constructed as simple particle dynamics in discretized space and time. Consequently, all particle velocities are also discrete. So we have particles that can move around but only within lattice nodes. Let's introduce a set of Boolean (true or false) variables

$$n_i\left(\vec{x},t\right),\quad i=0,..,M \tag{4.1}$$

These variables will describe the occupation of a particle. M is the number of directions of particle velocities at each node. For example, $i=0,..,6$ for hexagonal lattice is frequently used in LGA simulations. Boolean algebra has the advantage that it is without round-off errors. There can be either 0 or 1 particle at a lattice node.

$$\begin{aligned} n_i\left(\vec{x},t\right)&=0 \qquad \text{No particles at } \vec{x} \text{ and time } t \\ n_i\left(\vec{x},t\right)&=1 \qquad \text{A particle is present at site } \vec{x} \text{ and time } t \end{aligned} \tag{4.2}$$

Also, an exclusion rule applies, forbidding two particles sitting on the same node. We can write the evolution equation of the LGA as

$$n_i\left(\vec{x}+\vec{e}_i\,\delta t,t+1\right)=n_i\left(\vec{x},t\right)+\Omega_i\left(n\left(\vec{x},t\right)\right),\quad i=0,..,M, \tag{4.3}$$

where $\vec{e}_i$ are local particle velocities, Ω_i is the collision operator, and δt is time step. The collision operator contains all possible collisions. For example, in the case of a hexagonal lattice, two-, three- and four-body collisions are possible. There are scattering rules that bring proper dynamics to the system. Ω_i changes the occupation number due to the collision at site i and can have values of -1, 0, or 1. Number -1 means that the particle was destroyed, 0 leaves things unchanged, and 1 means a new particle is created. Boolean nature is preserved. It is important to stress that the interaction is completely local; neighboring sites do not interact. The configuration of particles at each time step evolves into two sequential substeps:

- *Streaming*: Each particle moves to the nearest node in the direction of its velocity.
- *Collision*: Particles arrive at a node and interact by changing their velocity directions according to scattering rules.

If we set the collision operator Ω_i to zero, then we obtain an equation for streaming alone.

In Fig. 4.2, we see an example of head-on collision of two particles at three sequential times (t − 1, t, t + 1). At t − 1, we have two incoming particles at sites 1 and 4. At site zero, all directions e1 to e6 are empty. After the streaming process, these two particles collide at time t. Directions e1 and e4 are now occupied. After collision at time t + 1, particles move according to scattering rules in directions e_2 and e_5. This is possible if e_2 and e_5 are empty. We can write the expression for this specific collision in form as

$$n_1 n_4 \bar{n}_2 \bar{n}_5 \tag{4.4}$$

where $\bar{n}_5$ means negation of n_5. This is one of the contributions to the Ω. Between n's is Boolean AND function. If the expression has a value of 1, then a collision is possible. For example, in Fig. 4.2, we get

$$11\bar{0}\bar{0} \rightarrow 1111 = 1. \tag{4.5}$$

Therefore, a collision is possible.

4.3 FROM LGA TO LBM

The main motivation for the transition from LGA to LBM was the desire to remove the statistical noise by replacing particle occupation variables n_i (Boolean variables) with single particle distribution functions

$$f_k = < n_k > \tag{4.6}$$

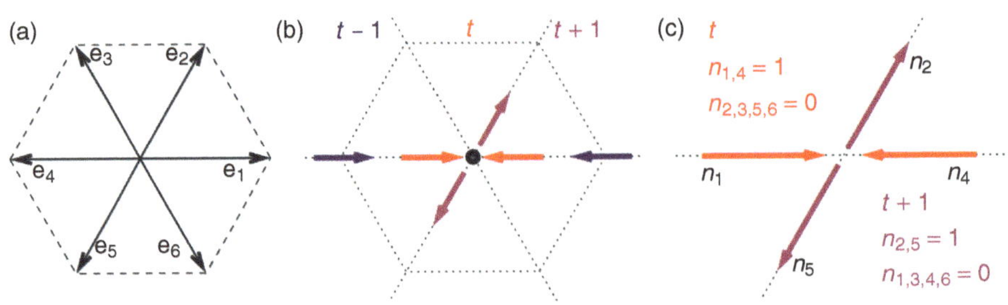

FIGURE 4.2

(a) FHP model with six velocity vectors. (b) Head-on collision of two particles at three sequential times. (c) Head-on collision with Boolean values.

These functions are an ensemble average of n_k and real variables [1]. n_k can be 0 or 1, and f_k can be any real number between 0 and 1. To obtain the macroscopic behavior of a system (streamlines) in the LGA, one has to average the state of each cell over a rather large patch of cells (e.g., a 32 × 32 square) and over several consecutive time steps [19]. With the replacement $n_k \rightarrow f_k$, noise is erased because f_k is by definition an average, a smooth quantity [20]. On the other hand, we get round-off errors. Now we will deal with the distribution function $f(\vec{x},\vec{e},t)$. It depends on the position vector $\vec{x}$, the velocity vector $\vec{e}$, and time t. $f(\vec{x},\vec{e},t)$ represents the number of particles with mass m at time t positioned between $\vec{x}+d\vec{x}$ that have velocities between $\vec{e}+d\vec{e}$. Now we apply force $\vec{F}$ on these particles. After time dt, position and velocity obtain new values (Fig. 4.3).

$$\begin{aligned}&\text{Position: } \vec{x} \rightarrow \vec{x}+\vec{e}\,dt\\&\text{Velocity: } \vec{e} \rightarrow \vec{e}+\frac{\vec{F}}{m}dt\end{aligned} \tag{4.7}$$

If there is no collision, the number of particles before and after applying force stays the same:

$$f\left(\vec{x}+\vec{e}dt,\vec{e}+\frac{\vec{F}}{m},t+dt\right)d\vec{x}d\vec{e} = f\left(\vec{x},\vec{e},t\right)d\vec{x}d\vec{e} \tag{4.8}$$

In case of a collision between particles, then all do not arrive at new positions. The rate of change between the final and initial status of the distribution function is called the collision operator Ω. The evolution equation with collisions is now written as

$$f\left(\vec{x}+\vec{e}dt,\vec{e}+\frac{\vec{F}}{m},t+dt\right)d\vec{x}d\vec{e} - f\left(\vec{x},\vec{e},t\right)d\vec{x}d\vec{e} = \Omega(f)\,d\vec{x}\,d\vec{e}\,dt \tag{4.9}$$

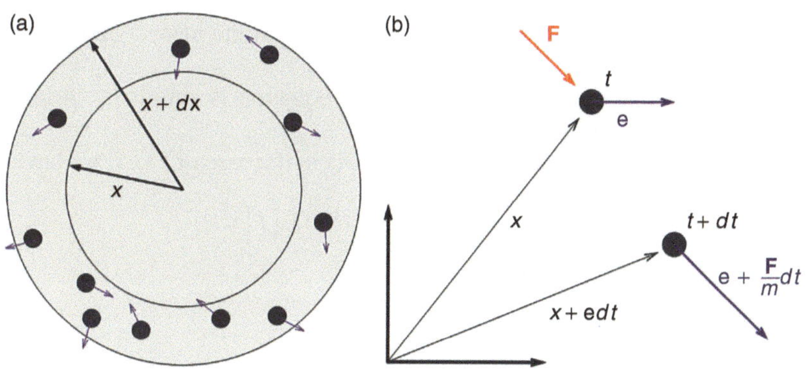

FIGURE 4.3

(a) Phase diagram. (b) Position and velocity vector before and after applying force [2].

By dividing the equation with $d\vec{x}\,d\vec{e}\,dt$ and in the limit $dt \to 0$, we can write:

$$\frac{Df}{dt} = \Omega(f) \tag{4.10}$$

The aforementioned equation states that the total rate of change of the distribution function is equal to the rate of the collision [2]. Let's expand

$$Df = \frac{\partial f}{\partial \vec{x}} d\vec{x} + \frac{\partial f}{\partial \vec{e}} d\vec{e} + \frac{\partial f}{\partial t} dt \tag{4.11}$$

and divide it by dt

$$\frac{Df}{dt} = \frac{\partial f}{\partial \vec{x}} \frac{d\vec{x}}{dt} + \frac{\partial f}{\partial \vec{e}} \frac{d\vec{e}}{dt} + \frac{\partial f}{\partial t} \tag{4.12}$$

Fraction $\frac{d\vec{x}}{dt}$ represents velocity $\vec{e}$, and fraction $\frac{d\vec{e}}{dt}$ represents acceleration $\vec{a}$, which can be written as $\vec{F}/m$. Finally, we obtain

$$\frac{\partial f}{\partial t} + \frac{\partial f}{\partial \vec{x}} \vec{e} + \frac{\vec{F}}{m} \frac{\partial f}{\partial \vec{e}} = \Omega(f) \tag{4.13}$$

We have derived the Boltzmann equation for a case without external forces:

$$\frac{\partial f}{\partial t} + \vec{e}.\vec{\nabla} f = \Omega(f) \tag{4.14}$$

Kinetic form stays the same as in LGA. We write it as

$$f_k\left(\vec{x} + \vec{e_k}\Delta t, t + \Delta t\right) = f_k\left(\vec{x}, t\right) + \Omega_k\left(f\left(\vec{x}, t\right)\right) \tag{4.15}$$

where Δt is a time increment and f_k is the particle velocity distribution along the kth direction, or in another words, f_k is the fraction of the particles having velocities in the interval $\vec{e_k}$ and $\vec{e_k} + d\vec{e_k}$. In the LBM, space is discretized in a way that is consistent with the kinetic equation, that is, the coordinates of the nearest neighbor nodes around $\vec{x}$ are $\vec{x} + \vec{e_k}\Delta t$ [2].

The main advantage of the aforementioned equation is simplicity. Adding the force term is also straightforward.

Eq. (4.2) is usually solved in two substeps. As in the case of LGA, we have

$$f_k\left(\vec{x} + \vec{e_k}\Delta t, t + \Delta t\right) = f_k\left(\vec{x}, t\right) + \Omega_k\left(f\left(\vec{x}, t\right)\right) \tag{4.16}$$

4.3.1 BGK MODEL EQUATION

The collision operator Ω_k mentioned earlier is in general a complex nonlinear integral. The idea is to linearize the collision term around its local equilibrium solution.

Ω_k is often replaced by the so-called BGK (Bhatnagar, Gross, Krook, 1954 [21]) collision operator

$$\Omega_k = -\frac{1}{\tau}\left(f_k - f_k^{\mathrm{eq}}\right) \tag{4.17}$$

where τ is the rate of relaxation toward local equilibrium and f_k^{eq} is the equilibrium distribution function. The collision operator is equivalent to the viscous term in the Navier–Stokes equation. The rate of relaxation τ is constant for all ks. This scheme is called the single-time relaxation scheme. All nodes relax with the same time scale τ [20]. The Navier–Stokes equation can be recovered from a Chapman–Enskog expansion [18]. This gives the kinematic viscosity υ in terms of the single relaxation time τ.

$$\upsilon = (\tau - 0.5\Delta t)c_s^2 \tag{4.18}$$

where $c_s = \frac{1}{\sqrt{3}}\frac{\Delta x}{\Delta t}$ is the sound speed in the lattice, Δx is the lattice spacing, and Δt is the time increment. Usually both Δx and Δt are set to unity. After introducing the BGK approximation, the Boltzmann equation (without external forces) can be written as

$$\frac{\partial f}{\partial t} + \vec{e}.\vec{\nabla} f = -\frac{1}{\tau}\left(f - f^{\mathrm{eq}}\right) \tag{4.19}$$

The form of the Boltzmann equation in a specific direction would be

$$\frac{\partial f_k}{\partial t} + \vec{e_k}.\vec{\nabla} f_k = -\frac{1}{\tau}\left(f_k - f_k^{\mathrm{eq}}\right) \tag{4.20}$$

We could say that this equation is the heart of LBM. It is the most popular kinetic model and replaces the Navier–Stokes equation in CFD simulations. The completely discretized equation with the time step Δt and space step $\Delta\vec{x_k} = \vec{e_k}\Delta t$ is

$$\begin{aligned}
&\text{Collision step:} \quad \overline{f_k}\left(\vec{x_k}, t+\Delta t\right) = f_k\left(\vec{x_k}, t\right) - \frac{\Delta t}{\tau}\left(f_k - f_k^{\mathrm{eq}}\right) \\
&\text{Streaming step:} \quad f_k\left(\vec{x_k} + \vec{e_k}\Delta t, t+\Delta t\right) = \overline{f_k}\left(\vec{x_k}, t+\Delta t\right)
\end{aligned} \tag{4.21}$$

Collision is not defined explicitly any more as in LGA. In the LBM, collision of the fluid particles is considered as relaxation toward a local equilibrium [22]. During the computation, there is no need to store both $f_k\left(\vec{x_k}, t+\Delta t\right)$ and $f_k\left(\vec{x_k}, t\right)$ [23] (Table 4.1).

4.3.2 EQUILIBRIUM DISTRIBUTION FUNCTION

The equilibrium distribution function f_k^{eq}, which appears in the BGK collision operator, is basically an expansion of the Maxwell's distribution function for a low Mach number Ma. We start with normalized Maxwell's distribution function

$$f = \frac{\rho}{2\pi/3} e^{-1.5\left(\vec{e}-\vec{u}\right)^2} = \frac{\rho}{2\pi/3} e^{-1.5\left(\vec{e}.\vec{e}\right)} e^{-1.5\left(2\vec{e}.\vec{u}-\vec{u}.\vec{u}\right)}. \tag{4.22}$$

Table 4.1 Comparison Between the Navier–Stokes Equation and Lattice Boltzmann Equation

Navier–Stokes equation	Lattice Boltzmann equation
$\rho\left(\frac{\partial \vec{u}}{\partial t}+\left(\vec{u}.\vec{\nabla}\right)\vec{u}\right)=-\vec{\nabla}p+\mu\nabla^2\vec{u}$	$\frac{\partial f}{\partial t}+\vec{e}.\vec{\nabla}f=-\frac{1}{\tau}\left(f_k-f_k^{eq}\right)$
Second-order PDE	First-order PDE
Need to treat the nonlinear convective term $\vec{u}.\vec{\nabla}\vec{u}$	Avoids convective term; convection becomes simple advection
Need to solve the Poisson equation for the pressure p	Pressure p is obtained from equation of state

where $\vec{u}$ is the macroscopic velocity of particles in a medium, $\vec{e}$ are velocity vectors in the specific lattice model, and ρ is macroscopic mass density. Now we expand Eq. (4.3) for small velocities $\mathrm{Ma}=u/c_s<<1$, where c_s represents the speed of sound.

$$f=\frac{\rho}{2\pi/3}e^{-1.5(\vec{e}.\vec{e})}\left[1+3\left(\vec{e}.\vec{u}\right)-1.5\left(\vec{u}.\vec{u}\right)+4.5\left(\vec{e}.\vec{u}\right)^2\right] \tag{4.23}$$

The expansion is done up to second order $O\left(\vec{u}.\vec{u}=u^2\right)$ to match the order of the Navier–Stokes equation. Here, the nonlinearity of the collision operator is hidden. The equilibrium distribution function is of the form

$$f_k^{eq}=\rho w_k\left[1+3\left(\vec{e}_k.\vec{u}\right)-1.5\left(\vec{u}.\vec{u}\right)+4.5\left(\vec{e}_k.\vec{u}\right)^2\right] \tag{4.24}$$

where k runs from 0 to M and represents available directions in the lattice and w_k are weighting factors. The complete procedure of deriving Eq. (4.5) can be seen in a paper by He and Luo [24].

4.4 MACROSCOPIC QUANTITIES

We live in a macroscopic world, and macroscopic quantities are of interest to us; distribution functions are not as meaningful. Macroscopic quantities (velocity $\vec{u}$, pressure p, mass density ρ, and momentum density $\rho\vec{u}$,...) are directly obtained by solving the Navier–Stokes equation. In the LBM, macroscopic quantities are obtained by evaluating the hydrodynamic moments of the distribution function f [23]. Connections of the distribution function to macroscopic quantities for the density and momentum flux are defined as follows:

$$\begin{aligned}\rho\left(\vec{x},t\right)&=m\int f\left(\vec{x},\vec{e},t\right)d\vec{e}\\ \rho\left(\vec{x},t\right)\vec{u}\left(\vec{x},t\right)&=m\int f\left(\vec{x},\vec{e},t\right)\vec{e}\,d\vec{e}\end{aligned} \tag{4.25}$$

In the discretized velocity space, the density and momentum fluxes can be evaluated as

$$\begin{aligned} &\text{Flow density}: \rho = \sum_k f_k, \\ &\text{Momentum}: \rho \vec{u} = \sum_k \vec{e}_k f_k, \end{aligned} \tag{4.26}$$

Macroscopic fluid density (6) is defined as the sum of the distribution functions. Macroscopic velocity (7) is an average of the microscopic velocities $\vec{e_k}$ weighted by the directional densities f_k. Mass m of particles is set to 1.

4.5 LATTICE ARRANGEMENTS

In this section, two-dimensional (2D) lattice arrangements used in the LBM are presented. The common terminology is to refer to the dimension of the problem and the number of velocity vectors D_nQ_m, where n represents the dimension of a problem and m refers to the speed model (number of linkages of a node) [2]. It is also worth mentioning that nodes in different positions regarding the central node have appropriate weighting factors w_k. Values of these factors depend on the distance from the central node, which has the highest value. Weighing factors are an important part of the calculation of equilibrium distribution function. The normalization condition for the weights is

$$\sum_k w_k = 1 \tag{4.27}$$

4.6 TWO-DIMENSIONAL LATTICE ARRANGEMENTS

D_2Q_9 has been widely and successfully used for simulations of 2D flows. We can describe it as nine-velocity square lattice model. It is a bit more demanding from a computational aspect than the hexagonal D_2Q_7 model, but it is more accurate. It has to be stressed that for fluid flow problems, in which the nonlinear term in the Navier–Stokes equation is important, the D_2Q_4 or D_2Q_5 model is not appropriate because of insufficient lattice symmetry (Fig. 4.4). It fails to achieve basic symmetry of the Navier–Stokes equation–rotational invariance. In plain words, a hexagon is "good enough" to replace a circle, but a square is not. Consequently, the lattice Boltzmann equation (LBE) cannot recover the correct Navier–Stokes equation on inappropriate lattices [2]. 2D lattice models with corresponding velocity vectors e_k, distribution functions f_k and weighting factors w_k are shown in Table 4.2.

4.7 BOUNDARY CONDITIONS

One of the important and crucial issues in LB simulation of flow is accurate modeling of the boundary conditions. Specifying boundary conditions for Navier–Stokes equations is somehow straightforward. This is not the case for LBM, in which the

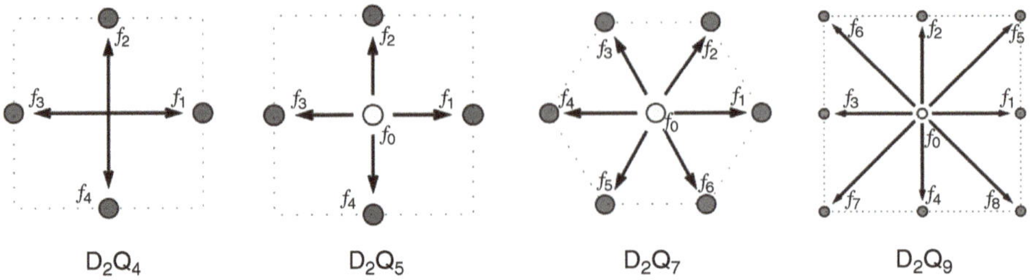

FIGURE 4.4 Two-Dimensional Lattice Models

Table 4.2 Two-Dimensional Lattice Models With Corresponding Velocity Vectors e_k, Distribution Functions f_k, and Weighting Factors w_k

Model	e_k	$f_k(w_k)$
D_2Q_4	$(\pm 1,0),(0,\pm 1)$	$f_{1,2,3,4}\left(\frac{1}{4}\right)$
D_2Q_5	$(0,0),(\pm 1,0),(0,\pm 1)$	$f_0\left(\frac{4}{6}\right), f_{1,2,3,4}\left(\frac{1}{12}\right)$
D_2Q_7	$(0,0),(\pm 1,0),\left(\pm\frac{1}{2},\pm\frac{\sqrt{3}}{3}\right)$	$f_0\left(\frac{1}{2}\right), f_{1,\ldots,4}\left(\frac{1}{12}\right)$
D_2Q_9	$(0,0),(\pm 1,0),(0,\pm 1),(\pm 1,\pm 1)$	$f_0\left(\frac{4}{9}\right), f_{1,2,3,4}\left(\frac{1}{9}\right), f_{5,6,7,8}\left(\frac{1}{36}\right)$

inward distribution functions to the integration domain need to be determined at the boundaries. Therefore, we need to determine appropriate equations for calculating the distribution functions at the boundaries for a given boundary condition. In the literature, different approaches are suggested and tested. In the following paragraphs, different boundary conditions are explained. The main idea is not to review different attempts but to discuss, in our point of view, the simple and more robust methods.

4.7.1 BOUNCE BACK

Bounce back is used to model a solid stationary or moving boundary condition, non-slip condition, or flow-over obstacles. The bounce-back method is quite simple and mainly implies that an incoming particle toward the solid boundary bounces back into the flow domain. In the literature, a few versions of the bounce-back scheme have been suggested. One of the schemes is to locate the wall at half the distance from the lattice sites, as shown in Fig. 4.5. The distribution functions f_4, f_7, and f_8

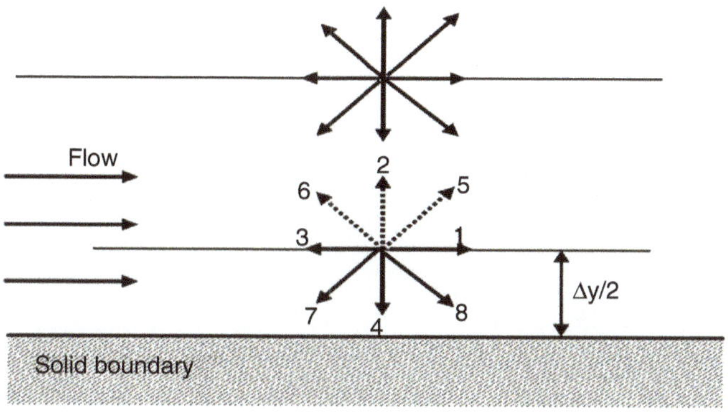

FIGURE 4.5 Bounce-Back Scheme [2]

are known from the streaming process. It is assumed that when these known distribution functions hit the wall, they bounce back to the solution domain. Therefore, $f_5 = f_7, f_2 = f_4,$ and $f_6 = f_8$: Notice that f_7 at node, say (i, j), is equal to f_7 of node $(i+1, j+1)$ after streaming. A similar argument is valid for f_8, which is coming from node $(i-1, j+1)$. Hence, the boundary condition must be applied after the streaming process.

4.7.2 BOUNDARY CONDITION WITH KNOWN VELOCITY

It is very common in practical applications to know the velocity component at the boundary (e.g., inlet velocity for a channel flow). Zhu and He described a method to calculate three unknown distribution functions based on Eqs (5.9) and (5.10) with equilibrium conditions assumption, normal to the boundary. Eq. (4.9) can be written as

$$\rho = f_0 + f_1 + f_2 + f_3 + f_4 + f_5 + f_6 + f_7 + f_8 \tag{4.28}$$

Eq. (4.10) can be written for x-component as

$$\rho u = f_1 + f_5 + f_8 - f_6 - f_7 - f_3 \tag{4.29}$$

and for y-component as

$$\rho v = f_2 + f_5 + f_6 - f_7 - f_8 \tag{4.30}$$

We have four unknowns, three distribution functions and an extra unknown, ρ, at the boundary. However, we have only three equations; we need to construct another equation or condition to solve for the four unknowns (three distribution functions and ρ). The condition comes from assumption of equilibrium condition, normal to the surface.

$$\rho v = f_2 + f_5 + f_6 - f_7 - f_8 \tag{4.31}$$

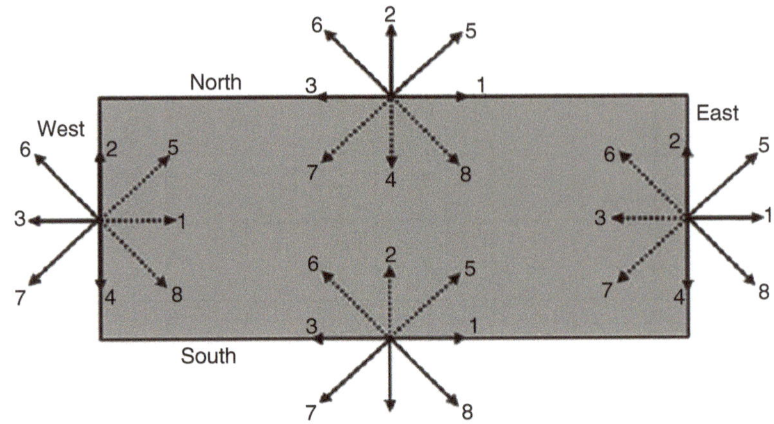

FIGURE 4.6 Distribution Functions at the Boundaries of a Domain [2]

Let us illustrate the aforementioned concept by an example. Fig. 4.6 shows a domain with east, west, north, and south boundaries. The velocity components are known at these boundaries.

In the figure, the unknown distribution functions are denoted by dotted lines. The known distribution functions after streaming are denoted by solid lines.

4.7.2.1 West boundary

The x-component velocity, $u = u_w$, and y-component velocity, $v = v_w$, are known and have given values, including zero (solid wall).

Then density can be written as

$$\rho_w = f_0 + f_1 + f_2 + f_3 + f_4 + f_5 + f_6 + f_7 + f_8 \tag{4.32}$$

The equation momentum can be written as

$$\rho_w u_w = f_1 + f_5 + f_8 - f_6 - f_7 - f_3 \tag{4.33}$$

And momentum along the vertical direction can be written as

$$\rho_w v_w = f_2 + f_5 + f_6 - f_7 - f_8 \tag{4.34}$$

The equilibrium condition normal to the boundary yields

$$f_1 - f_1^{eq} = f_3 - f_3^{eq} \tag{4.35}$$

where f^{eq} can be calculated as follows.

$$f_1^{eq} = \frac{1}{9}\rho_w\left[1 + 3u_w + \frac{9}{3}u_w^2 - \frac{3}{2}(u_w^2 + v_w^2)\right] \tag{4.36}$$

$$f_3^{eq} = \frac{1}{9}\rho_w\left[1 - 3u_w + \frac{9}{3}u_w^2 - \frac{3}{2}(u_w^2 + v_w^2)\right] \tag{4.37}$$

Substituting the aforementioned equations in Eq. (4.35) yields

$$f_1 = f_5 + \frac{2}{3}\rho_w u_w \tag{4.38}$$

For the west boundary, the unknown distribution functions are f_1, f_5, and f_8. Solving Eqs (4.32)–(4.38) for four unknowns, ρ_w, f_1, f_5, and f_8 yields the following equations:

$$\rho_w = \frac{1}{1-u_w}[f_0 + f_2 + f_4 + 2(f_3 + f_6 + f_7)] \tag{4.39}$$

$$f_5 = f_7 - \frac{1}{2}(f_4 - f_2) + \frac{1}{6}\rho_w u_w + \frac{1}{2}\rho_w u_w \tag{4.40}$$

$$f_8 = f_6 - \frac{1}{2}(f_2 - f_4) + \frac{1}{6}\rho_w u_w + \frac{1}{2}\rho_w u_w \tag{4.41}$$

Hence, four unknowns are specified at the west boundary; a similar method can be applied for other boundaries. The conditions for other boundaries are given in the following equations without derivations following the same procedure as earlier.

4.7.3 OPEN BOUNDARY CONDITION

In some problems, the outlet velocity is not known. The normal practice is to use extrapolation for the unknown distribution functions. For instance, if the east boundary condition on Fig. 4.6 represent the outlet condition, then f_3, f_6, and f_7 need to be calculated at the east boundary, $i = n$. A second-order polynomial can be used, as:

$$f_{3,n} = 2f_{3,n-1} - f_{3,n-2} \tag{4.42}$$

$$f_{6,n} = 2f_{6,n-1} - f_{6,n-2} \tag{4.43}$$

$$f_{7,n} = 2f_{7,n-1} - f_{7,n-2} \tag{4.44}$$

4.7.4 PERIODIC BOUNDARY CONDITION

The periodic boundary condition becomes necessary to apply to isolate repeating flow conditions (e.g., flow over bank of tubes as shown in Fig. 4.7). The flow conditions above the line a and below the line b are the same. Hence, it is sufficient to model the flow between these two lines and use periodic boundary conditions alone within these boundaries. The distribution functions that are leaving line a are the same as the distribution functions entering from line band and vice versa.

The distributions functions, f_4, f_7, and f_8 are unknown on the line a, and f_2, f_5, and f_6 are unknown on the line b. The periodic boundary is as follows:

Along line a:

$$f_{4,a} = f_{4,b}, f_{7,a} = f_{7,b}, f_{8,a} = f_{8,b} \tag{4.45}$$

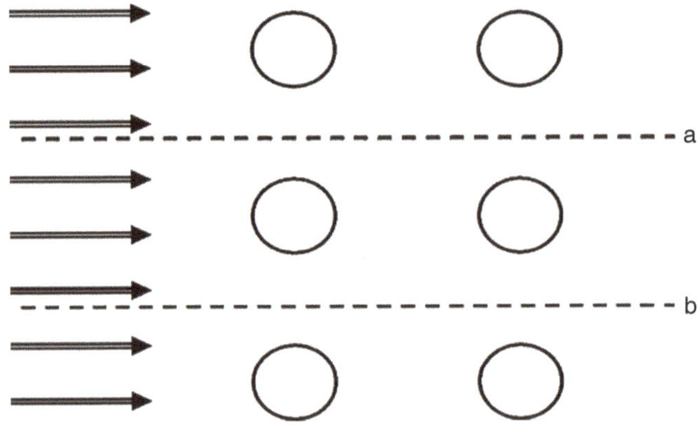

FIGURE 4.7 The Periodic Boundary Conditions [2]

Along line b:

$$f_{2,b} = f_{4,b}, f_{5,b} = f_{5,a}, f_{6,b} = f_{6,a} \tag{4.46}$$

4.7.5 SYMMETRY CONDITION

Many practical problems show symmetry along a line or a plane. In these cases, it is beneficial to find a solution for only one part of the domain, which saves computer resources. An example is flow in a channel (Fig. 4.8). The flow above the symmetry line is a mirror image of the flow below the symmetry line. Therefore, the integration should be carried only for one part of the domain, and the symmetry condition needs to be applied along the symmetry line.

The distribution functions f_5, f_2, and f_6 are unknowns. The way to construct these functions is to set them equal to their mirror images, that is, $f_5 = f_8$, $f_2 = f_4$, and $f_6 = f_7$.

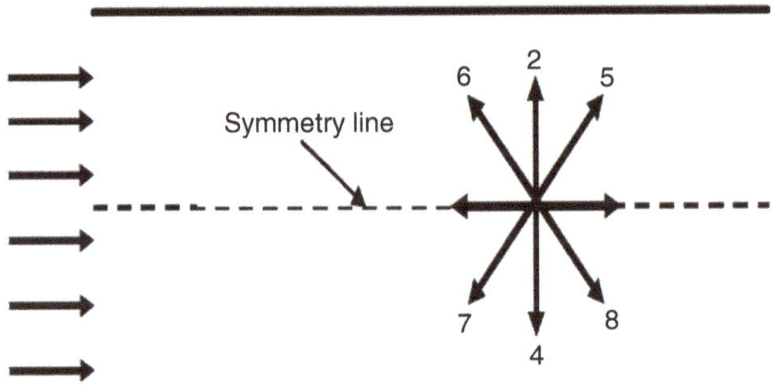

FIGURE 4.8 The Symmetry Boundary Conditions [2]

4.8 LATTICE BOLTZMANN METHOD FOR TEMPERATURE

For scalar function (temperature), the following equation needs to be solved.

$$g_k(x+\Delta x, t+\Delta t) = g_k(x,t)\left[1-\omega_s\right] + \omega_s g_k^{eq}(x,t) \tag{4.47}$$

The equilibrium distribution function is as in the advection problem:

$$g_k^{eq}(x,t) = w_k \phi(x,t)\left[1 + \frac{c_k.\vec{u}}{c_s^2}\right] \tag{4.48}$$

The relaxation time for temperature is

$$\omega_s = \frac{1}{3\alpha + 0.5} \tag{4.49}$$

where α is the thermal diffusion coefficient.

4.9 THERMAL BOUNDARY CONDITION

4.9.1 CONSTANT TEMPERATURE BOUNDARY CONDITION

For walls with a constant temperature, the following equations should be used. For example, if the left wall temperature equals T_w, we have (see Fig. 4.6):

$$\begin{aligned} g_1(0,y) &= T_w(w(1)+w(3)) - g_3(0,y) \\ g_5(0,y) &= T_w(w(5)+w(7)) - g_7(0,y) \\ g_8(0,y) &= T_w(w(8)+w(6)) - g_6(0,y) \end{aligned} \tag{4.50}$$

4.9.2 ADIABATIC BOUNDARY CONDITION

If the bottom boundary is adiabatic, the temperature gradient is zero. Therefore, the following boundary conditions are used (see Fig. 4.6):

$$\begin{aligned} g_2(x,0) &= g_2(x,1) \\ g_5(x,0) &= g_5(x,1)] \\ g_6(x,0) &= g_6(x,1) \end{aligned} \tag{4.51}$$

4.9.3 CONSTANT HEAT FLUX BOUNDARY CONDITION

If the bottom boundary is exposed to a constant heat flux, the boundary conditions are defined as follows (see Fig. 4.6):

$$\begin{aligned} g_2(x,0) &= g_2(x,1) + \frac{2}{3}q''\left(\frac{dy}{k}\right) \\ g_5(x,0) &= g_5(x,1) + \frac{1}{6}q''\left(\frac{dy}{k}\right) \\ g_6(x,0) &= g_6(x,1) + \frac{1}{6}q''\left(\frac{dy}{k}\right) \end{aligned} \tag{4.52}$$

4.10 SAMPLE FORTRAN CODE FOR LBM

4.10.1 LID-DRIVEN CAVITY

After introducing BGK approximation, the general form of the LBM equation for flow filed is as follow [2]:

$$f_i(x+c_i\Delta t, t+\Delta t) = f_i(x,t) + \frac{\Delta t}{\tau_v}[f_i^{\text{eq}}(x,t) - f_i(x,t)]$$

$$f_i^{\text{eq}} = w_i\rho\left[1 + \frac{c_i.u}{c_s^2} + \frac{1}{2}\frac{(c_i.u)^2}{c_s^4} - \frac{1}{2}\frac{u^2}{c_s^2}\right] \tag{4.53}$$

where Δt denotes the lattice time step, c_i is the discrete lattice velocity in direction i, and τ_v denotes the lattice relaxation time for the flow field. The kinetic viscosity υ is defined in terms of its respective relaxation times, that is, $\upsilon = c_s^2(\tau_v - 1/2)$.

Finally, macroscopic variables are calculated with the following formula:

$$\text{Flow density}: \rho = \sum_i f_i,$$

$$\text{Momentum}: \rho \text{u} = \sum_i \text{c}_i f_i, \tag{4.54}$$

FORTRAN code for this problem is presented in Appendix A8. Streamline and velocity profiles at the midsection and at the midheight of the cavity are shown in Fig. 4.9.

4.10.2 NATURAL CONVECTION IN A CAVITY

The density and distribution functions (i.e., f and g) are calculated by solving the LBE, which is a special discretization of the kinetic Boltzmann equation. After introducing BGK approximation, the general form of the LBE with external force is as follows:

For the flow field:

$$f_i(x+c_i\Delta t, t+\Delta t) = f_i(x,t) + \frac{\Delta t}{\tau_v}[f_i^{\text{eq}}(x,t) - f_i(x,t)] + \Delta t c_i F_k \tag{4.55}$$

For the temperature field:

$$g_i(x+c_i\Delta t, t+\Delta t) = g_i(x,t) + \frac{\Delta t}{\tau_C}[g_i^{\text{eq}}(x,t) - g_i(x,t)] \tag{4.56}$$

where Δt denotes the lattice time step, c_i is the discrete lattice velocity in direction i, F_k is the external force in direction of lattice velocity, and τ_v and τ_C denote the lattice relaxation time for the flow and temperature fields. The kinetic viscosity υ and the thermal diffusivity α are defined in terms of their respective relaxation times, that is, $\upsilon = c_s^2(\tau_v - 1/2)$ and $\alpha = c_s^2(\tau_C - 1/2)$, respectively. Note that the limitation $0.5 < \tau$ should be satisfied for both relaxation times to ensure that viscosity and thermal

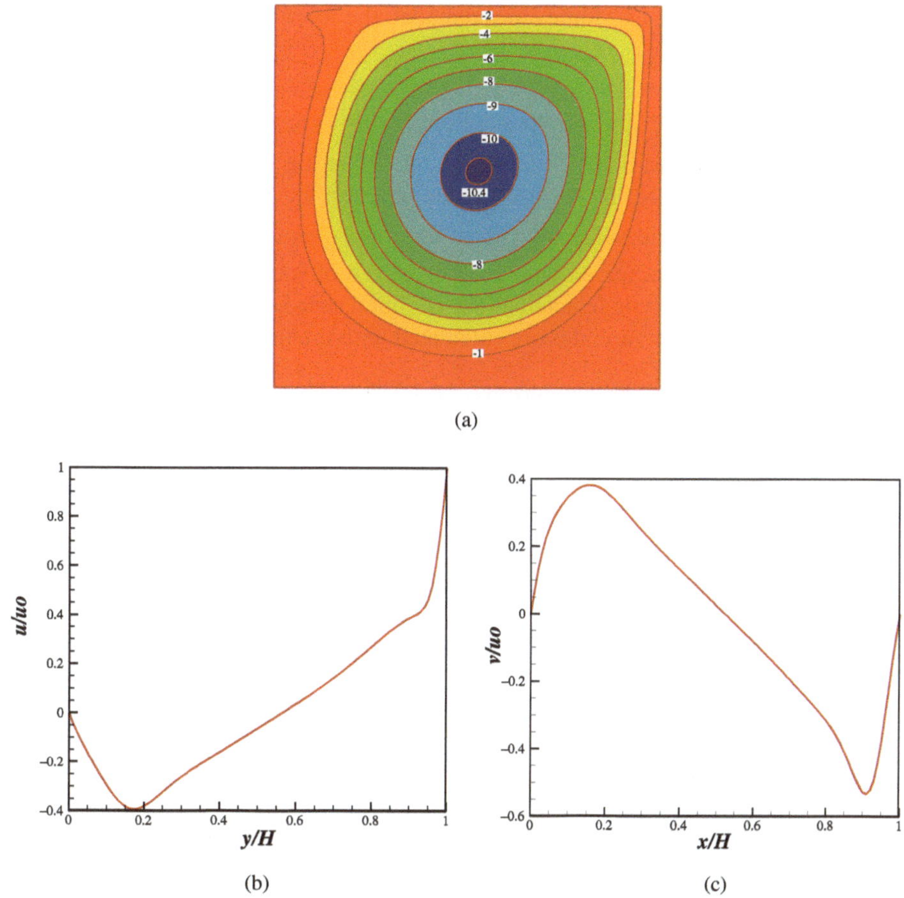

FIGURE 4.9

Streamline (a) velocity profiles (b, c) at the midsection and at the midheight of the cavity.

diffusivity are positive. Furthermore, the local equilibrium distribution function determines the type of problem that needs to be solved. It also models the equilibrium distribution functions, which are calculated with Eqs (4.55) and (4.56) for the flow and temperature fields, respectively.

$$f_i^{eq} = w_i \rho \left[1 + \frac{c_i.u}{c_s^2} + \frac{1}{2}\frac{(c_i.u)^2}{c_s^4} - \frac{1}{2}\frac{u^2}{c_s^2} \right] \tag{4.57}$$

$$g_i^{eq} = w_i T \left[1 + \frac{c_i.u}{c_s^2} \right] \tag{4.58}$$

where w_i is a weighting factor and ρ is the lattice fluid density.

To incorporate buoyancy forces in the model, the force term in Eq. (4.56) needs to be calculated as [2]

$$F = 3w_i\rho\left[g_y\beta(T - T_m)\right] \tag{4.59}$$

For natural convection, the Boussinesq approximation is applied, and radiation heat transfer is negligible. To ensure that the code works in a near incompressible regime, the characteristic velocity of the flow for the natural ($V_{\text{natural}} \equiv \sqrt{\beta g_y \Delta T H}$) regime must be small compared with the fluid speed of sound. In the present study, the characteristic velocity is selected as 0.1 of sound speed.

Finally, macroscopic variables are calculated with the following formula:

$$\begin{aligned}&\text{Flow density:} && \rho = \sum_i f_i,\\ &\text{Momentum:} && \rho \mathbf{u} = \sum_i \mathbf{c}_i f_i,\\ &\text{Temperature:} && T = \sum_i g_i.\end{aligned} \tag{4.60}$$

FORTRAN code of this problem is presented in Appendix A9. Streamlines and isotherms for different Rayleigh numbers are shown in Fig. 4.10.

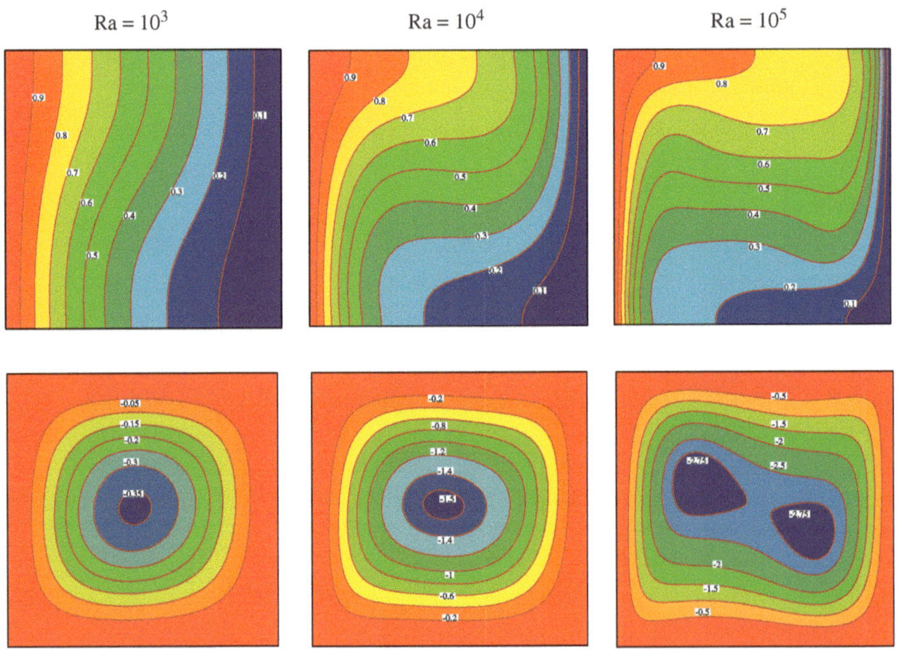

FIGURE 4.10 Streamline and Isotherms for Different Rayleigh Numbers

4.11 CURVED BOUNDARY CONDITIONS

4.11.1 CURVED BOUNDARY TREATMENT FOR VELOCITY

For treating velocity and temperature fields with curved boundaries, the method proposed by Yan and Zu [25] has been used. An arbitrary curved wall separating the solid region from fluid is shown in Fig. 4.11. The link between the fluid node x_f and the wall node x_w intersects the physical boundary at x_b. The fraction of the intersected link in the fluid region is $\Delta = |x_f - x_w| / |x_f - x_b|$. To calculate the postcollision distribution function $\tilde{f}_{\bar{\alpha}}(x_b, t)$ based on the surrounding node information, a Chapman–Enskog expansion for the postcollision distribution function on the right-hand side of Eq. (4.21) is conducted as

$$\tilde{f}_{\bar{\alpha}}(x_b, t) = (1-\chi)\tilde{f}_{\alpha}(x_f, t) + \chi f_{\alpha}^{*}(x_b, t) + 2w_{\alpha}\rho\frac{3}{c^2}e_{\bar{\alpha}}.u_w \tag{4.61}$$

where

$$\begin{aligned} &f_{\alpha}^{*}(x_b, t) = f_{\alpha}^{eq}(x_f, t) + w_{\alpha}\rho(x_f, t)\frac{3}{c^2}e_{\alpha}\cdot(u_{bf} - u_f), \\ &u_{bf} = u_{ff} = u(x_{ff}, t), \qquad \chi = \frac{(2\Delta - 1)}{\tau - 2}, \qquad \text{if } 0 \le \Delta \le \frac{1}{2} \\ &u_{bf} = \frac{1}{2\Delta}(2\Delta - 3)u_f + \frac{3}{2\Delta}u_w, \quad \chi = \frac{(2\Delta - 1)}{\tau - 1/2}, \qquad \text{if } \frac{1}{2} \le \Delta \le 1 \end{aligned} \tag{4.62}$$

In the aforementioned, $e_{\bar{\alpha}} \equiv -e_{\alpha}$; u_f is the fluid velocity near the wall; u_w is the velocity of solid wall; and u_{bf} is an imaginary velocity for interpolations.

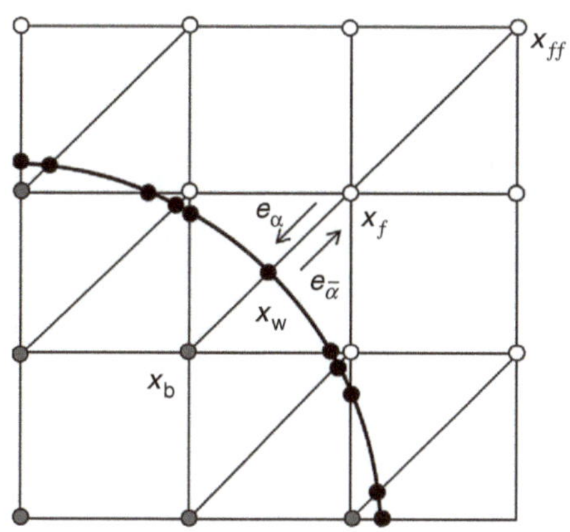

FIGURE 4.11

Discrete velocity set of two-dimensional nine-velocity (D_2Q_9) model and curved boundary and lattice nodes.

4.11.2 CURVED BOUNDARY TREATMENT FOR TEMPERATURE

Following the work of Yan and Zu [25], the nonequilibrium parts of the temperature distribution function can be defined as:

$$g_{\bar{\alpha}}(\mathrm{x}_b,t) = g_{\bar{\alpha}}^{\mathrm{eq}}(\mathrm{x}_b,t) + g_{\bar{\alpha}}^{\mathrm{neq}}(\mathrm{x}_b,t) \quad (4.63)$$

Substituting Eq. (4.63) into the Boltzmann equation leads to:

$$\tilde{g}_{\bar{\alpha}}(\mathrm{x}_\mathrm{b},t+\Delta t) = g_{\bar{\alpha}}^{\mathrm{eq}}(\mathrm{x}_\mathrm{b},t) + (1-\frac{1}{\tau_\mathrm{s}})g_{\bar{\alpha}}^{\mathrm{neq}}(\mathrm{x}_\mathrm{b},t) \quad (4.64)$$

Obviously, both $g_{\bar{\alpha}}^{\mathrm{eq}}(\mathrm{x}_b,t)$ and $g_{\bar{\alpha}}^{\mathrm{neq}}(\mathrm{x}_b,t)$ are needed to calculate the value of $\tilde{g}_{\bar{\alpha}}(\mathrm{x}_\mathrm{b},t+\Delta t)$. In Eq. (4.64), the equilibrium part is defined as

$$g_{\bar{\alpha}}^{eq}(\mathrm{x}_\mathrm{b},t) = w_{\bar{\alpha}} T_\mathrm{b}^* \left(1+\frac{3}{c^2}\mathrm{e}_{\bar{\alpha}}.\mathrm{u}_\mathrm{b}^*\right) \quad (4.65)$$

where T_b^* is defined as a function of $T_{\mathrm{b}1} = [T_\mathrm{w} + (\Delta-1)T_\mathrm{f}]/\Delta$ and $T_{\mathrm{b}2} = [2T_\mathrm{w} + (\Delta-1)T_\mathrm{ff}]/(1+\Delta)$

$$\begin{aligned} T_\mathrm{b}^* &= T_{\mathrm{b}1}, && \text{if } \Delta \geq 0.75 \\ T_\mathrm{b}^* &= T_{\mathrm{b}1} + (1-\Delta)T_{\mathrm{b}2}, && \text{if } \Delta \leq 0.75 \end{aligned} \quad (4.66)$$

and u_b^* is defined as a function of $u_{\mathrm{b}1} = [u_\mathrm{w} + (\Delta-1)u_\mathrm{f}]/\Delta$ and $u_{\mathrm{b}2} = [2u_\mathrm{w} + (\Delta-1)u_\mathrm{ff}]/(1+\Delta)$.

$$\begin{aligned} u_\mathrm{b}^* &= u_{\mathrm{b}1}, && \text{if } \Delta \geq 0.75 \\ u_\mathrm{b}^* &= u_{\mathrm{b}1} + (1-\Delta)u_{\mathrm{b}2}, && \text{if } \Delta \leq 0.75 \end{aligned} \quad (4.67)$$

The nonequilibrium part in Eq. (4.64) is defined as:

$$g_{\alpha}^{\mathrm{neq}}(\mathrm{x}_\mathrm{b},t) = \Delta g_{\alpha}^{\mathrm{neq}}(x_\mathrm{f},t) + (1-\Delta)g_{\alpha}^{\mathrm{neq}}(x_\mathrm{ff},t) \quad (4.68)$$

4.12 THE LATTICE BOLTZMANN METHOD FOR MHD

4.12.1 LBM FOR MHD WITH TWO DISTRIBUTION FUNCTIONS

The LBM used here is the same as that used by Kao and Yang [26] and Barrios et al. [27]. The thermal LBM uses two distribution functions, f and g, for the flow and temperature fields, respectively. It uses modeling of movement of fluid particles to capture macroscopic fluid quantities such as velocity, pressure, and temperature. In this approach, the fluid domain is discretized to uniform Cartesian cells. Each cell holds a fixed number of distribution functions, which represent the number of fluid particles moving in these discrete directions. The D_2Q_9 model was used, and values of $w_0 = 4/9$ for $|c_0| = 0$ (for the static particle), $w_{1-4} = 1/9$ for $|c_{1-4}| = 1$, and $w_{5-9} = 1/36$ for $|c_{5-9}| = \sqrt{2}$ are assigned in this model. The density and distribution functions, that is, the f and g, are calculated by solving the LBE, which is a special discretization of the kinetic Boltzmann equation. After introducing BGK approximation, the general form of the LBE with external force is as follows:

For the flow field:

$$f_i(x + c_i\Delta t, t + \Delta t) = f_i(x,t) + \frac{\Delta t}{\tau_v}[f_i^{eq}(x,t) - f_i(x,t)] + \Delta t c_i F_k \tag{4.69}$$

For the temperature field:

$$g_i(x + c_i\Delta t, t + \Delta t) = g_i(x,t) + \frac{\Delta t}{\tau_C}[g_i^{eq}(x,t) - g_i(x,t)] \tag{4.70}$$

where Δt denotes lattice time step, c_i is the discrete lattice velocity in direction i, F_k is the external force in direction of lattice velocity, and τ_v and τ_C denote the lattice relaxation time for the flow and temperature fields. The kinetic viscosity υ and the thermal diffusivity α are defined in terms of their respective relaxation times, that is, $\upsilon = c_s^2(\tau_v - 1/2)$ and $\alpha = c_s^2(\tau_C - 1/2)$, respectively. Note that the limitation $0.5 < \tau$ should be satisfied for both relaxation times to ensure that viscosity and thermal diffusivity are positive. Furthermore, the local equilibrium distribution function determines the type of problem that needs to be solved. It also models the equilibrium distribution functions, which are calculated with Eqs (4.69) and (4.70) for the flow and temperature fields, respectively.

$$f_i^{eq} = w_i\rho\left[1 + \frac{c_i.u}{c_s^2} + \frac{1}{2}\frac{(c_i.u)^2}{c_s^4} - \frac{1}{2}\frac{u^2}{c_s^2}\right] \tag{4.71}$$

$$g_i^{eq} = w_i T\left[1 + \frac{c_i.u}{c_s^2}\right] \tag{4.72}$$

where w_i is a weighting factor and ρ is the lattice fluid density.

To incorporate buoyancy forces and magnetic forces in the model, the force term in Eq. (4.69) needs to be calculated as

$$\begin{aligned} &F = F_x + F_y \\ &F_x = 3w_i\rho\left[A\left(v\sin(\theta_M)\cos(\theta_M)\right) - \left(u\sin^2(\theta_M)\right)\right], \\ &F_y = 3w_i\rho\left[g_y\beta(T - T_m) + A\left(u\sin(\theta_M)\cos(\theta_M)\right) - \left(v\cos^2(\theta_M)\right)\right] \end{aligned} \tag{4.73}$$

where A is $A = Ha^2\mu/L^2$, $Ha = LB\sqrt{\sigma/\mu}$ is the Hartmann number, and θ_M is the direction of the magnetic field. For natural convection, the Boussinesq approximation is applied, and radiation heat transfer is negligible. To ensure that the code works in a near incompressible regime, the characteristic velocity of the flow for natural ($V_{natural} \equiv \sqrt{\beta g_y \Delta TH}$) regime must be small compared with the fluid speed of sound. In the present study, the characteristic velocity is selected as 0.1 of sound speed.

Finally, macroscopic variables are calculated with the following formula:

$$\begin{aligned}&\text{Flow density:} && \rho=\sum_i f_i,\\ &\text{Momentum:} && \rho \mathrm{u}=\sum_i \mathrm{c}_i f_i,\\ &\text{Temperature:} && T=\sum_i g_i.\end{aligned} \tag{4.74}$$

4.12.2 LBM FOR MHD WITH THREE DISTRIBUTION FUNCTIONS

There are two other Lattice Boltzmann models for MHD flows: the multispeed (MS) model and multidistribution function (MDF) model. In the MS model, the equilibrium distribution is transformed in order not only to include the magnetic field force but also to be equal to the magnetic field vector on a different vector base. This is achieved by allowing an extra degree of freedom and by defining supplementary vectors on each base vector. Combining the base and the supplementary vectors, the two vector bases are defined for the momentum and the magnetic field. In the MDF model presented by Dellar [28], the Lorentz force can be introduced as a point-wise force. The induction equation is also solved using an Lattice Bhatnagar-Gross-Krook (LBGK) equation by introducing an independent distribution function. MDF models can improve the numerical stability. The LBM used here is the same as that used by Kao and Yang [29]. The thermal LBM uses two distribution functions, *f*, *g*, and *B*, for the flow, temperature, and magnetic fields, respectively. It uses modeling of movement of fluid particles to capture macroscopic fluid quantities such as velocity, pressure, temperature, and magnetic field. In this approach, the fluid domain is discretized to uniform Cartesian cells. Each cell holds a fixed number of distribution functions, which represent the number of fluid particles moving in these discrete directions. The D2Q9 model was used and values of $w_0 = 4/9$ for $|c_0| = 0$ (for the static particle), $w_{1-4} = 1/9$ for $|c_{1-4}| = 1$, and $w_{5-9} = 1/36$ for $|c_{5-9}| = \sqrt{2}$ are assigned in this model. The density and distribution functions, that is, the *f*, *g*, and *B* are calculated by solving the LBE, which is a special discretization of the kinetic Boltzmann equation. After introducing BGK approximation, the general form of LBE with external force is

For the flow field:

$$f_i(x+c_i\Delta t, t+\Delta t) = f_i(x,t) + \frac{\Delta t}{\tau_v}[f_i^{\mathrm{eq}}(x,t) - f_i(x,t)] + \Delta t c_i F_k \tag{4.75}$$

For the temperature field:

$$g_i(x+c_i\Delta t, t+\Delta t) = g_i(x,t) + \frac{\Delta t}{\tau_C}[g_i^{\mathrm{eq}}(x,t) - g_i(x,t)] \tag{4.76}$$

where Δt denotes lattice time step, c_i is the discrete lattice velocity in direction i, F_k is the external force in direction of lattice velocity, and τ_v and τ_c denote the lattice relaxation time for the flow and temperature fields. The kinetic viscosity υ and the thermal diffusivity α are defined in terms of their respective relaxation times, that is, $\upsilon = c_s^2(\tau_v - 1/2)$ and $\alpha = c_s^2(\tau_C - 1/2)$, respectively. Note that the limitation $0.5 < \tau$

should be satisfied for both relaxation times to ensure that viscosity and thermal diffusivity are positive. Furthermore, the local equilibrium distribution function determines the type of problem that needs to be solved. It also models the equilibrium distribution functions for flow and temperature fields, respectively. In this study, the density distribution function, f_i^{eq}, was modified to consider the magnetic effect:

$$f_i^{\text{eq}} = w_i\rho\left[1+\frac{c_i.u}{c_s^2}+\frac{1}{2}\frac{(c_i.u)^2}{c_s^4}-\frac{1}{2}\frac{u^2}{c_s^2}\right]+\frac{w_i}{2c_s^2}\left[\frac{B^2c^2}{2}-(c.B)^2\right] \tag{4.77}$$

$$g_i^{\text{eq}} = w_iT\left[1+\frac{c_i.u}{c_s^2}\right] \tag{4.78}$$

where B is the magnetic field, w_i is weighting factor, and c_s is the speed of sound and defined by $c_s = c/\sqrt{3}$. Similar to the density equilibrium function (f_i^{eq}), for calculating the magnetic field, the magnetic equilibrium function considered as [28]

$$h_{ix}^{\text{eq}} = \lambda_i\left[B_x+\frac{1}{c_s^2}e_{ix}\left(u_yB_x-u_xB_y\right)\right] \tag{4.79}$$

$$h_{iy}^{\text{eq}} = \lambda_i\left[B_y+\frac{1}{c_s^2}e_{iy}\left(u_xB_y-u_yB_x\right)\right] \tag{4.80}$$

where λ_i is the weighting factor of the magnetic field and defined in the fifth direction by [28]

$$\lambda_i = \begin{cases} \frac{1}{3} & \text{for} \quad i=0 \\ \frac{1}{6} & \text{for} \quad i=1-4 \end{cases} \tag{4.81}$$

For solving the velocity and magnetic field, the bellow equation must considered [28].

$$h_i\left(x+c_i\Delta t, t+\Delta t\right) = h_i\left(x,t\right)+\frac{\Delta t}{\tau_m}\left[h_i^{\text{eq}}\left(x,t\right)-h_i\left(x,t\right)\right] \tag{4.82}$$

The magnetic resistivity, like kinetic viscosity υ and the thermal diffusivity α, is defined in terms of its respective relaxation time $\eta = c_s^2(\tau_m - 1/2)$.

To incorporate buoyancy force in the model, the force term in the Eq. (4.75) needs to be calculated as below in the vertical direction (y):

$$F = 3w_ig_y\beta\theta \tag{4.83}$$

For natural convection, the Boussinesq approximation is applied, and radiation heat transfer is negligible. To ensure that the code works in a near incompressible regime, the characteristic velocity of the flow for natural ($V_{\text{natural}} \equiv \sqrt{\beta g_y \Delta TH}$) regime must be small compared with the fluid speed of sound. In the present study, the characteristic velocity selected as 0.1 of the sound speed.

Finally, macroscopic variables are calculated with the following formula:

$$\rho = \sum_i f_i, \quad \rho\text{u} = \sum_i \text{c}_i f_i, \quad T = \sum_i g_i, \quad B_x = \sum_i h_{ix}, \quad B_y = \sum_i h_{iy} \tag{4.84}$$

4.13 THE LATTICE BOLTZMANN MODEL FOR NANOFLUID

To simulate the nanofluid by the LBM, because of the interparticle potentials and other forces on the nanoparticles, the nanofluid behaves differently from the pure liquid from the mesoscopic point of view and is of higher efficiency in energy transport as well as better stabilization than the common solid-liquid mixture. For modeling the nanofluid because of changing in the fluid thermal conductivity, density, heat capacitance, and thermal expansion, some of the governed equations should change. The effective density, effective heat capacity, thermal expansion, and electrical conductivity of the nanofluid are defined as mentioned in Chapter 1.

4.14 APPLICATION OF LBM FOR NANOFLUID HYDROTHERMAL BEHAVIOR IN PRESENCE OF MAGNETIC FIELD

4.14.1 VALIDATION

A validation test was carried for natural convection in an enclosure filled with Cu–water for different Grashof numbers with the results of Khanafer et al. [30] in Fig. 4.12. The present numerical solution is validated by comparing the present code results against the results of Davis [31] for Pr = 0.71.This comparison is presented in Table 4.3. Another test for validation of the current code was performed for the case

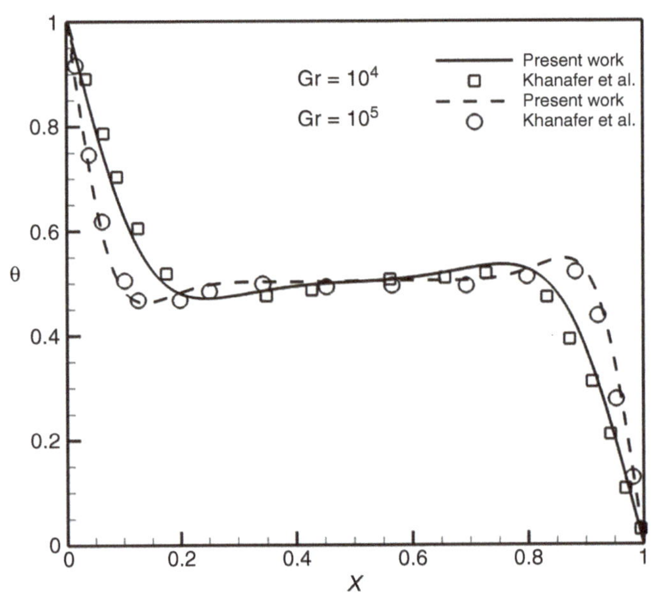

FIGURE 4.12 Comparison of the Temperature on Axial Midline Between the Present Results and Numerical Results by Khanafer et al. [30] $\phi = 0.1$ and Pr = 6.8 (Cu–Water)

Table 4.3 Comparison of the Present Solution With Previous Works for Different Rayleigh Numbers When Pr = 0.7

Ra	Present	Davis [31]
10^3	1.1432	1.118
10^4	2.2749	2.243
10^5	4.5199	4.519

Table 4.4 Average Nusselt Number Versus at Different Grashof Number Under Various Strengths of the Magnetic Field at Pr = 0.733

	$Gr = 2 \times 10^4$		$Gr = 2 \times 10^5$	
Ha	Present	Rudraiah et al. [32]	Present	Rudraiah et al. [32]
0	2.5665	2.566894	5.093205	5.151789
10	2.26626	2.261644	4.9047	4.952211
50	1.09954	1.083047	2.67911	2.689937
100	1.02218	1.008833	1.46048	1.456019

of natural convection in a square enclosure in the presence of a magnetic field. In this test case, the average Nusselt number using different Gr and Ha numbers has been compared with those obtained by Rudraiah et al. [28] as shown in Table 4.4. All of the previous comparisons indicate the accuracy of the present LBM code.

4.14.2 MAGNETIC FIELD EFFECTS ON NATURAL CONVECTION AROUND A HORIZONTAL CIRCULAR CYLINDER INSIDE A SQUARE ENCLOSURE FILLED WITH NANOFLUID

4.14.2.1 Problem definition

The physical model used in this work is shown in Fig. 4.13. The model consists of a concentric annulus between a cold outer square cylinder with temperature T_c and a heated inner circular cylinder with temperature T_h, ($T_h > T_c$). Also, it is assumed that the walls are insulating magnetic field. In Fig. 4.13, $\lambda = L/2r$ denotes the aspect ratio, and γ is measured counterclockwise from the upward vertical plane through the center of outer cylinders [33].

4.14.2.2 Effect of active parameters

In this example, the effective thermal conductivity and viscosity of Cu–water nanofluid are calculated using the Maxwell–Garnetts and Brinkman models, respectively. Also, the MDF model was used for simulating the effect of uniform magnetic field [33]. Fig. 4.14 shows the effect of nanoparticles on the streamlines and isotherms. As seen, the velocity components of nanofluid are increased because of an increase

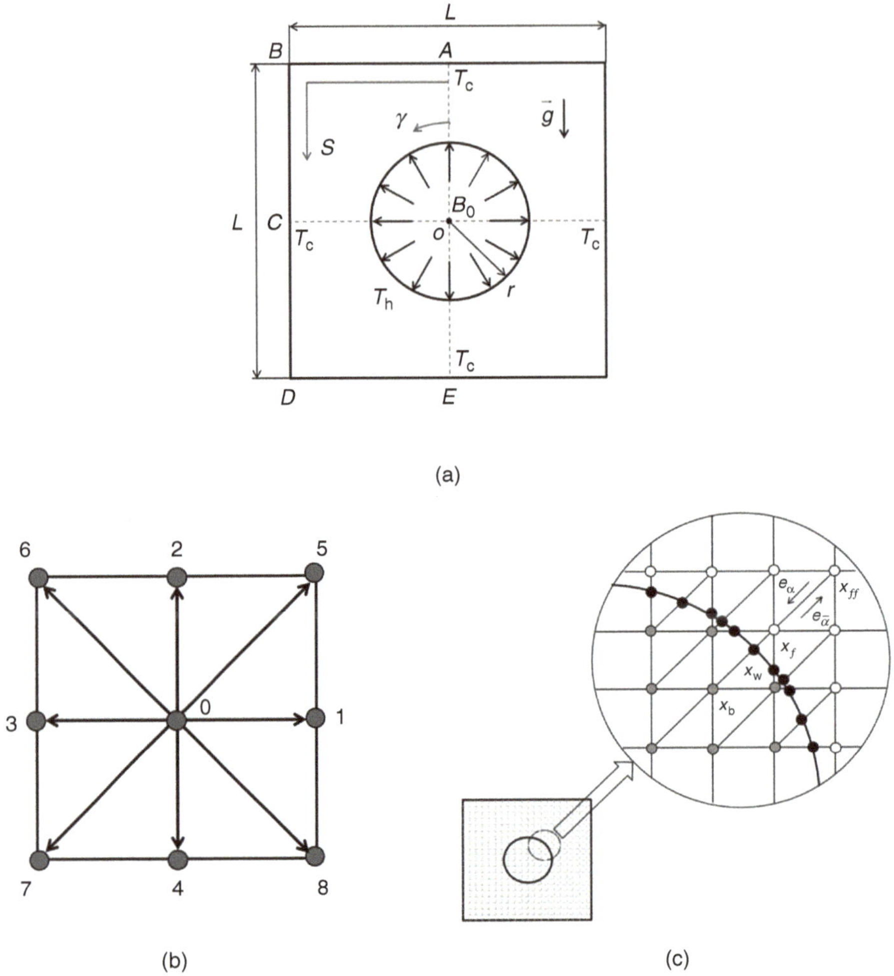

FIGURE 4.13

(a) Geometry of the problem. (b) Discrete velocity set of two-dimensional nine-velocity (D_2Q_9) model. (c) Curved boundary and lattice nodes.

in the energy transport in the fluid with the increasing of volume fraction. Thus, the absolute values of stream functions that indicate the strength of flow increase with increasing the volume fraction of nanofluid. Effects of Rayleigh number and Hartmann number on the isotherms and streamlines contours when $\phi = 0.06$ and $\lambda = 2.5$ is shown in Fig. 4.15. As seen, for all cases, an increase of the Hartmann number decreases the maximum value of absolute stream function this indicates that the convection mechanism is suppressed because of the presence of magnetic field. At Ra = 10^4, the conduction heat transfer mechanism is more pronounced, and the isotherms are parallel to each other and take the enclosure shape. At this Rayleigh number, two eddies exist within the cavity, and the upper one is stronger than the

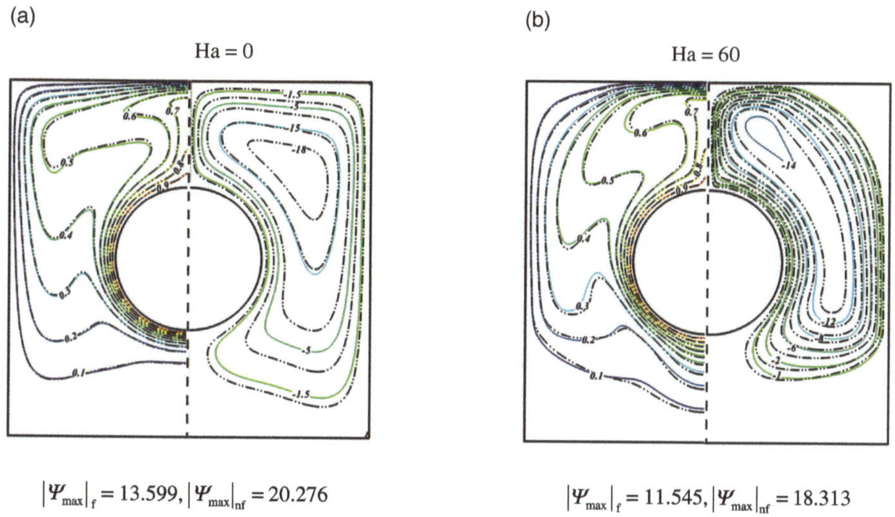

FIGURE 4.14

Comparison of the Isotherm (a) and Streamline (b) Contours Between Nanofluid ($\phi = 0.06$) (—) and Pure Fluid ($\phi = 0$) (–…–) for Different Values of λ at Ra = 10^6, $\lambda = 2.5$, and Pr = 6.8.

lower one. As the Hartmann number increases up to Ha = 40, the size of the lower vortex decreases and then increases after Ha = 40. At Ra = 10^5, the convective heat transfer becomes comparable with conduction. Moreover, a thermal plume appears over the hot circular cylinder, and the thermal boundary layer becomes stratified in the lower part of the enclosure, which is because of the existence of a hot wall of the circular cylinder over the enclosure at this area. It is an interesting point that increasing the magnetic field enhances the thickness of the thermal boundary layer under the hot cylinder decreases. The streamlines depict a single cell moves upward and approaches the vertical centerline of the enclosure as the Hartmann number enhances. At Ra = 10^6, the thermal plume that is formed over the hot circular cylinder, which causes the hot fluid, impinges on the cold wall of the enclosure. Hence, the thermal boundary layer decreases over the cold wall of the enclosure. The flow pattern inside the enclosure is similar to that of Ra = 10^5 except the value of stream function, which enhances is greater than its value at Ra = 10^5, this increment is the characteristic of flow at high Rayleigh numbers. The distribution of the local Nusselt numbers only over the left half of the enclosure and inner cylinder are shown in Figs 4.16 and 4.17. As a general observation, an increase of the Rayleigh number and nanoparticle volume fraction enhance the local Nusselt number value, but the effect of Hartmann number on the local Nusselt number is in contrast with the effect of Ra and ϕ.

The effects of the Rayleigh numbers, Hartmann number, and nanoparticle volume fraction on the local Nusselt number along the walls of the enclosure (Nu_c) is shown in Fig. 4.16. The local Nusselt number has a maximum value at point A, which is the stagnation point on the top wall of the enclosure. When we move from point A to B along the top wall of the enclosure, the local Nusselt number decreases and reaches

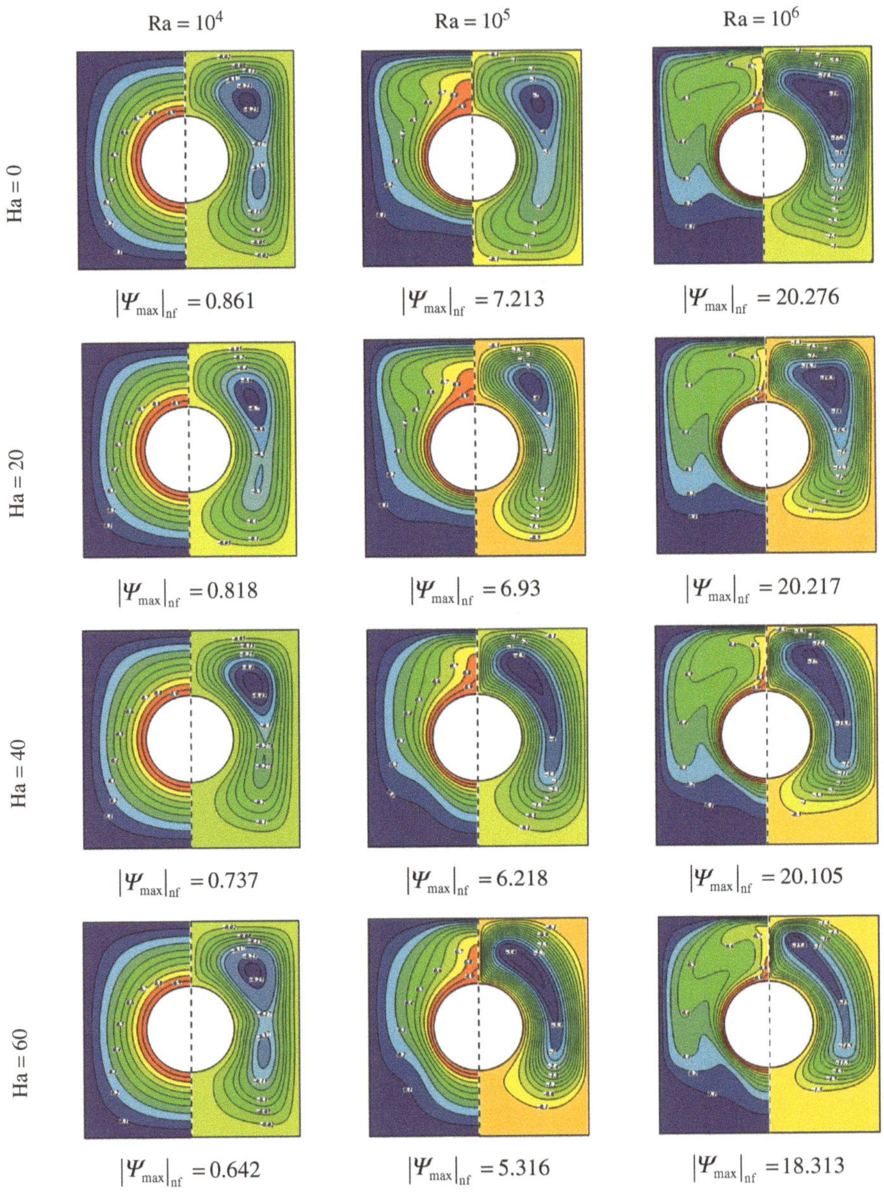

FIGURE 4.15

Effects of Rayleigh Number (Ra) and Hartmann Number (Ha) on Isotherm (Left) and Streamline (Right) Contours When $\phi = 0.06$ and $\lambda = 2.5$.

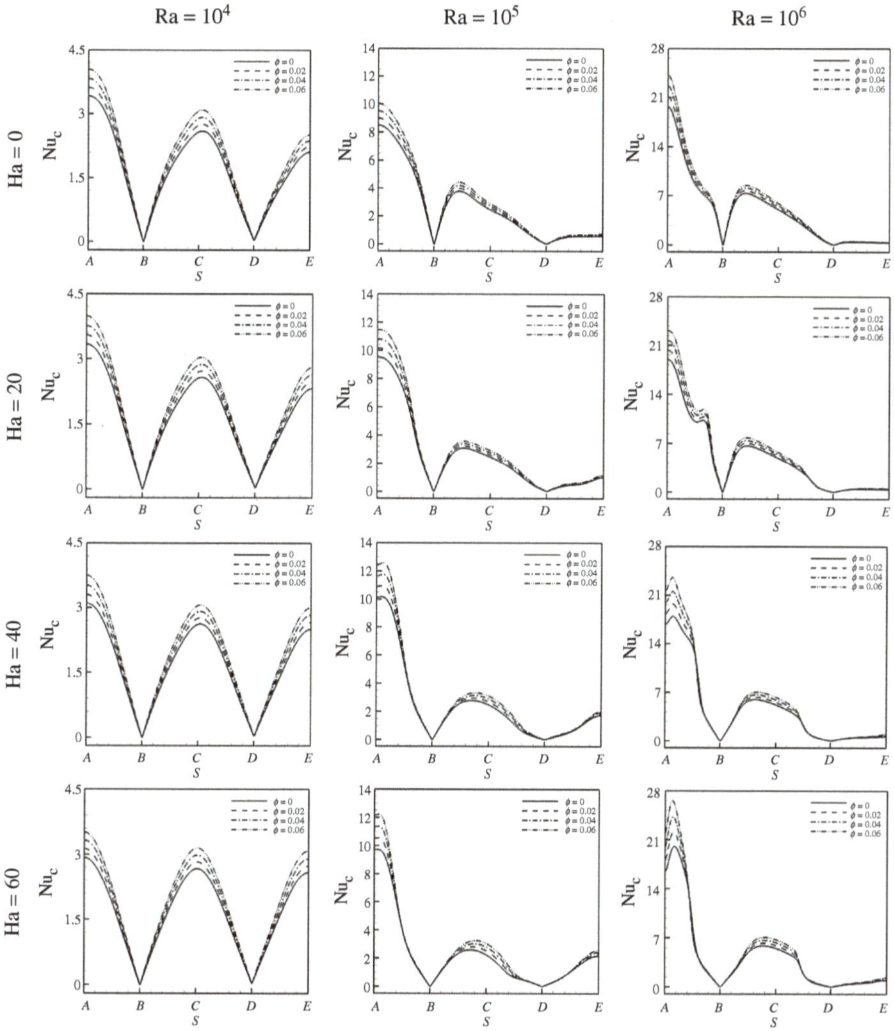

FIGURE 4.16 Effects of Rayleigh Number (Ra), Hartmann Number (Ha), and Nanoparticle Volume Fraction (ϕ) on the Local Nusselt Number Along the Walls of the Enclosure (Nu_χ) When $\lambda = 2.5$

a local minimum value close to zero at point B. When we move from point B to C along the right wall of the enclosure, the local Nusselt number increases, reaches a local maximum value, and decreases again until it has a local minimum value at point D. When we move farther from point D to E, the local Nusselt number increases slightly again. At $Ra = 10^4$, the Nu_c profile is nearly symmetry respect to the horizontal center line. As the Rayleigh number enhances (e.g., $Ra = 10^5$ and $Ra = 10^6$), the Nu_c profile is no longer in symmetry, and the local Nusselt number is considerably small over the bottom wall of the enclosure. In addition, the local maximum Nusselt number near the vertical centerline occurs where the flow impinges on the cold wall. It is worthwhile mentioning that with an increase of the Rayleigh number, the local maximum of Nu_c between B and D approaches point B because of the increase in convective forces. At $Ra = 10^5$ and 10^6 increases of the Hartmann number, the value of the local Nusselt number decreases near E over the cold wall of the enclosure.

The local Nusselt number distribution on the surface of the inner circular cylinder is depicted in Fig. 4.17. Generally, as we move from $\lambda = 0$ to 180°, Nu_h enhances.

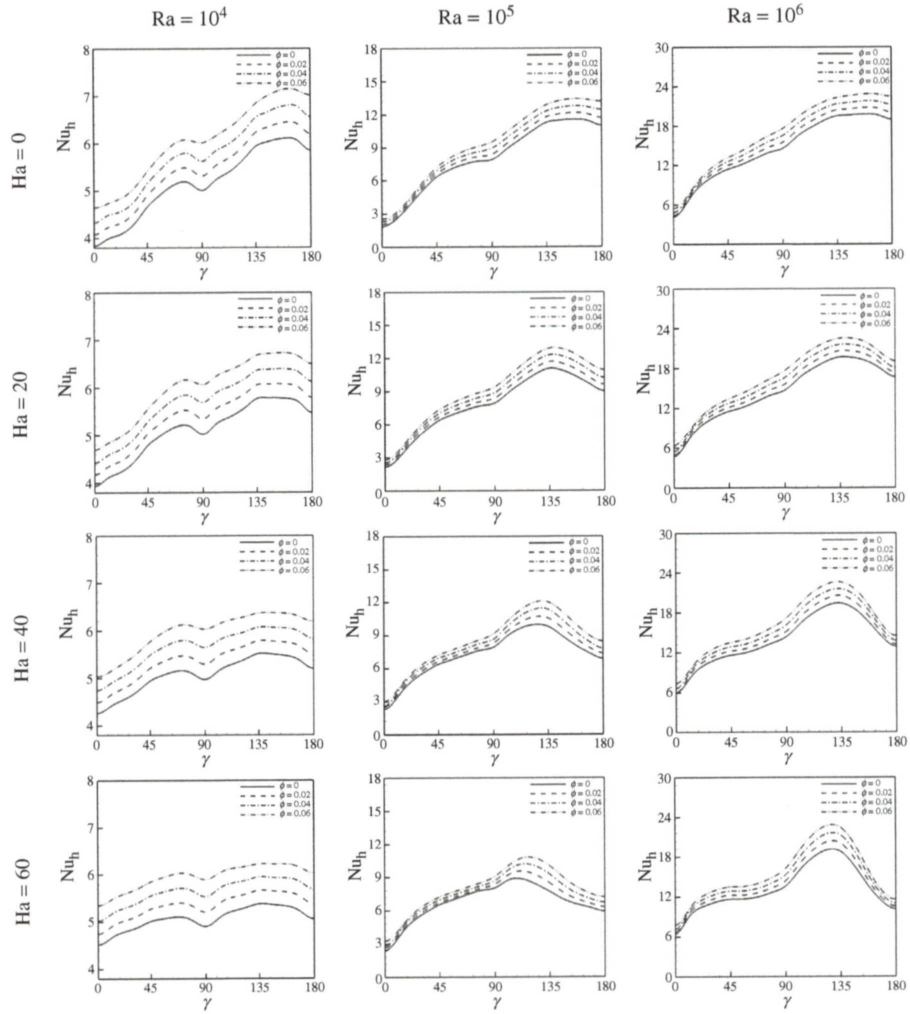

FIGURE 4.17 Effects of Rayleigh Numbers (Ra), Hartmann Number (Ha) and Nanoparticle Volume Fraction (ϕ) on the Local Nusselt Number Along the Surface of the Inner Cylinder (Nu_h) When $\lambda = 2.5$

At Ra = 10^5 and 10^6, an increase of the Hartmann number leads to an increase in the average Nusselt number at $\gamma = 180°$, which is due to an increase of the thermal boundary layer thickness in the presence of a magnetic field. Fig. 4.18a,b shows the averaged Nusselt number over the surface of the enclosure $\left(\overline{\mathrm{Nu}_c}\right)$ and the averaged Nusselt number over the inner cylinder $\left(\overline{\mathrm{Nu}_h}\right)$ as a function of the Hartmann number for different Rayleigh numbers. As it can be seen, the average Nusselt number on the both inner and outer cylinders increases as the Rayleigh number enhances or the Hartmann number decreases. Also, this figure shows that the effect of the Hartmann number is more pronounced at a high Rayleigh number. To estimate the enhancement of heat transfer between the case of $\phi = 0.06$ and the pure fluid (base fluid) case, the heat transfer enhancement is defined as

$$\mathrm{En} = \frac{\mathrm{Nu}(\phi = 0.06) - \mathrm{Nu}(\mathrm{basefluid})}{\mathrm{Nu}(\mathrm{basefluid})} \times 100 \tag{4.85}$$

The heat transfer enhancement ratios due to the addition of nanoparticles for different values of Ha and Ra are shown in Fig. 4.18c. The figure shows that the ratio of enhancement has different behavior for various values of Rayleigh number. When Ra = 10^4 and 10^5, by increasing the Hartmann number from 0 to 40, heat transfer enhancement increases; by increasing the Hartmann number, the heat transfer enhancement decreases. This change is more pronounced for Ra = 10^5. When Ra = 10^6, the minimum of En is obtained at Ha = 20. Moreover, it can be found that for Ha < 20, the enhancement in heat transfer at Ra = 10^4 is greater than at other Rayleigh numbers. Also, for Ha > 20, the maximum of En is obtained for Ra = 10^5.

4.14.3 MHD FREE CONVECTION IN AN ECCENTRIC SEMI-ANNULUS FILLED WITH NANOFLUID

4.14.3.1 Problem definition

The physical model used in this work is shown in Fig. 4.19. In this figure, $\lambda = L/2r$ denotes aspect ratios, and γ is measured counterclockwise from the upward vertical plane through the center of the outer cylinders [6]. The system consists of a square enclosure with sides of length L, within which a circular cylinder with $\lambda = 3.5$ is located and moves along the vertical centerline in the range from $\delta/L = 0.2$ to 0.8. The walls of the square enclosure were kept at a constant low temperature of T_c, and the cylinder was kept at a constant high temperature of T_h, ($T_h > T_c$). Also, it is also assumed that the uniform magnetic field ($\vec{B} = B_x \vec{e}_x + B_y \vec{e}_y$) of constant magnitude $B = \sqrt{B_x^2 + B_y^2}$ is applied, where $\vec{e}_x$ and $\vec{e}_y$ are unit vectors in the Cartesian coordinate system. The orientation of the magnetic field form an angle θ_M with the horizontal axis such that $\theta_M = \cot^{-1}\left(B_x/B_y\right.$. The electric current J and the electromagnetic force F are defined by $J = \sigma\left(\vec{V} \times \vec{B}\right)$ and $F = \sigma\left(\vec{V} \times \vec{B}\right) \times \vec{B}$, respectively. In this study, we considered $\theta_M = 0$.

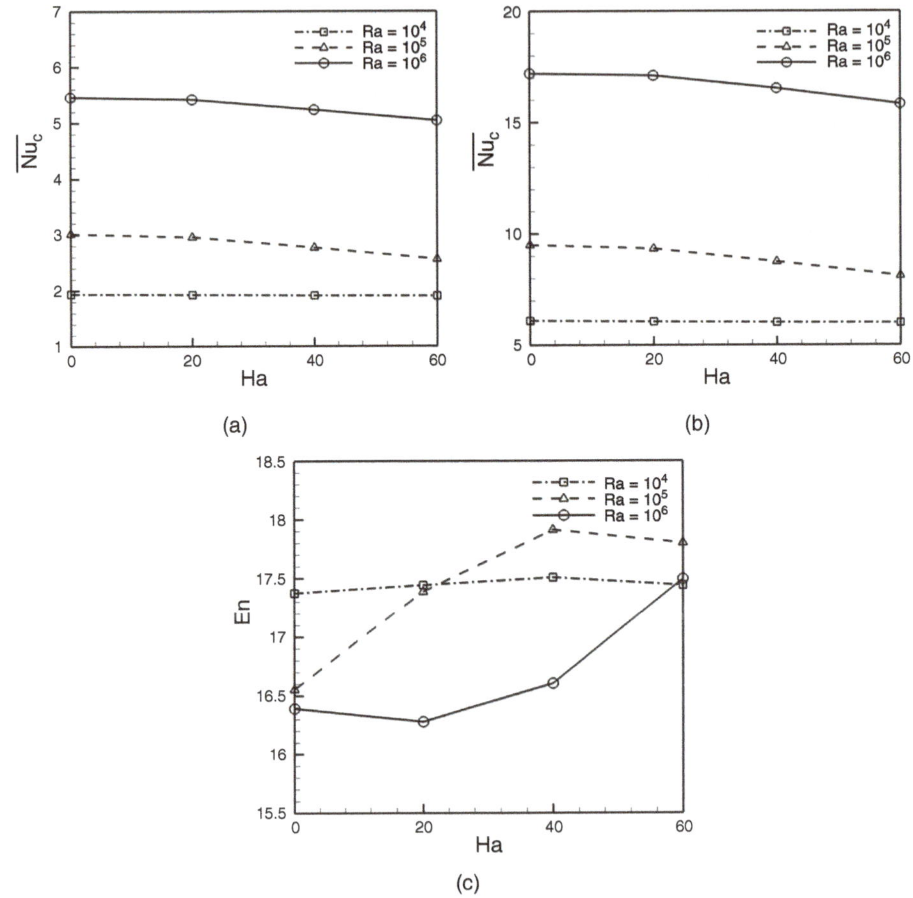

FIGURE 4.18

Effects of Rayleigh number (Ra) and Hartmann number (Ha) on Local Nusselt number along (a) the walls of the enclosure $\left(\overline{Nu_c}\right)$; (b) the surface of the inner cylinder $\left(\overline{Nu_h}\right)$ when $\lambda = 2.5$ and $\phi = 0.06$. (c) Effects of Rayleigh numbers (Ra) and Hartmann number (Ha) on the ratio of enhancement of heat transfer caused by the addition of nanoparticles when $\lambda = 2.5$.

4.14.3.2 Effect of active parameters

LBM is applied to solve the problem of the MHD effect on free convection in an eccentric annulus filled with nanofluid. Calculations are made for various values of Hartmann number (Ha = 0 − 40), volume fraction of nanoparticle ($\phi = 0, 0.02, 0.04$, and 0.06), Rayleigh number (Ra = 10^4, 10^5, and 10^6), and position of the inner circular cylinder ($\delta/L = 0.2, 0.4, 0.6$ to 0.8) when Prandtl number ($Pr = 6.8$) and aspect ratio ($\lambda = 3.5$) are fixed.

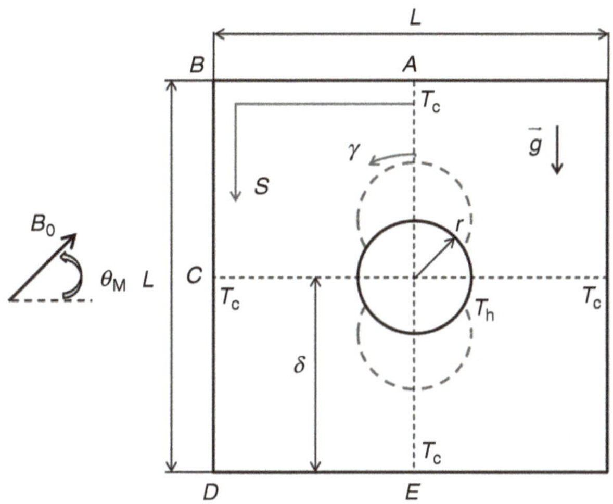

FIGURE 4.19 Geometry of the Problem

The effects of Cu nanoparticles on the streamlines and isotherms in the presence and absence of the magnetic field are shown in Fig. 4.20.The flow strength (absolute values of stream function) increases when nanoparticle are added in to the base fluid. The thermal boundary layer thickness increases as the volume fraction of the nanofluid increases. It is interesting to note from this figure that the effect of the nanofluid volume fraction is more pronounced when the hot cylinder is located at the upper location in the presence of a magnetic field. This is owing to domination of the conduction heat transfer mechanism below the hot cylinder. Also, the presence of the magnetic field helps the conduction mechanism to be dominated.

Fig. 4.21 shows the effects of Ha, Ra, and δ/L on isotherm and streamline contours. At 10^4, increasing δ/L up to 0.6 leads to a decrease in the maximum value of stream function as δ/L enhances further, $|\psi_{max}|_{nf}$ increases. At high Raleigh numbers (e.g., Ra = 10^5 and 10^6), the values of $|\psi_{max}|_{nf}$ continuously decrease by augmentation of δ/L. As seen, $|\psi_{max}|_{nf}$ has an inverse relationship with the Hartmann number.

At Ra = 10^4, the distribution of the flow and thermal fields for $\delta/L = 0.2$ and 0.4 shows the symmetric shapes about the horizontal center line compared with that of corresponding $\delta/L = 0.8$ and 0.6, respectively, because conduction is the dominant mode of heat transfer at this low Rayleigh number. The circulation of flow shows one rotating eddy for the streamlines. When $\delta/L < 0.5$, the core of the main vortex is located in the upper half of the enclosure, but when $\delta/L > 0.5$, the main vortex moves downward. Also, a thermal plume starts to appear on the top of the inner cylinder, and as a result, the isotherms move upward. The thermal plume diminishes as the inner cylinder moves upward.

As the Rayleigh number increases up to 10^5, the role of convection in heat transfer becomes more significant, and consequently, the thermal boundary layer thicknesses

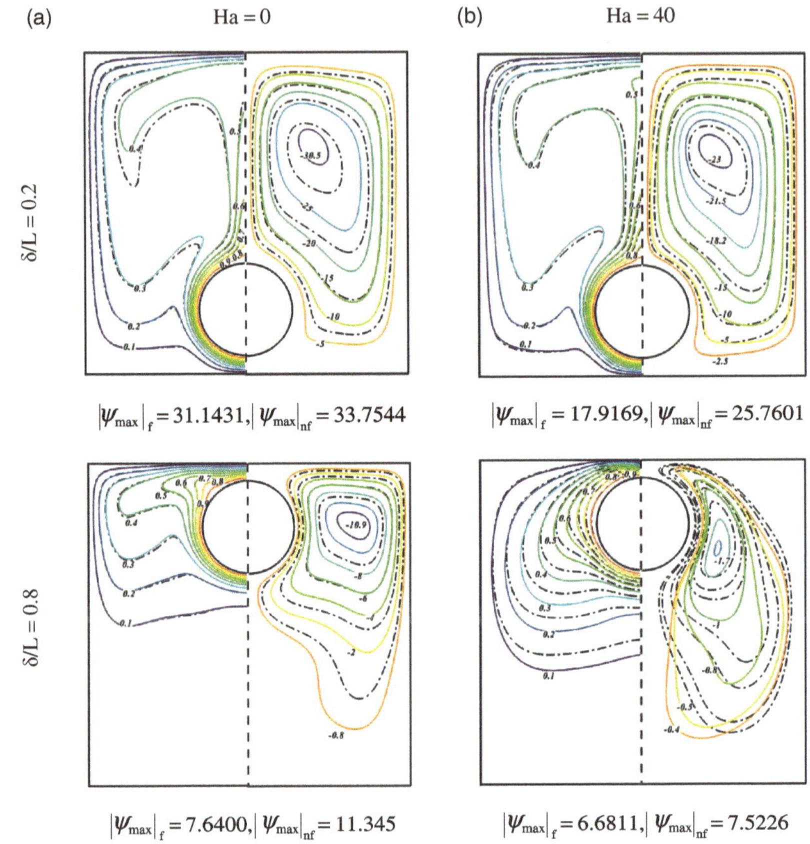

FIGURE 4.20

Comparison of the Isotherm (a) and Streamline (b) Contours Between Nanofluid ($\phi = 0.06$) (—) and Pure Fluid (= 0) (– · · –) for Different Values of Ha and d/*L* at I = 3.5, Ra = 10^6, and Pr = 6.8.

on the surface of the inner cylinder become thinner. It is an interesting to notice that at Ra = 10^5, the core of main vortex is located in the upper half of the enclosure for all positions of the inner cylinder. Also, this figure shows that at $\delta/L = 0.8$, the thermal plume forms at the top of the left cylinder. A similar flow pattern is observed at Ra = 10^6, but the value of $|\psi_{max}|_{nf}$ and the thermal gradient near the hot inner cylinder is greater at this Raleigh number.

When the magnetic field is imposed on the enclosure, the velocity field is suppressed owing to the retarding effect of the Lorenz force. Therefore, the intensity of convection weakens significantly. The braking effect of the magnetic field is observed from the maximum stream function value. Also, applying the magnetic field makes the isotherms parallel to each other because of domination of the conduction mode of heat transfer. Also, the effect of the position of the inner cylinder is more

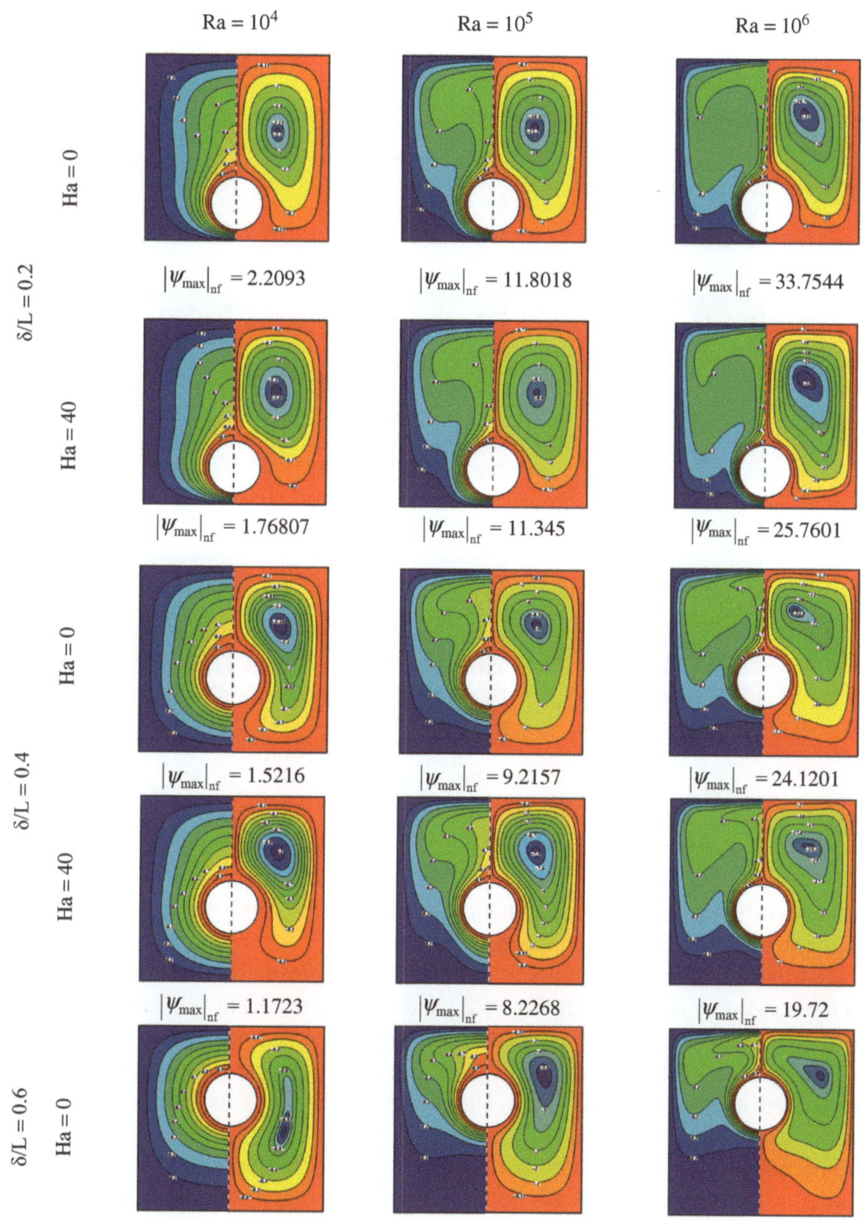

FIGURE 4.21

Effects of Ha, Ra, and δ/L on Isotherm (Left) and Streamline (Right) Contours for Cu–Water ($\phi = 0.06$) When $\lambda = 3.5$, and Pr = 6.8.

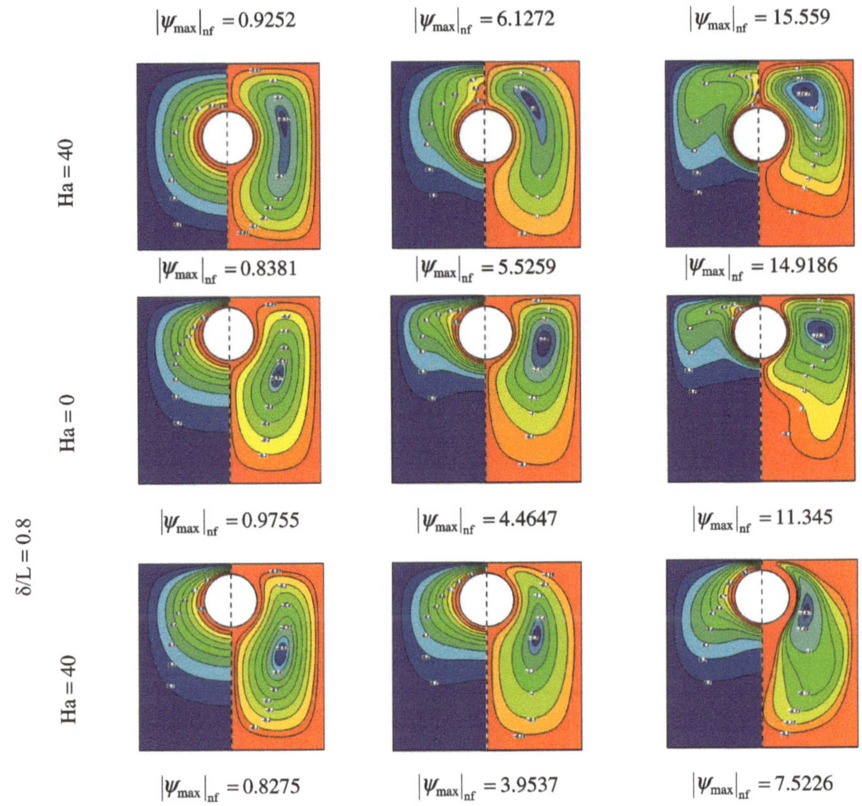

FIGURE 4.21 ***(Continued)***

obvious in the presence of a magnetic field because the magnetic field boosts the conduction heat transfer mechanism.

Distribution of the local Nusselt numbers only over the left half of the enclosure and inner cylinder are shown in Figs 4.22 and 4.23 for different values of Ha, Ra, δ/L, and ϕ. The Nusselt number is expected to decrease with an increase of the volume fraction because of the decrement in the temperature gradient, but the opposite trend is observed in Figs 4.22 and 4.23. This observation is due to the fact that Nusselt number is a multiplication of the temperature gradient and the thermal conductivity ratio (conductivity of the nanofluid to the conductivity of the base fluid). Because the reduction in the temperature gradient due to the presence of nanoparticles is much smaller than the thermal conductivity ratio, an enhancement in the Nusselt takes place by increasing the volume fraction of the nanoparticles. At Ra = 10^4, when the center of inner cylinder located in the lower half of the enclosure, the minimum value of Nu_h is obtained at $\gamma = 0°$ and its maximum value occurs at $\gamma = 180°$, but the opposite trend is observed for $\delta/L > 0.5$. At Ra = 10^5 and 10^6, the Nu_h profile shows two different behaviors at δ/L greater and smaller than 0.6. This means that for $\delta/L < 0.6$,

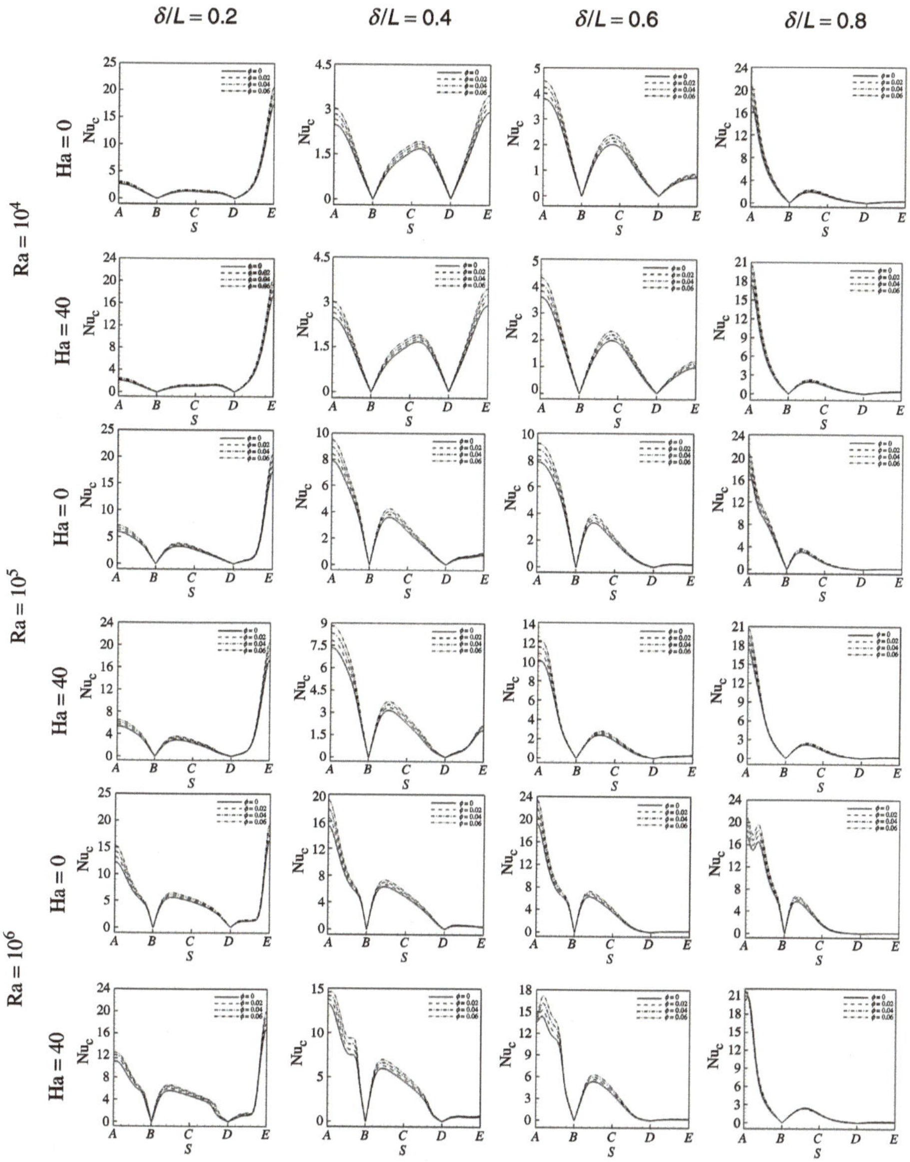

FIGURE 4.22 Effects of Ha, Ra, δ/L, and ϕ on the Local Nusselt Number Along the Walls of the Enclosure (Nu_c) for Cu–Water When $\lambda = 3.5$ and Pr = 6.8

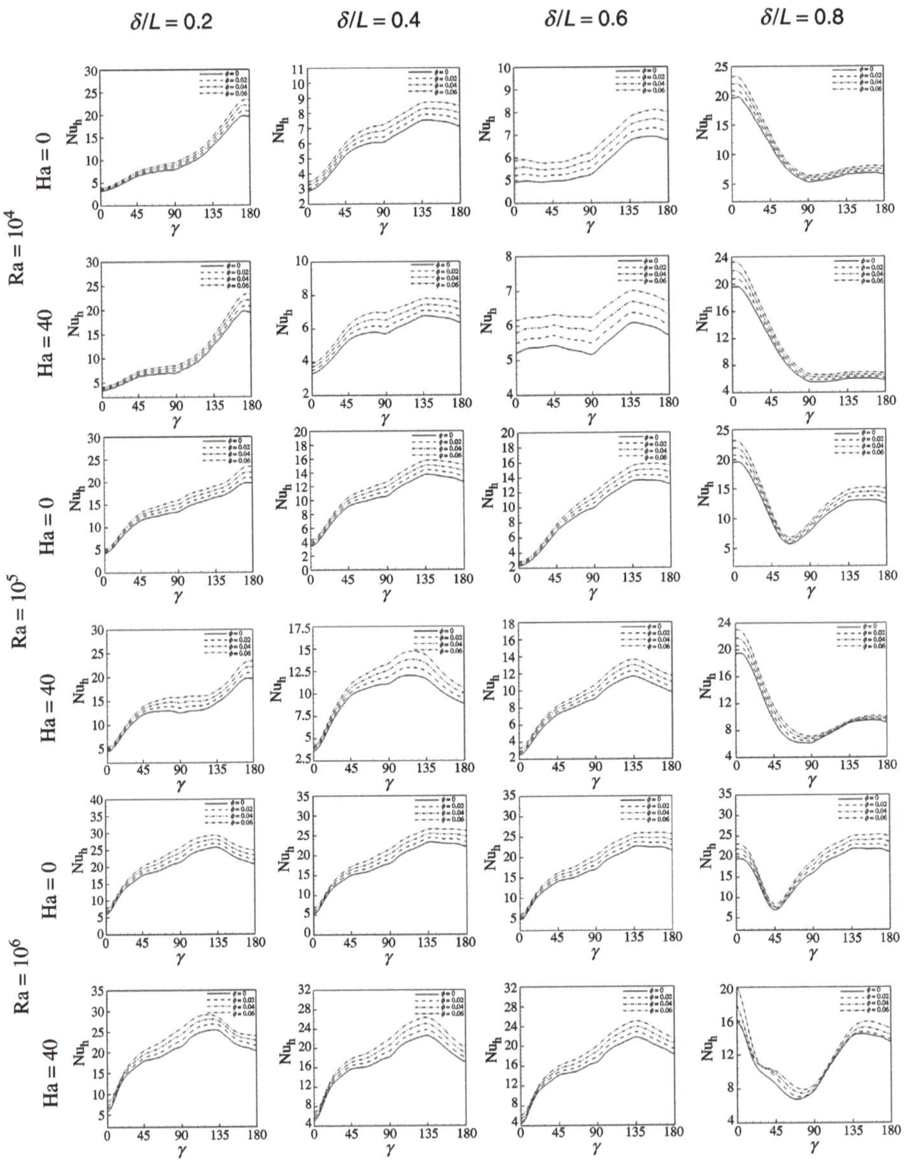

FIGURE 4.23 Effects of Ha, Ra, δ/L, and ϕ on the Local Nusselt Number Along the Surface of the Inner Cylinder (Nu_h) for Cu–Water When $\lambda = 3.5$ and $Pr = 6.8$

the local Nusselt number increases with an augmentation of γ, but for $\delta/L > 0.6$, at first the Nu_h decreases and then increases. The minimum values of Nu_h for these cases correspond to the existence of a thermal plume over the heated cylinder. It can be seen that at a high Rayleigh number, the magnetic field effect on the local Nusselt number is more obvious. As the inner cylinder is situated at the upper position, the Nu_c profiles become smoother because of the type of heat transfer mechanism. When $\delta/L = 0.2$, maximum values of the local Nusselt number are obtained at $\gamma = 180°$ because the minimum distance between the cold and hot cylinders exists at this region.

The corresponding polynomial representations of such a model for the Nusselt number along the walls of the enclosure $\left(\overline{Nu_c}\right)$ and along the surface of the inner cylinder $\left(\overline{Nu_h}\right)$ are as follows:

$$\begin{aligned}\overline{Nu_c} &= b_{13} + b_{23}Y_1 + b_{33}Y_2 + b_{43}Y_1^2 + b_{53}Y_2^2 + b_{63}Y_1Y_2 \\ Y_1 &= b_{11} + b_{21}\,\mathrm{Ra}^* + b_{31}\delta^* + b_{41}\,\mathrm{Ra}^{*2} + b_{51}\delta^{*2} + b_{61}\,\mathrm{Ra}^*\,\delta^* \\ Y_2 &= b_{12} + b_{22}\,\mathrm{Ha}^* + b_{32}\phi + b_{42}\,\mathrm{Ha}^{*2} + b_{52}\phi^2 + b_{62}\,\mathrm{Ha}^*\,\phi\end{aligned} \tag{4.86}$$

$$\begin{aligned}\overline{Nu_h} &= a_{13} + a_{23}Y_1 + a_{33}Y_2 + a_{43}Y_1^2 + a_{53}Y_2^2 + a_{63}Y_1Y_2 \\ Y_1 &= a_{11} + a_{21}\,\mathrm{Ra}^* + a_{31}\delta^* + a_{41}\,\mathrm{Ra}^{*2} + a_{51}\delta^{*2} + a_{61}\,\mathrm{Ra}^*\,\delta^* \\ Y_2 &= a_{12} + a_{22}\,\mathrm{Ha}^* + a_{32}\phi + a_{42}\,\mathrm{Ha}^{*2} + a_{52}\phi^2 + a_{62}\,\mathrm{Ha}^*\,\phi\end{aligned} \tag{4.87}$$

where $\mathrm{Ra}^* = \log(\mathrm{Ra})$, $\delta^* = \delta/L$ and $\mathrm{Ha}^* = \mathrm{Ha}/100$. Also, a_{ij} and b_{ij} can be found in Tables 4.5 and 4.6. The effects of Ha, Ra, and δ/L on the average Nusselt number along the walls of the enclosure and the surface of the inner cylinder are depicted in Figs 4.24 and 4.25. Variation of the average Nusselt number over the hot surface is similar to that of cold surface. In addition, the value of the Nusselt number over the inner cylinder is considerably greater than the cold one. Because the conduction heat transfer mechanism is dominant at $\mathrm{Ra} = 10^4$, the profile of the average Nusselt number is almost symmetric about $\delta/L = 0.5$. It can be found that the minimum value of the average Nusselt number is obtained at $\delta/L = 0.6$ when $\mathrm{Ra} = 10^5$. For $\mathrm{Ra} = 10^6$, the

Table 4.5 Constant Coefficient for Using Eq. (4.86)

b_{ij}	$i = 1$	$i = 2$	$i = 3$	$i = 4$	$i = 5$	$i = 6$
$j = 1$	4.983644	−1.8922	−1.86745	0.355825	5.82276	−1.02723
$j = 2$	2.777641	−0.49729	7.766133	−0.19892	9.476536	−1.33354
$j = 3$	−0.85206	0.598477	0.000487	−0.09365	−0.00053	0.345664

Table 4.6 Constant Coefficient for Using Eq. (4.87)

a_{ij}	$i = 1$	$i = 2$	$i = 3$	$i = 4$	$i = 5$	$i = 6$
$j = 1$	21.5249	−8.15581	−8.09223	1.553241	25.67044	−4.56336
$j = 2$	12.40464	−2.22502	31.3948	−0.89001	43.90568	−5.46778
$j = 3$	−3.82584	0.602369	0.003993	−0.02115	−0.00012	0.077697

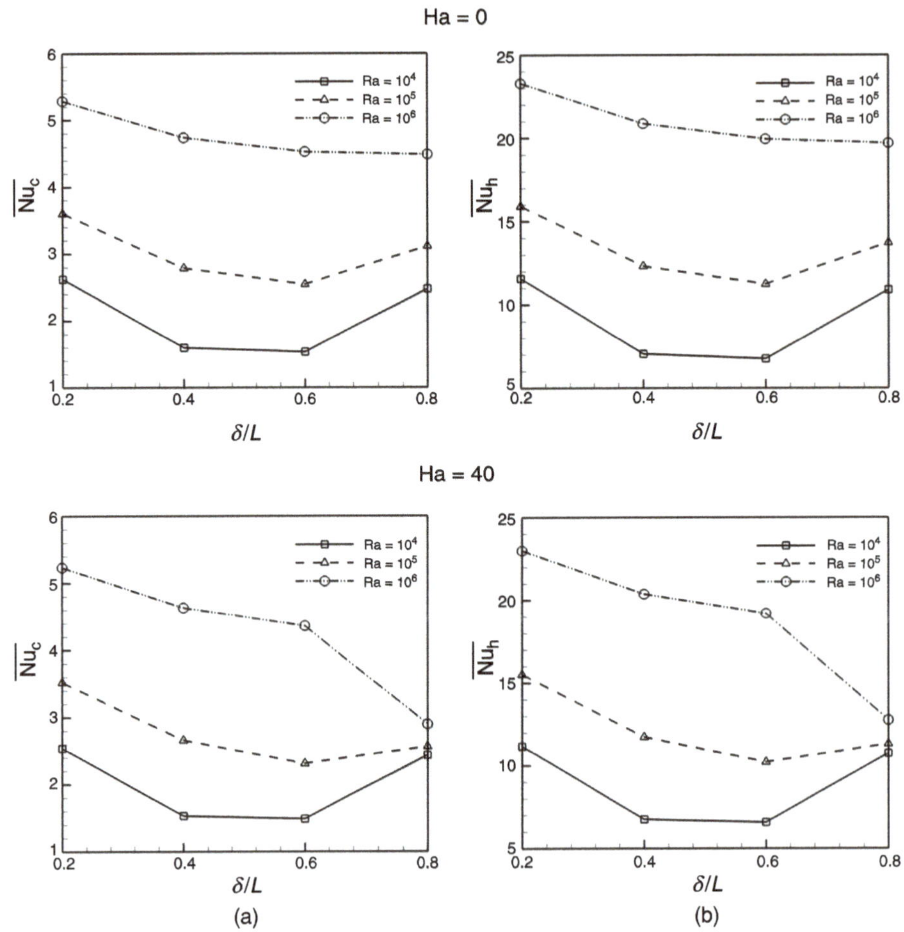

FIGURE 4.24

Effects of Ha, Ra, and δ/L on the Average Nusselt Number Along the Walls of the Enclosure $(\overline{Nu_c})$ (a) and the Surface of the Inner Cylinder $(\overline{Nu_h})$ (b) for Cu–Water When $\phi = 0.06$ and $\lambda = 3.5$.

average Nusselt number decreases continuously with increases of δ/L. An increasing Hartmann number causes the Lorenz force to increase and leads to a substantial suppression of the convection. Therefore, the Nusselt number has a reverse relationship with the Hartmann number.

The enhancement of heat transfer between the case of $\phi = 0.06$ and the pure fluid (base fluid) case is defined as

$$\text{En} = \frac{\overline{\text{Nu}_\text{h}}(\phi = 0.06) - \overline{\text{Nu}_\text{h}}(\text{basefluid})}{\overline{\text{Nu}_\text{h}}(\text{basefluid})} = 100 \qquad (25)$$

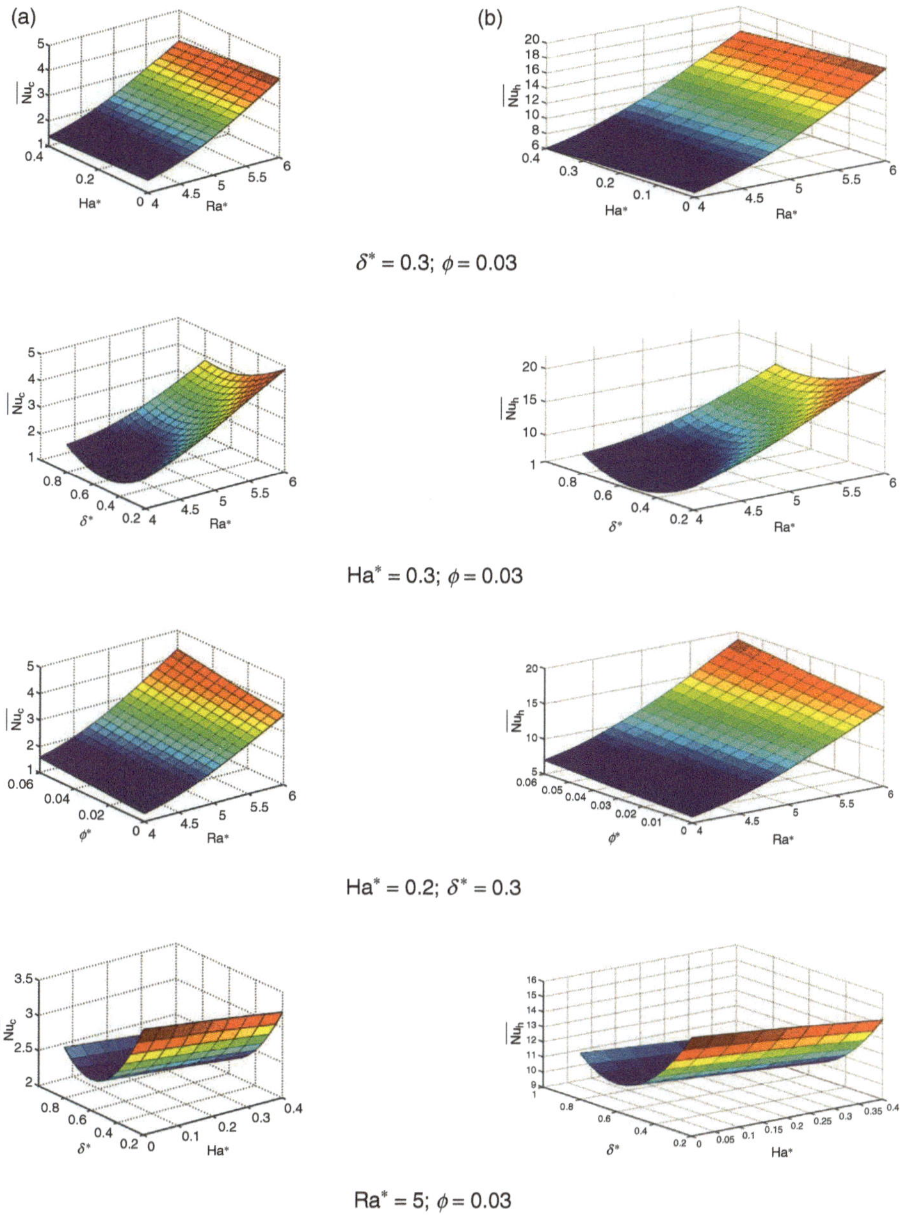

FIGURE 4.25

Variation of the Average Nusselt Number Along (a) the Walls of the Enclosure $\left(\overline{Nu_c}\right)$ and (b) the Surface of the Inner Cylinder $\left(\overline{Nu_h}\right)$ for Various Input Parameters.

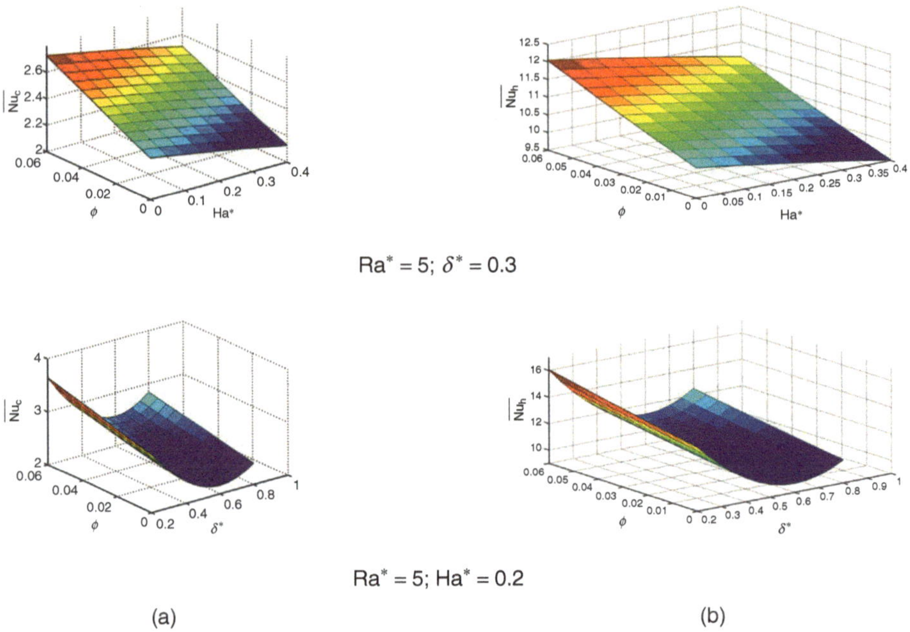

FIGURE 4.25 ***(Continued)***

Fig. 4.26 depicts the effects of the Raleigh number, Hartmann number, and position of the inner cylinder on the enhancement of heat transfer. As seen, the effect of nanoparticles is more pronounced at low Rayleigh numbers than at high Rayleigh numbers because of greater amount of rate of enhancement, and an increasing Rayleigh number leads to decrease in the ratio of enhancement of heat transfer. This

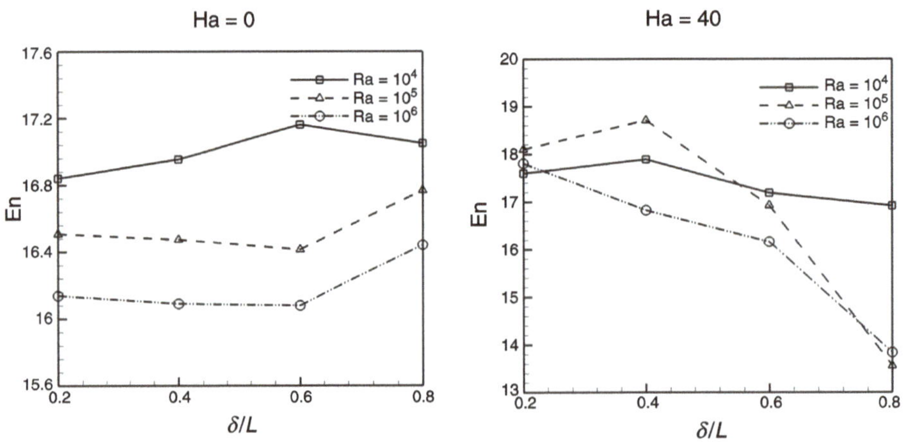

FIGURE 4.26 Effects of Ha, Ra, and δ/L on the Ratio of Enhancement in Heat Transfer Caused by the Addition of Nanoparticles for Cu–Water When $\lambda = 3.5$

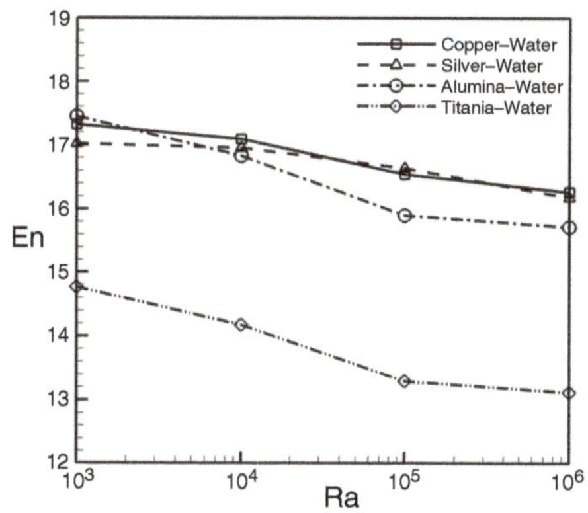

FIGURE 4.27 Effects of Types of Nanoparticle on Enhancement in Heat Transfer Caused by the Addition of Nanoparticles When $\lambda = 3.5$, Pr = 6.8, and Ha = 0

observation can be explained by noting that at low Rayleigh numbers, the heat transfer is dominant by conduction. Therefore, the addition of high thermal conductivity nanoparticles will increase the conduction and therefore make the enhancement more effective. Also, it can be seen that whereas the maximum value of En is obtained at $\delta/L = 0.4$ for Ra = 10^4, the minimum value of heat transfer enhancement at Ra = 10^5 and 10^6 is obtained at $\delta/L = 0.6$ and 0.4, respectively. Also, this figure shows that enhancement increases with an augmentation of the Hartmann number. Fig. 4.27 shows the effects of types of nanoparticles on enhancement in heat transfer caused by the addition of nanoparticles. It can be seen that maximum of enhancement occurs for Cu–water.

4.14.4 MAGNETOHYDRODYNAMIC NATURAL CONVECTION HEAT TRANSFER OF AL_2O_3–WATER NANOFLUID IN A HORIZONTAL CYLINDRICAL ENCLOSURE WITH AN INNER TRIANGULAR CYLINDER

4.14.4.1 Problem definition

The physical model used in this work is shown in Fig. 4.28. The inner cylinder is maintained at a constant high temperature T_h, and the temperature of the outer cylinder is set at a constant low temperature T_c. λ is the aspect ratio, which is defined as the ratio of the radius of the outer cylinder r_0 to the radius of the circumscribed circle r_i of the inner equilateral triangle, $\lambda = r_o / r_i$ [34]. The Rayleigh number Ra is defined as

$$Ra = \frac{g\beta\Delta T (r_o - r_i)^3}{\alpha \upsilon}. \tag{4.88}$$

In this study, we consider λ as equal to 2. Also, we consider a uniform magnetic field $(\vec{B} = B_x \vec{e_x} + B_y \vec{e_y})$ of constant magnitude $B = \sqrt{B_x^2 + B_y^2}$, where $\vec{e_x}$ and $\vec{e_y}$ are unit vectors in the Cartesian coordinate system. Let the orientation of

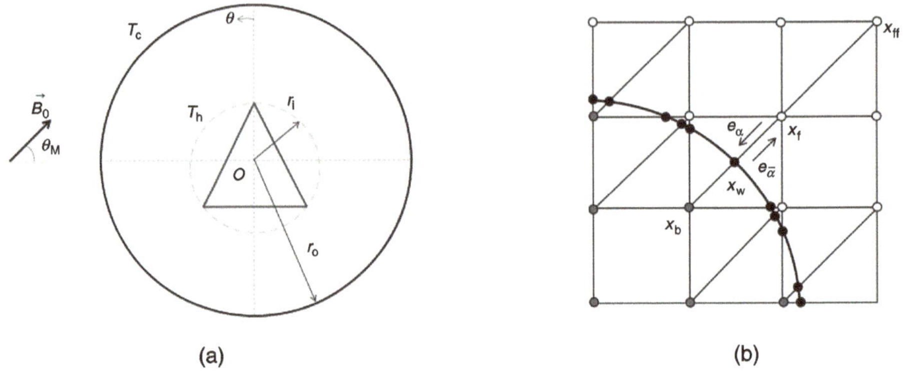

FIGURE 4.28

(a) Geometry of the problem. (b) Curved boundary and lattice nodes.

the magnetic field form an angle θ_M with horizontal axis such that $\cot^{-1}\theta_M = B_x/B$. The electric current $\vec{J}$ and the electromagnetic force $\vec{F}$ are defined by $\vec{J} = \sigma\left(\vec{V}\times\vec{B}\right)$ and $\vec{F} = \sigma\left(\vec{V}\times\vec{B}\right)\times\vec{B}$, respectively. In this study, λ and θ_M are equal to 2 and 0, respectively.

4.14.4.2 Effect of active parameters

Fig. 4.29 shows the effects of the Rayleigh number and Hartmann number on isotherms and streamlines. For Ra = 10^3, conduction is the dominant mechanism for heat transfer. However, by increasing the Rayleigh number, one can increase the buoyancy forces and overcome the viscous forces. Hence, convection currents dominate at high Rayleigh numbers. Moreover, the isotherms are distorted at higher Rayleigh numbers because of the stronger convection effects with increasing magnitude of the velocity circulating in the enclosure.

When the magnetic field is imposed on the enclosure, the velocity field is being suppressed owing to the retarding effect of the Lorenz force. Therefore, the intensity of convection weakens significantly. The braking effect of the magnetic field is observed from the maximum stream function value. The core vortex is shifted upward vertically as the Hartmann number increases. Also, the imposition of the magnetic field leads to avoidance of the thermal plume over the inner wall. At high Hartmann numbers, the conduction heat transfer mechanism is more pronounced. For this reason, the isotherms are parallel to each other.

The effects of the Rayleigh number and the Hartmann number on the local Nusselt number along the surface of the outer cylinder are shown in Fig. 4.30. It can be observed that the local Nusselt number profiles are generally symmetrical about the vertical center line of the outer circular cylinder. The local Nusselt number decreases when we increase θ from 0 to 180° along the outer cylinder. An increase in the magnetic field leads to a decrease in the local Nusselt number.

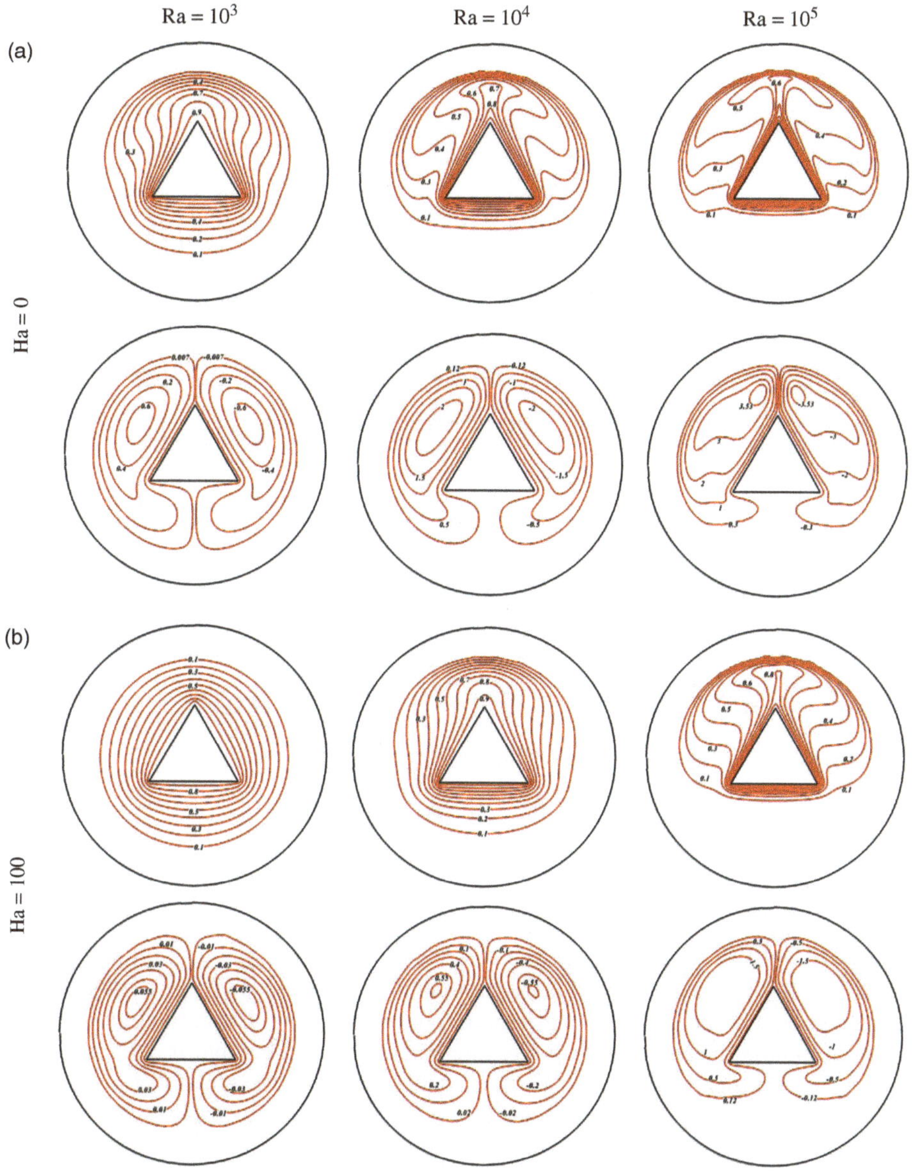

FIGURE 4.29

Effects of the Rayleigh Number Ra and Hartmann Number Ha on Isotherms (a) and Streamlines (b) for Al_2O_3–Water for $\phi = 0.04$.

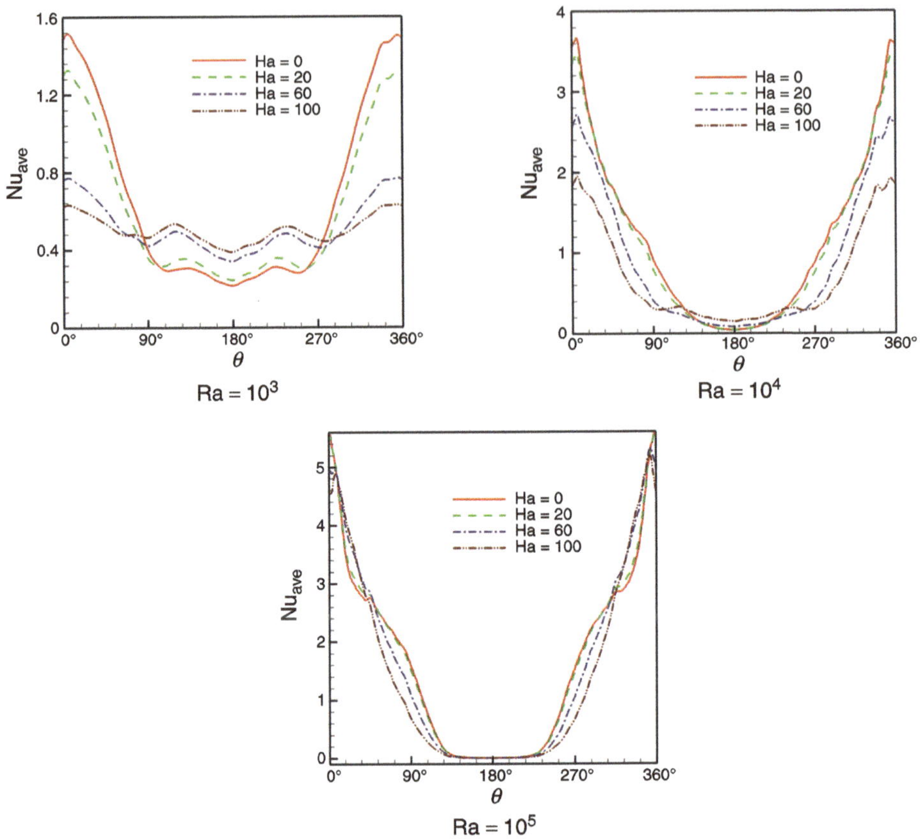

FIGURE 4.30 Effects of Ra and Ha on the Local Nusselt Number Along the Surface of the Outer Cylinder (Nu_{loc}) When $\phi = 0.04$

Fig. 4.31 depicts the effects of the Hartmann number and Rayleigh number on the average Nusselt number over the inner and outer cylinders. It can be found that the Nusselt number increases with increasing Rayleigh numbers, but it decreases with increasing Hartmann numbers. Also, Fig. 4.32 shows that the effect of the magnetic field on the average Nusselt number is more pronounced for higher values of the Rayleigh number. The corresponding polynomial representation of such a model for the Nusselt number is as follows.

$$\begin{aligned} \mathrm{Nu} &= a_{12} + a_{22} Y_1 + a_{32}\,\mathrm{Ha}^* + a_{42} Y_1^2 + a_{52}\,\mathrm{Ha}^{*2} + a_{62} Y_1\,\mathrm{Ha}^*, \\ Y_1 &= a_{11} + a_{21}\,\mathrm{Ra}^* + a_{31}\phi + a_{41}\,\mathrm{Ra}^{*2} + a_{51}\phi^2 + a_{61}\,\mathrm{Ra}^*\phi, \end{aligned} \tag{4.89}$$

where $\mathrm{Ra}^* = \log(\mathrm{Ra})$ and $\mathrm{Ha}^* = \mathrm{Ha}/100$. Also, values of a_{ij} can be found in Table 4.7 (e.g., a_{21} equals o.52675).

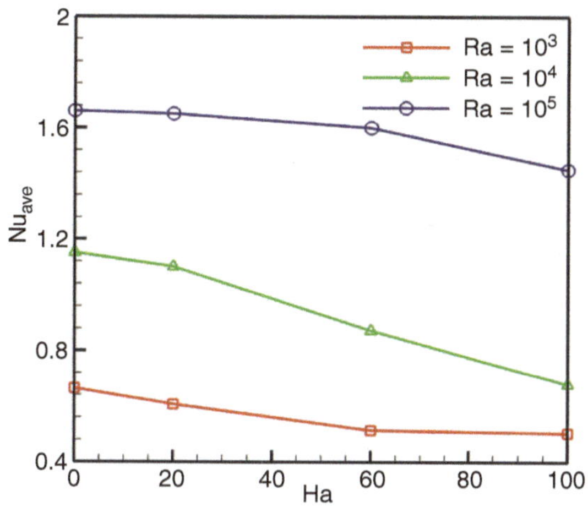

FIGURE 4.31 Effects of Ra and Ha on the Average Nusselt Number Along the Surface of the Outer Cylinder (Nu_{ave}) When $\phi = 0.04$

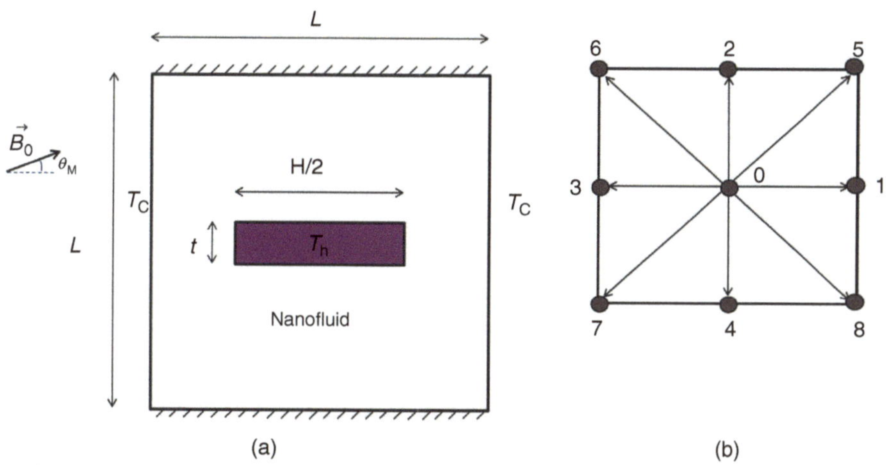

FIGURE 4.32

(a) Geometry of the problem. (b) Discrete velocity set of two-dimensional nine-velocity (D_2Q_9) model.

Table 4.7 Values of the Coefficients in Eq. (4.89)

a_{ij}	$i = 1$	$i = 2$	$i = 3$	$i = 4$	$i = 5$	$i = 6$
$j = 1$	0.989708	−0.52675	−2.57505	0.123062	−0.103	1.290625
$j = 2$	0.12906	0.98409	−0.24789	0.005106	−0.03631	0.011477

4.14.5 MAGNETIC FIELD EFFECTS ON NATURAL CONVECTION AROUND A HORIZONTAL CIRCULAR CYLINDER INSIDE A SQUARE ENCLOSURE FILLED WITH NANOFLUID

4.14.5.1 Problem definition

The considered physical geometry with related parameters and coordinates is shown in Fig. 4.32. A rectangular body with height t and width $H/2$ is placed in the center of the enclosure; it is supposed to be isothermal at higher temperatures than the two vertical isothermal walls, and the top and bottom walls are insulated [17]. Also, it is also assumed that the uniform magnetic field ($\vec{B} = B_x \vec{e_x} + B_y \vec{e_y}$) of constant magnitude $B = \sqrt{B_x^2 + B_y^2}$ is applied, where $\vec{e_x}$ and $\vec{e_y}$ are unit vectors in the Cartesian coordinate system. The orientation of the magnetic field form an angle θ_M with the horizontal axis such that $\theta_M = B_x / B_y$. The electric current J and the electromagnetic force F are defined by $J = \sigma\left(\vec{V} \times \vec{B}\right)$ and $F = \sigma\left(\vec{V} \times \vec{B}\right) \times \vec{B}$, respectively. In this study, θ_M equals zero.

According to Sheikholeslami and Ganji [17], one can find the volumetric entropy generation rate as

$$S^{\bullet}_{\text{gen}} = \text{HTI} + \text{FFI} \tag{4.90}$$

where HTI is the irreversibility caused by heat transfer in the direction of finite temperature gradients and FFI is the contribution of fluid friction irreversibility to the total generated entropy.

In terms of the primitive variables, HTI and FFI become

$$\text{HTI} = \frac{k(\nabla T.\nabla T)}{T^2} \qquad \text{FFI} = \frac{\mu\phi}{T} \tag{4.91}$$

One can also define the Bejan number, Be, as

$$\text{Be} = \frac{\text{HTI}}{\text{HTI} + \text{FFI}} \tag{4.92}$$

Note that a Be value more or less than 0.5 shows that the contribution of HTI to the total entropy generation is higher or lower than that of FFI. The limiting value of Be = 1 shows that the only active entropy generation mechanism is HTI; Be = 0 represents no HTI contribution.

The dimensionless form of entropy generation rate, Ns, is defined as

$$\text{Ns} = \frac{\left(\frac{H}{\Omega}\right)^2 S^{\bullet}_{\text{gen}}}{k} \tag{4.93}$$

One finds that

$$\text{Ns} = \frac{\left(\frac{\partial\theta}{\partial x}\right)^2 + \left(\frac{\partial\theta}{\partial y}\right)^2}{\Omega^2\left(\Omega^{-1} + \theta\right)^2} + \frac{\text{Ge}\varphi}{\text{Ra}\Omega^2\left(\Omega^{-1} + \theta\right)} \tag{4.94}$$

where the dimensionless temperature difference is defined as

$$\Omega = \frac{T_h - T_c}{T_c} \tag{4.95}$$

$$\varphi = 2\left(\frac{\partial u}{\partial x}\right)^2 + 2\left(\frac{\partial v}{\partial x}\right)^2 + \left(\frac{\partial u}{\partial x} + \frac{\partial v}{\partial x}\right)^2 \tag{4.96}$$

$$\mathrm{Ge} = \frac{g\beta H}{C_p} \tag{4.97}$$

The average Ns is denoted by <Ns>, where the angle brackets show an average taken over the area, as

$$< \mathrm{Ns} >= \int_A \frac{\mathrm{Ns}}{A} dA \tag{4.98}$$

Selecting the fluid, trapped between the heated plate and the cavity, as the thermodynamic system, one observes that the amount of heat entered through the heated plate is equal to the one transferred to the surroundings via the isothermal walls. Moreover, one notes that the total volumetric entropy generation rate is obtainable as

$$< S^{\bullet}_{\mathrm{gen}} >= \frac{q''}{H}\left(\frac{1}{T_c} - \frac{1}{T_H}\right) \tag{4.99}$$

where, in terms of Nu, it reads

$$< S^{\bullet}_{\mathrm{gen}} >= 4\mathrm{Nu}\frac{k}{H^2}\frac{\Omega^2}{1+\Omega} \tag{4.100}$$

Applying perturbation techniques for small values of Ω, say $\Omega << 1$, one has

$$< S^{\bullet}_{\mathrm{gen}} >\approx \frac{4\,\mathrm{Nu}\,\Omega^2(1-\Omega)k}{H^2} \tag{4.101}$$

The dimensionless entropy generation number can be obtained as

$$< \mathrm{Ns} >= \frac{4\,\mathrm{Nu}}{1+\Omega} \tag{4.102}$$

To compare the total heat transfer rate, the Nusselt number is used. The average Nusselt numbers are defined as follows:

$$\mathrm{Nu}_{\mathrm{ave}} = \frac{k_{\mathrm{nf}}}{k_{\mathrm{f}}}\int_0^1 \frac{1}{4}\left(\left.\frac{\partial T}{\partial x}\right|_{X=0} + \left.\frac{\partial T}{\partial x}\right|_{X=1}\right) dy. \tag{4.103}$$

4.14.5.2 Effect of active parameters

The effects of nanoparticles on streamlines and isotherms are shown in Fig. 4.33. By adding nanoparticles, the absolute values of stream functions indicate that the strength of flow decreases. Although thermal boundary layer thickness decreases with adding nanoparticle in the base fluid, the Nusselt number increases with increases of the nanofluid volume fraction because of an increment in thermal conductivity. Fig. 4.34

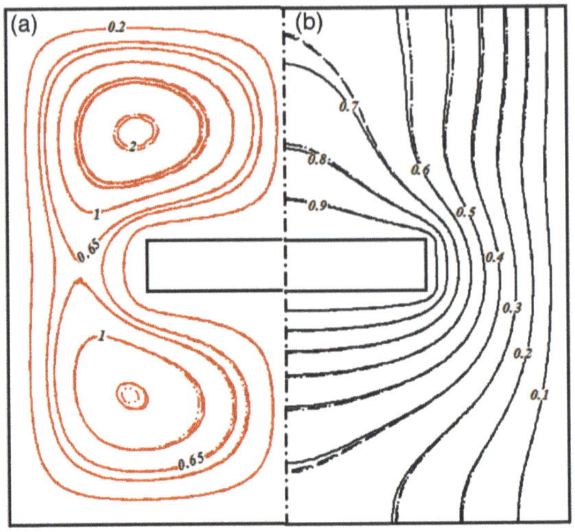

FIGURE 4.33

Streamline (a) and Isotherm (b) Contours Between CuO–Water Nanofluid ($\phi = 0.04$) (–··–) and Pure Fluid ($\phi = 0$) (—) When Ra = 10^4, $L/t = 10$, and Ha = 0.

shows the isotherms and streamlines for different Rayleigh numbers. In general, the heated lighter fluid is lifted and moves upward along the hot surface of the body and the vertical symmetry line until it encounters the cold top wall. Then the fluid becomes gradually colder and denser while it moves horizontally outward in contact with the cold top wall. Consequently, the cooled denser fluid descends along the cold side walls. For Ra = 10^3, the heat transfer in the enclosure is mainly dominated by the conduction mode. The circulation of the flow shows two overall rotating symmetric eddies with two inner vortices, respectively, as shown in Fig. 4.34 for the streamlines. At Ra = 10^4, the patterns of the isotherms and streamlines are about the same as those for Ra = 10^3. However, a careful observation indicates that the thermal boundary layer on the bottom part of the body is thinner than that on the opposite side, and the inner lower vortex becomes slightly smaller in size and weaker in strength compared with the upper one because the effect of convection on heat transfer and flow increases with increasing the Rayleigh number. As the Rayleigh number increases up to 10^5, the role of convection in heat transfer becomes more significant, and consequently, the thermal boundary layer on the surface of the body becomes thinner. Also, a plume starts to appear on the top of the body, and as a result, the isotherms move upward, giving rise to a stronger thermal gradient in the upper part of the enclosure and a much lower thermal gradient in the lower part. In consequence, the dominant flow is in the upper half of the enclosure, and correspondingly, the core of the recirculating eddies is located only in the upper half. At this Rayleigh number, the flow filed undergoes a bifurcation where two inner vortices merge. The flow at the bottom of the enclosure is very weak compared with that at the middle and top regions, which suggests stratification effects in the lower regions of the enclosure.

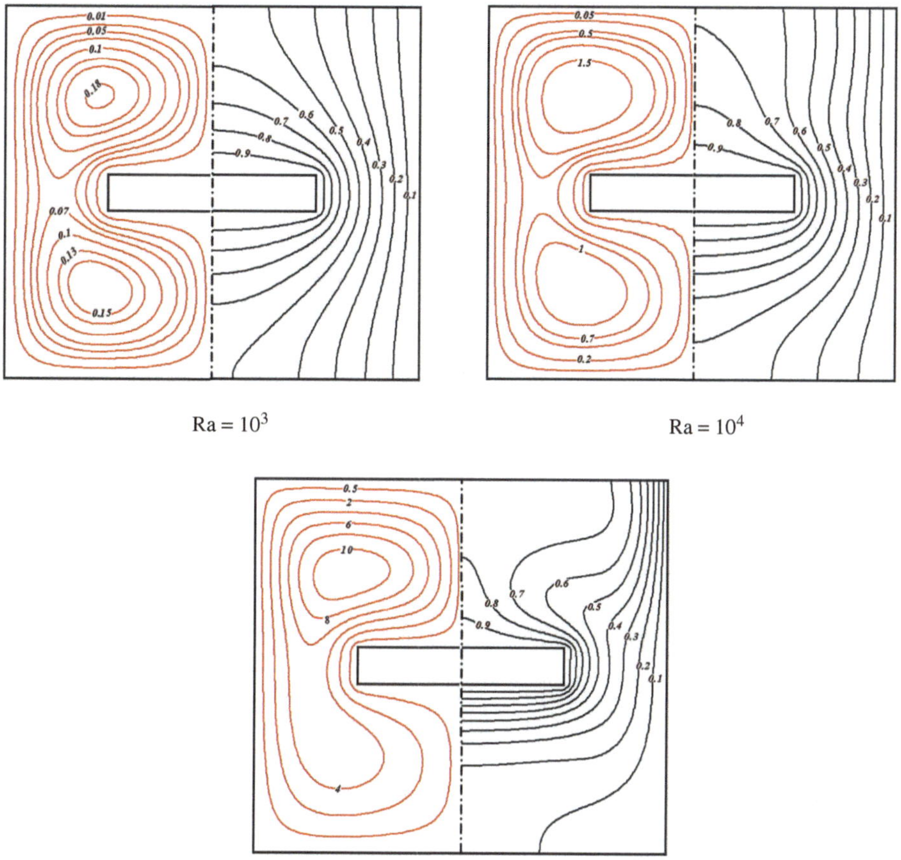

Ra = 10^5

FIGURE 4.34 Effects of Rayleigh Numbers on Streamline (Red) and Isotherm (Black) Contours When $L/t = 10$, Ha = 0, and $\phi = 0.04$

The effects of the Hartmann number on the streamlines and isotherms are shown in Fig. 4.35. The Hartmann number is the ratio of electromagnetic force to the viscous force. Variation of Hartmann number leads to the variation of the Lorentz force caused by the magnetic field, and the Lorentz force produces more resistance to the transport phenomena. An increase of the Hartmann number causes the flow strength to decrease considerably. As the Hartmann number increases, the primary eddy divides into two secondary eddies, which rotate in the same direction. The pattern of the isotherms is affected strongly by the changing intensity of the magnetic field. There is high-temperature gradient at the bottom of the body in the absent of a magnetic field. With an increasing Hartmann number, thermal boundary layer thickness increases at the bottom of the body. The convection is suppressed at the higher Hartmann number. It causes the plume on the top of the body wall to disappear, and the

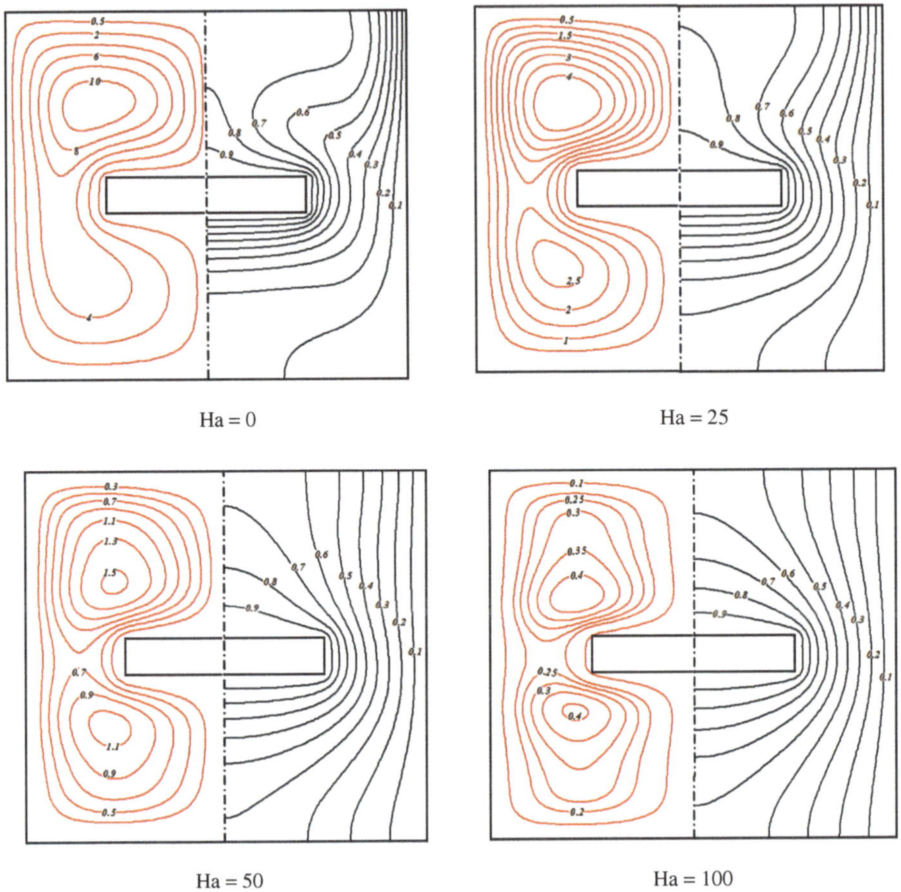

FIGURE 4.35 Effect of Hartmann Number on the Streamlines (Red) and Isotherms (Black) When $L/t = 10$, Ra = 10^5, and $\phi = 0.04$

isothermal lines become concentric and parallel between the body and the enclosure. It is shown that the convection heat transfer becomes weaker and causes the heat transfer mostly dominated by conduction between the cylinders.

As shown in Table 4.8, at Ra = 10^3, for all values of Hartmann number, the absolute values of stream function decrease with an increase of the nanoparticle volume fraction, but for higher values of Rayleigh number (e.g., Ra = 10^4 and Ra = 10^5), the effect of the nanoparticle volume fraction on the absolute values of the stream function depend on the value of the Hartmann number. In these Rayleigh numbers, when Ha ≤ 50, the absolute values of the stream function decrease as the nanoparticle volume fraction decreases, but the opposite trend is observed when Ha > 50.

Fig. 4.36 shows the effects of the nanoparticle volume fraction, Rayleigh number, and Hartmann number for Cu–water nanofluids on the average Nusselt number ratio

Table 4.8 Effects of the Nanoparticle Volume Fraction and Hartmann Number on $|\psi_{max}|$ When $L/t = 10$

		Ra		
Ha	ϕ	10^3	10^4	10^5
0	0	0.188952	2.05299	11.48
0	0.04	0.18672	2.0787	11.4451
50	0	0.018267	0.142036	1.57514
50	0.04	0.016833	0.145037	1.55233
100	0	0.016523	0.046635	0.444481
100	0.04	0.01025	0.044569	0.446956

$\left(\text{Nu}^* = \left(\text{Nu}_{\text{ave}}\big|_{\phi=0.04}\right) / \left(\text{Nu}_{\text{ave}}\big|_{\phi=0}\right)\right)$. It can be found that the effect of nanoparticles is more pronounced at low Rayleigh numbers than at high Rayleigh numbers because of greater values of average Nusselt number ratios. This observation can be explained by noting that at low Rayleigh numbers, the heat transfer is dominant by conduction. Therefore, the addition of high thermal conductivity nanoparticles will increase the conduction and therefore make the enhancement more effective. When Ha $<$ 75, minimum values of the average Nusselt number ratio are obtained for Ra = 10^5, but when Ha $>$ 75, minimum values of this ratio occur for Ra = 10^4.

The effects of the Hartmann number, the nanoparticle volume fraction, Rayleigh number, and dimensionless temperature difference for CuO–water on the dimensionless entropy generation number are shown in Fig. 4.37. The dimensionless entropy generation number increases with an increasing nanoparticle volume fraction and Rayleigh number. As seen in Fig. 4.37, <Ns> decreases with an increase of Ω.

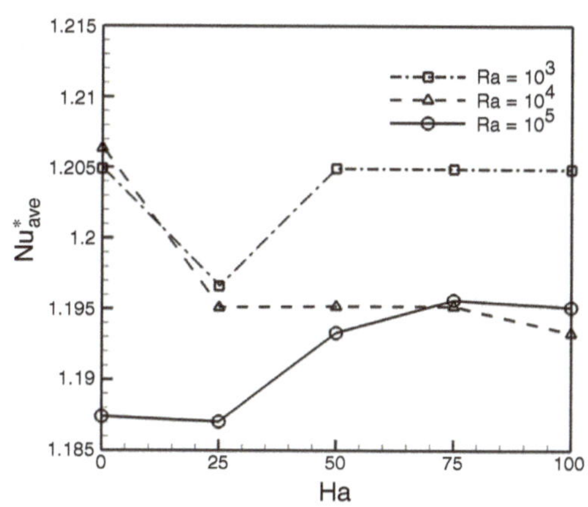

FIGURE 4.36 Effects of the Nanoparticle Volume Fraction, Rayleigh Number, and Hartmann Number for CuO–Water Nanofluids on the Average Nusselt Number Ratio When $L/t = 10$

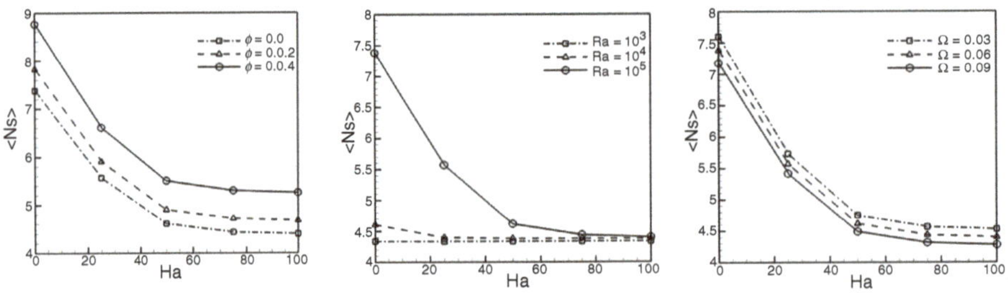

(a) (b) (c)

FIGURE 4.37

Effects of the Nanoparticle Volume Fraction, Hartmann Number, Rayleigh Number, and Dimensionless Temperature Difference for CuO–Water on the Dimensionless Entropy Generation Number (Ns) When Ra = 10^5 and $\Omega = 0.06$ (a); $\phi = 0.04$ and $\Omega = 0.06$ (b); and $\phi = 0.04$, Ra = 10^5, and $L/t = 10$ (c).

This fact is in line with the predictions of Eq. (4.15). This also makes physical sense because a s $\Omega = T_h/T_c - 1$, higher values of Ω imply a greater temperature difference (leading to higher heat transfer rates) and consequently boosted HTI values. So a decrease in <Ns>, according to $S^{\bullet}_{gen} = \mathrm{Ns}.k(\Omega/L)^2$, leads to an increase in the total entropy generation. Also, these figures indicate that an increasing Hartmann number leads to decreases in the dimensionless entropy generation number.

4.14.6 FREE CONVECTION OF FERROFLUID IN A CAVITY HEATED FROM BELOW IN THE PRESENCE OF AN EXTERNAL MAGNETIC FIELD

4.14.6.1 Problem definition

The numerical model consists in a 2D square cavity with side equal to H which represents the characteristic dimension of the problem (see Fig. 4.38). The heat source is centrally located on the bottom surface and its length 1 varied from 1/5 to 4/5 of H; the ratio *1/H* is called ε. The cooling is achieved by the two vertical walls. The heat source has a temperature T_h while the cooling walls have a temperature $T_c (T_h > T_c)$; all the other surfaces are adiabatic.

The cavity is filled with a mixture of kerosene and nanoscale ferromagnetic particle of cobalt. The ferrofluid in the cavity is Newtonian, incompressible, and laminar. The thermophysical properties of the nanoscale ferromagnetic particle are assumed to be constant (Tables 4.9 and 4.10). The density variation of the ferrofluid is approximated by the standard Boussinesq model. The external uniform magnetic source with a constant magnitude is applied at $x = 0.5$ H and $y = 1.2$ H. Moreover, the inter actions between nanoscale ferromagnetic particles of cobalt are neglected [10].

To compare the total heat transfer rate, the Nusselt number is used. The local and average Nusselt numbers are defined as follows:

$$\mathrm{Nu} = \frac{k_{\mathrm{nf}}}{k_{\mathrm{f}}} \frac{\partial T}{\partial Y}\bigg|_{Y=0} \quad \text{and} \quad \mathrm{Nu}_{\mathrm{ave}} = \int_{\frac{1-\varepsilon}{2}}^{\frac{1+\varepsilon}{2}} \mathrm{Nu}\, dX \tag{4.104}$$

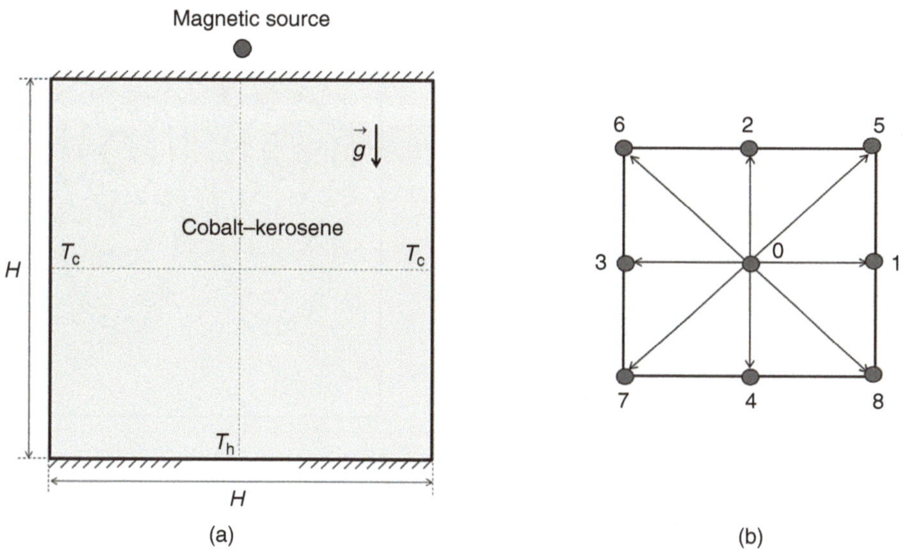

FIGURE 4.38

(a) Geometry of the problem. (b) Discrete velocity set of two-dimensional nine-velocity (D_2Q_9) model.

Table 4.9 Thermophysical Properties of Kerosene and Cobalt

	ρ (kg/m^3)	C_p (j/kgk)	K (W/m.k)	μ (kg/ms)
Kerosene	780	2090	0.149	0.00164
Cobalt	8900	420	100	–

Table 4.10 The Saturation Magnetization of Different Particle Sizes and the Magnetization Coefficient

Particle size (nm)	18	45	110
Saturation magnetization (emu/g)	39.1	29.5	19.9
Magnetization coefficient	1.325	1	0.675

4.14.6.2 Effect of active parameters

Fig. 4.39 shows the effect of volume fraction of cobalt on streamlines and isotherms. As seen, the velocity components increase with an increase of the nanoparticle volume fraction, which enhances the energy transport within the fluid. Thus, the absolute values of stream functions, indicate that the strength of flow increases with increasing the volume fraction of cobalt. The sensitivity of the thermal boundary layer thickness to the volume fraction of the nanoparticles is related to the increased thermal conductivity of the nanofluid. In fact, higher values of thermal conductivity

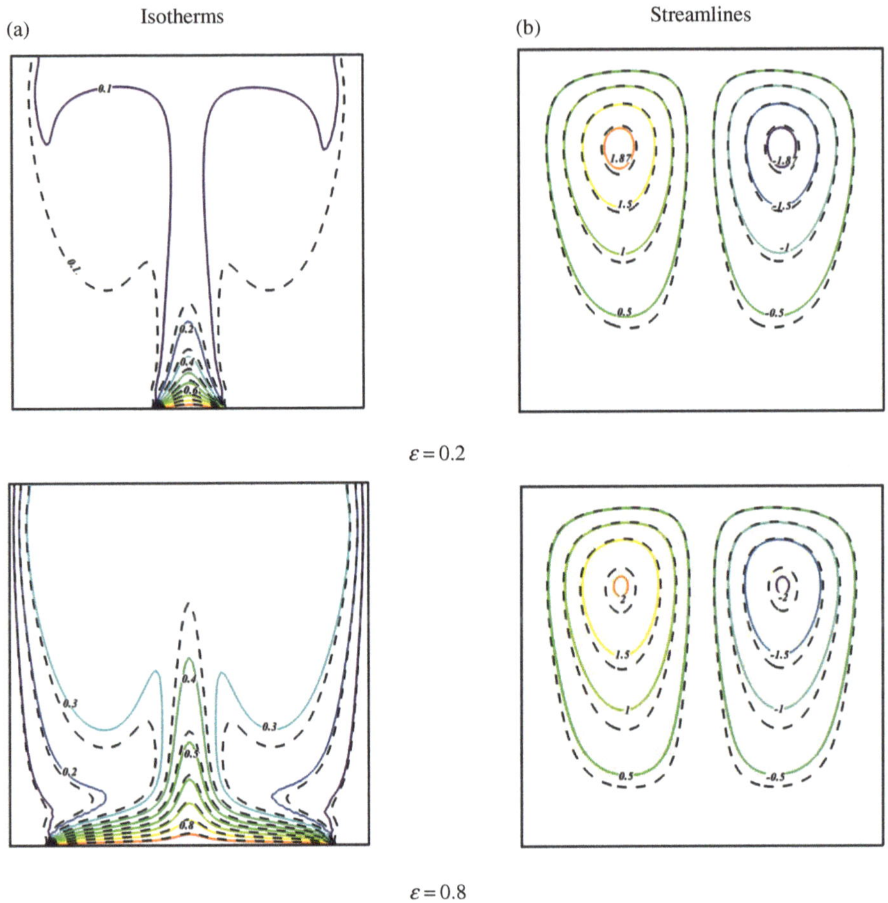

FIGURE 4.39

Comparison of the Isotherms (a) and Streamlines (b) Contours Between Nanofluid ($\phi = 0.04$) (- - -) and Pure Fluid ($\phi = 0$) (—) When $A = 1.325$ and Ra = 10^5.

are accompanied by higher values of thermal diffusivity. The high value of thermal diffusivity causes a drop in the temperature gradients and accordingly increases the boundary layer thickness.

The effects of the Rayleigh number, magnetic coefficient, and heat source length on streamlines and isotherms are shown in Figs 4.40 and 4.41. For low Rayleigh numbers, conduction is the dominant mechanism for heat transfer compared with the convection mechanism. However, by increasing the Rayleigh number, the buoyancy forces increase and overcome the viscous forces, and the heat transfer is dominated by convection at high Rayleigh numbers. Moreover, the isotherms are more distorted at higher Rayleigh numbers because of stronger convection effects.

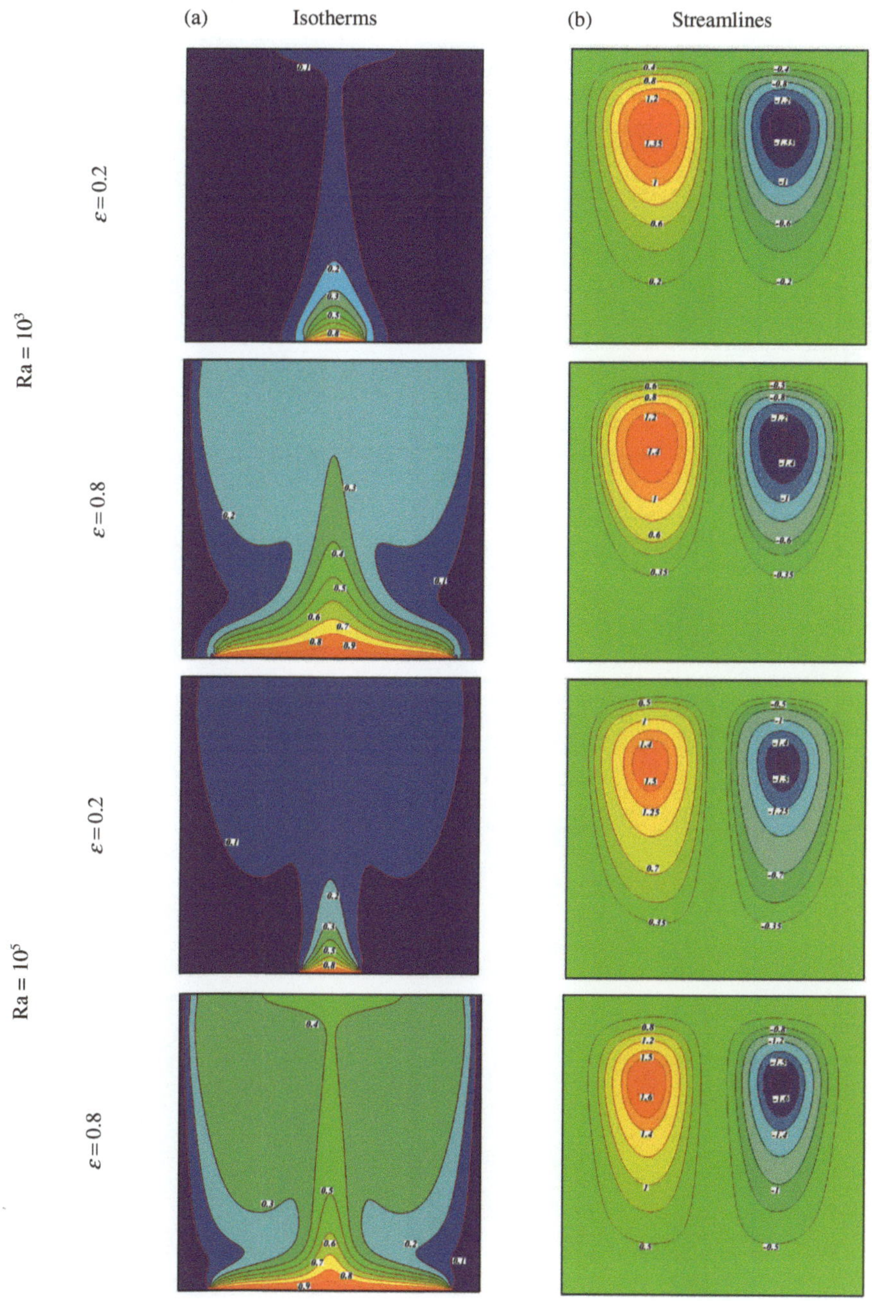

FIGURE 4.40

Effects of Ra and ε on Isotherms (a) and Streamlines (b) When $\phi = 0.04$ and $A = 1$.

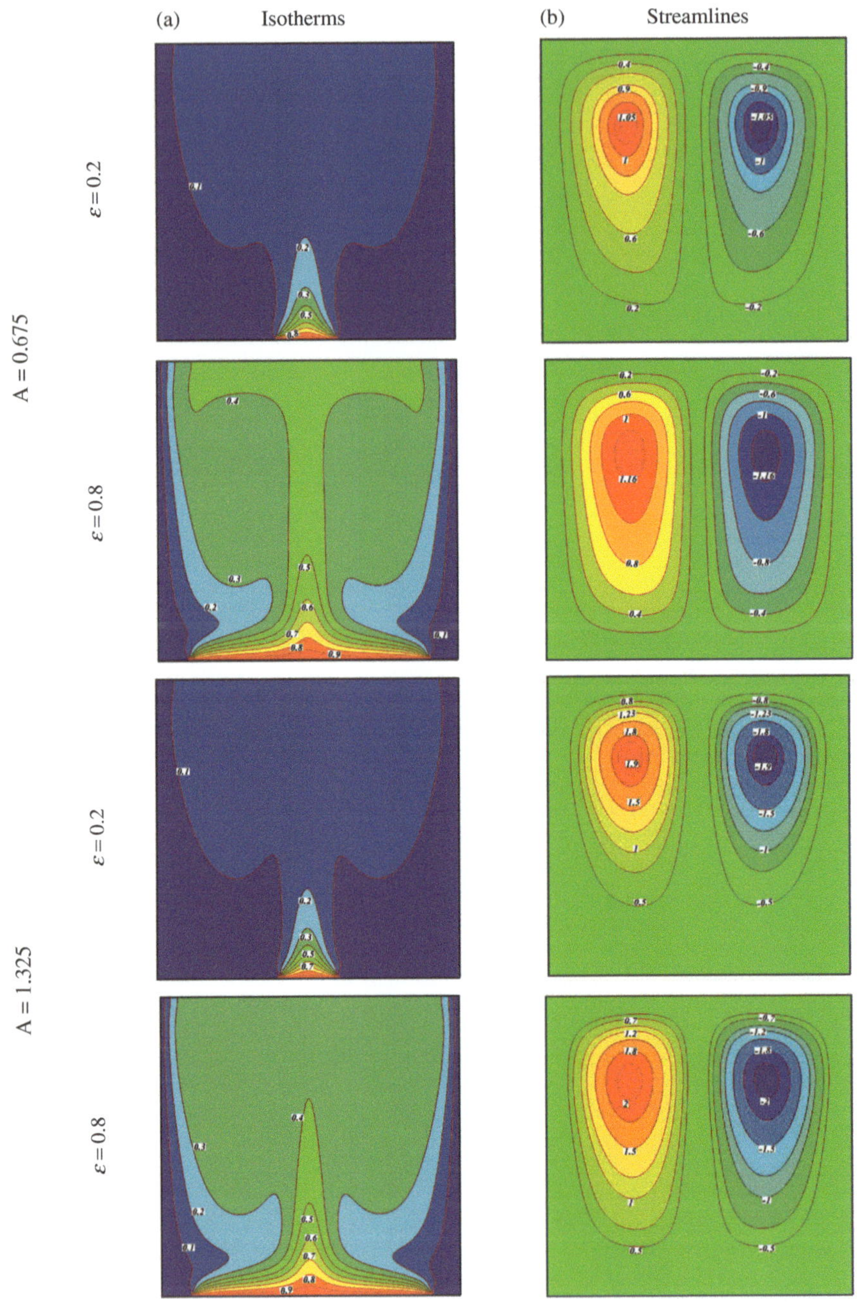

FIGURE 4.41

Effects of A and ε on Isotherms (a) and Streamlines (b) when Ra = 10^5 and $\phi = 0.04$.

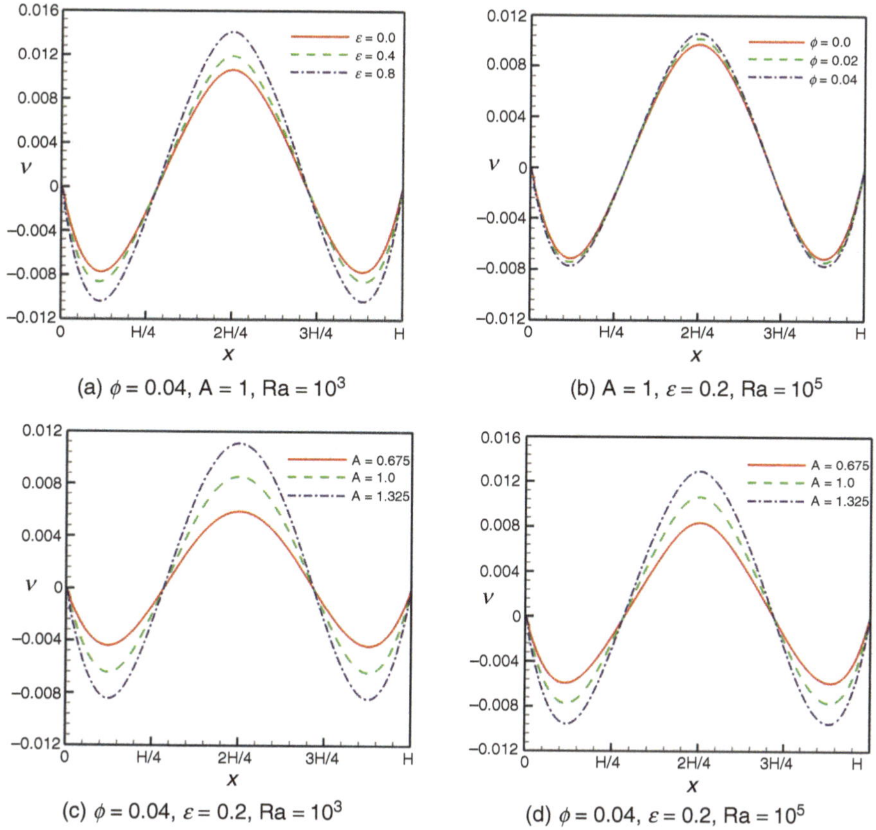

FIGURE 4.42 Effects of ϕ, ε, Ra, and A on Velocity Profile at $x = 0.5$ H

The magnetic force, which depends on the magnetization coefficient in our model, is affected by the size and volume fraction of particles and the additional magnetic source. To understand the relation between the intensity of the magnetic force and the flow pattern, three magnetic coefficients (0.675, 1, and 1.325) were used here. The effect of the magnetization coefficient on flow and heat transfer is similar to that of the Rayleigh number. Increasing the heat source length leads to an increase in the absolute values of the stream function and a decrease in the thermal boundary layer thickness.

The effects of the Rayleigh number, magnetic coefficient, heat source length, and volume fraction of cobalt on velocity and temperature profile are shown in Figs 4.42 and 4.43, respectively. Both profiles increase with an increase of the Rayleigh number, heat source length, and volume fraction of cobalt. Also, it can be found that the temperature profile decreases with an increase of the magnetic coefficient, and the velocity profile decreases with an increase of A. This implies that the smaller particle has better heat dissipation ability. Fig. 4.44 shows the effects of ϕ, ε, Ra, and

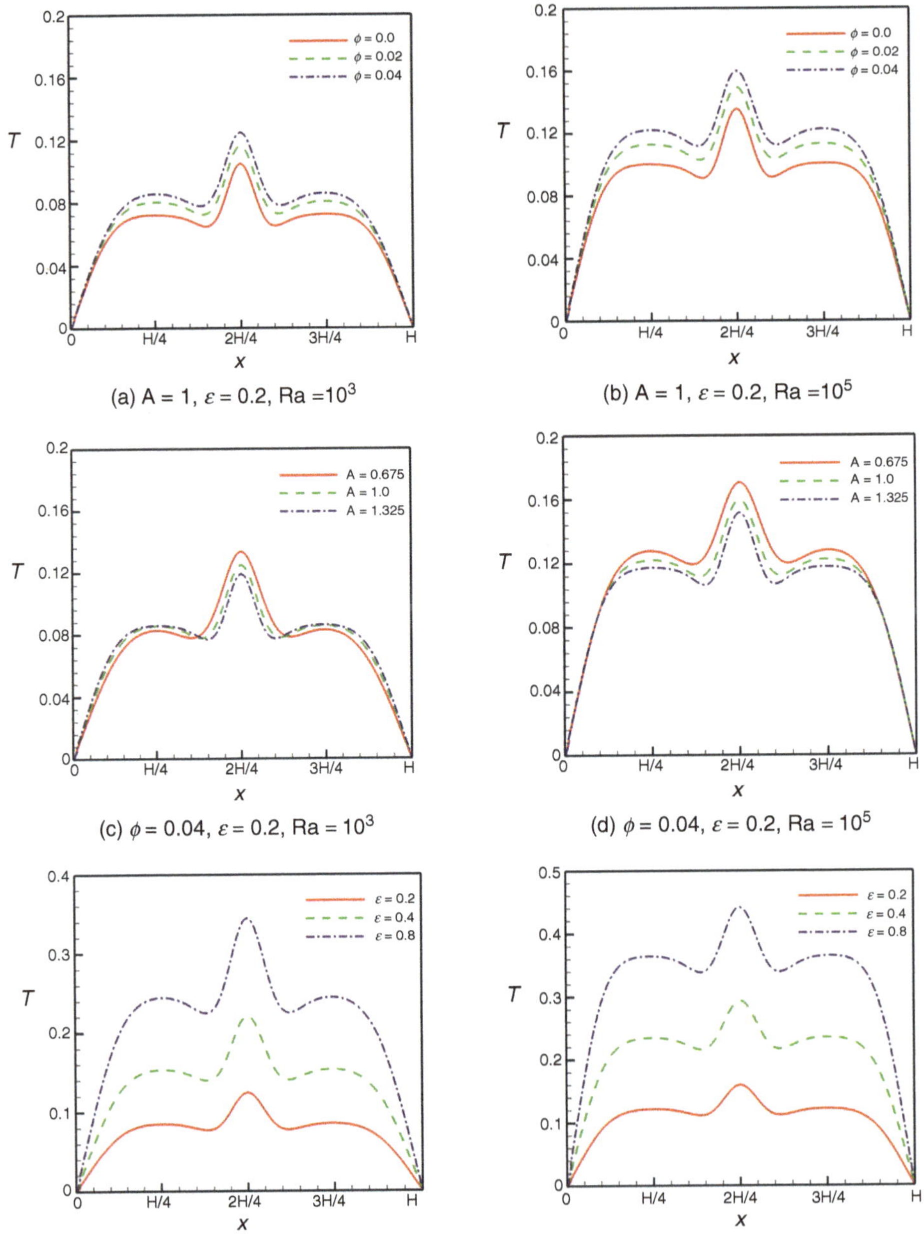

(a) A = 1, ε = 0.2, Ra = 10^3

(b) A = 1, ε = 0.2, Ra = 10^5

(c) ϕ = 0.04, ε = 0.2, Ra = 10^3

(d) ϕ = 0.04, ε = 0.2, Ra = 10^5

(e) ϕ = 0.04, A = 1, Ra = 10^3

(f) ϕ = 0.04, A = 1, Ra = 10^5

FIGURE 4.43 Effects of ϕ, ε, Ra, and A on the Temperature Profile at x = 0.5 H

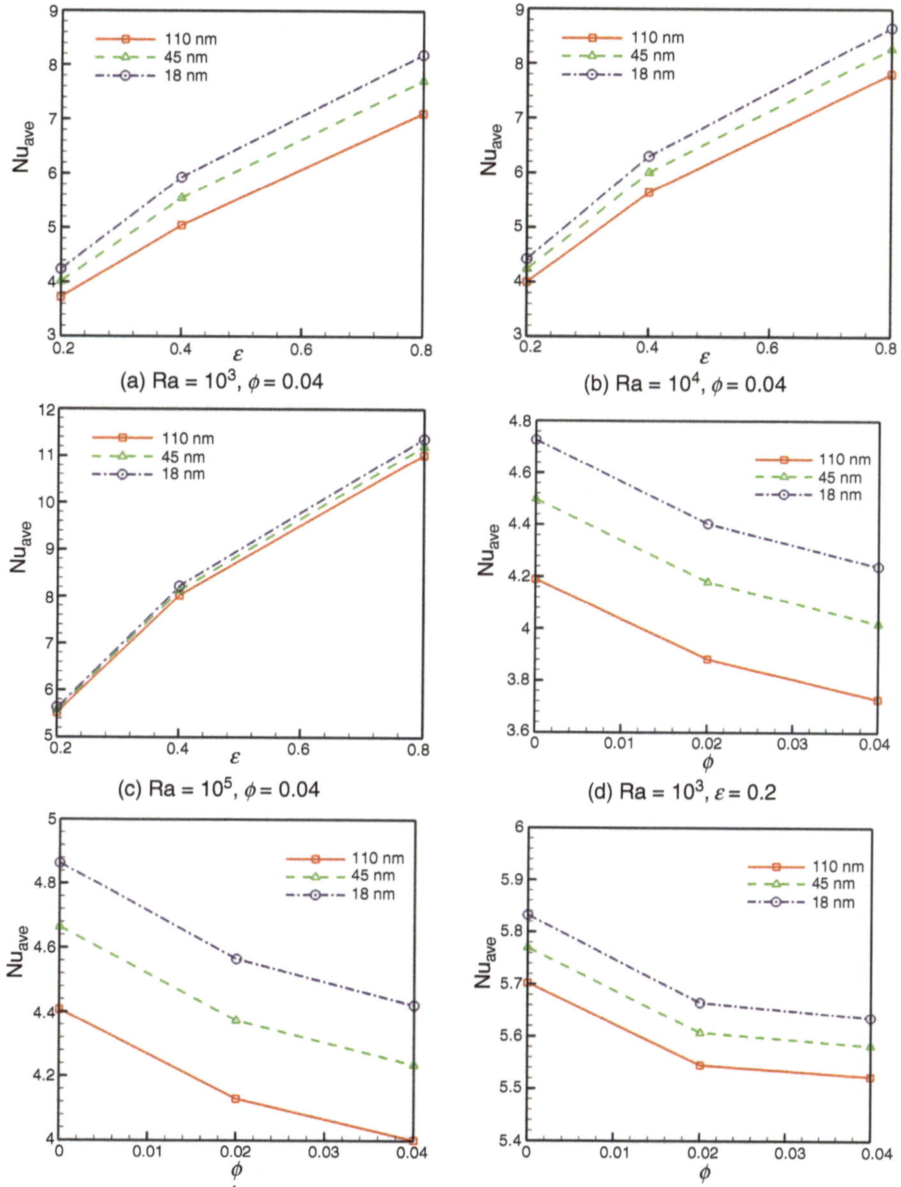

(a) Ra = 10^3, $\phi = 0.04$

(b) Ra = 10^4, $\phi = 0.04$

(c) Ra = 10^5, $\phi = 0.04$

(d) Ra = 10^3, $\varepsilon = 0.2$

(e) Ra = 10^4, $\varepsilon = 0.2$

(f) Ra = 10^5, $\varepsilon = 0.2$

FIGURE 4.44 Effects of ϕ, ε, Ra, and A on the Average Nusselt Number

A on the Nusselt number. The simulation result shown in this figure indicates that the Nusselt number is an increasing function of the Rayleigh number, magnetic coefficient, and heat source length, and it is a decreasing function of the volume fraction of cobalt.

4.14.7 LATTICE BOLTZMANN METHOD FOR SIMULATION OF THE MAGNETIC FIELD EFFECT ON THE HYDROTHERMAL BEHAVIOR OF NANOFLUID IN A CUBIC CAVITY

4.14.7.1 Problem definition

The numerical model consists of a three-dimensional square cavity with side equal to L, which represents the characteristic dimension of the problem (Fig. 4.45). The hot and cold wall are located at $X = x/L = 1$ and $X = x/L = 0$, respectively [35]. All of the other surfaces are adiabatic. Also, it is assumed that the uniform magnetic field ($\vec{B} = B_x \vec{e_x} + B_y \vec{e_y} + B_z \vec{e_z}$) of constant magnitude $B = \sqrt{B_x^2 + B_y^2 + B_z^2}$ is applied, where $\vec{e_x}, \vec{e_y}$, and $\vec{e_z}$ are unit vectors in the Cartesian coordinate system. The orientation of the magnetic field forms an angle θ_z with z-axis and θ_x with x-axis. The electric current J and the electromagnetic force F are defined by $J = \sigma\left(\vec{V} \times \vec{B}\right)$ and $F = \sigma\left(\vec{V} \times \vec{B}\right) \times \vec{B}$, respectively. In this study, we assume that $\theta_x = \theta_z = 90°$.

The D_3Q_{19} model was used (Fig. 4.46). These 19 velocities are given as follows.

$$c_i = \begin{pmatrix} 0 & 0 & 0 & 0 & -1 & -1 & -1 & -1 & 1 & 1 & 1 & 0 & 0 & 0 & 0 & -1 & 1 & 0 \\ -1 & -1 & 1 & 1 & 0 & -1 & -1 & 1 & 0 & -1 & 1 & 0 & 0 & -1 & 1 & 0 & 0 & 0 \\ -1 & 1 & -1 & 1 & -1 & 0 & 0 & 0 & 1 & 0 & 0 & -1 & 1 & 0 & 0 & 0 & 0 & 0 \end{pmatrix} \quad (4.105)$$

Weighting factor (w_i) is defined as follows:

$$w_i = \begin{cases} 1/3 & i = 0 \\ 1/18 & i = 1:6 \\ 1/36 & i = 7:18 \end{cases} \quad (4.106)$$

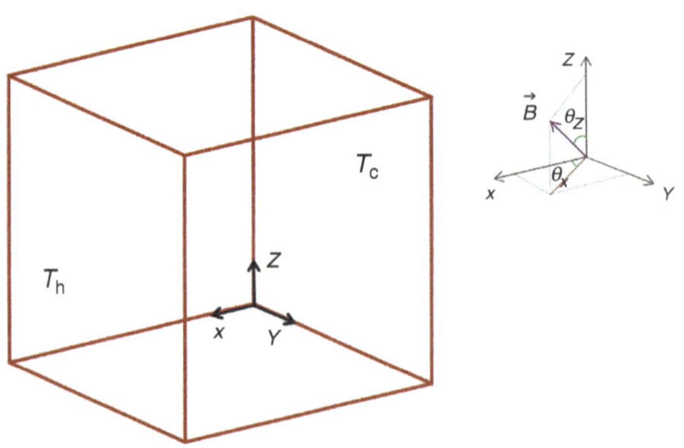

FIGURE 4.45 Geometry of the Problem

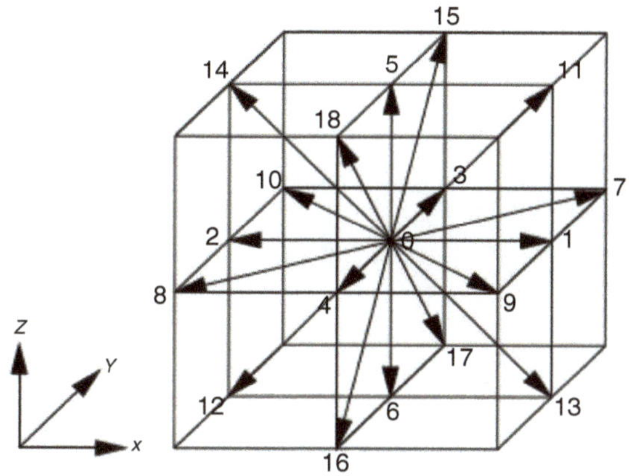

FIGURE 4.46 Discrete Velocity for the D_3Q_{19} Model

To incorporate buoyancy forces and magnetic forces in the model, the force term in Eq. (4.69) needs to be calculated as below [2]:

$$
\begin{aligned}
&F = F_x + F_y + F_z \\
&F_x = 3w_i\rho A\big[-\cos(\theta_z)\big(u\cos(\theta_z) + w\sin(\theta_z)\cos(\theta_x)\big) \\
&\quad -\sin(\theta_z)\sin(\theta_x)\big(u\sin(\theta_z)\sin(\theta_x) - v\sin(\theta_z)\cos(\theta_x)\big)\big], \\
&F_y = 3w_i\rho A\big[\sin(\theta_z)\cos(\theta_x)\big(u\sin(\theta_z)\sin(\theta_x) - v\sin(\theta_z)\cos(\theta_x)\big) \\
&\quad -\cos(\theta_z)\big(v\cos(\theta_z) - w\sin(\theta_z)\sin(\theta_x)\big)\big] \\
&F_z = 3w_i\rho\big[g_z\beta(T-T_m) + A\sin(\theta_z)\sin(\theta_x)\big(v\cos(\theta_z) - w\sin(\theta_z)\sin(\theta_x)\big) \\
&\quad +A\sin(\theta_z)\cos(\theta_x)\big(u\cos(\theta_z) - w\sin(\theta_z)\cos(\theta_x)\big)\big]
\end{aligned}
\tag{4.107}
$$

where A is $A = \dfrac{\mathrm{Ha}^2\mu}{L^2}$; $\mathrm{Ha} = LB_0\sqrt{\dfrac{\sigma}{\mu}}$ is the Hartmann number; θ_z and θ_x are the orientation of the magnetic field with z- and x-axis, respectively; and u, v, and w are velocity components in the x, y, and z directions, respectively.

4.14.7.2 Effect of active parameters

Three-dimensional nanofluid flow and heat transfer in the presence of a magnetic field are studied. The LBM scheme was used to obtain the numerical simulation in a cubic cavity filled with Al_2O_3–water. The effect of active parameters such as Rayleigh number (Ra = 10^3, 10^4, and 10^5), Hartmann number (Ha = 0, 20, 60, and 100), and volume fraction of nanoparticle ($\phi = 0$ and 0.04) on flow and heat transfer are examined.

The effects of the Rayleigh number and Hartmann number on isotherms and streamlines are shown in Figs 4.47 and 4.48. An increasing Rayleigh number causes

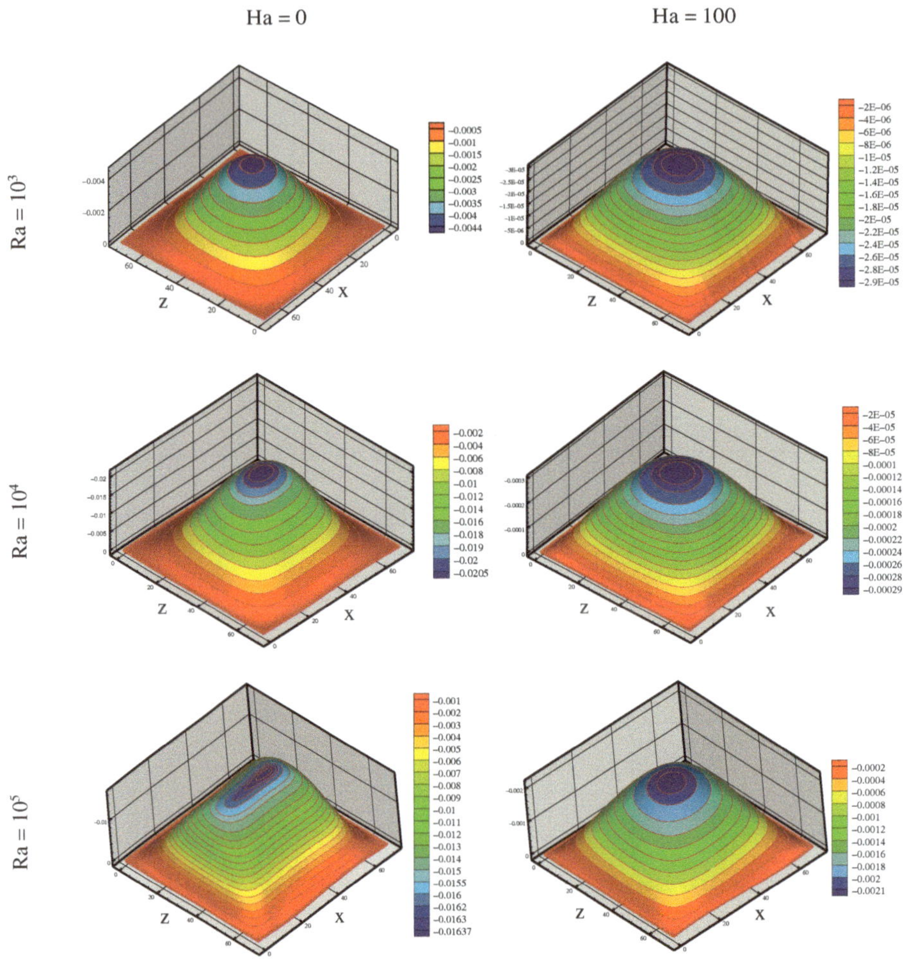

FIGURE 4.47 Effects of the Rayleigh Number and Hartmann Number on Streamlines at $Y = y/L = 0.5$ When $\phi = 0.04$

the buoyancy forces to enhance so that convective heat transfer becomes stronger for high Rayleigh numbers. The maximum value of stream function increases with an increase of the Rayleigh number. The Lorentz force increases with an increase of the Hartmann number. An increasing Lorentz force leads to domination of the conduction heat transfer mechanism. Therefore, isotherms become parallel to each other with an increase of the Hartmann number.

The effects of the Rayleigh number and Hartmann number on the local and average Nusselt number are depicted in Figs 4.49 and 4.50. As the Rayleigh number increases, the thermal boundary layer thickness decreases, and the opposite trend is

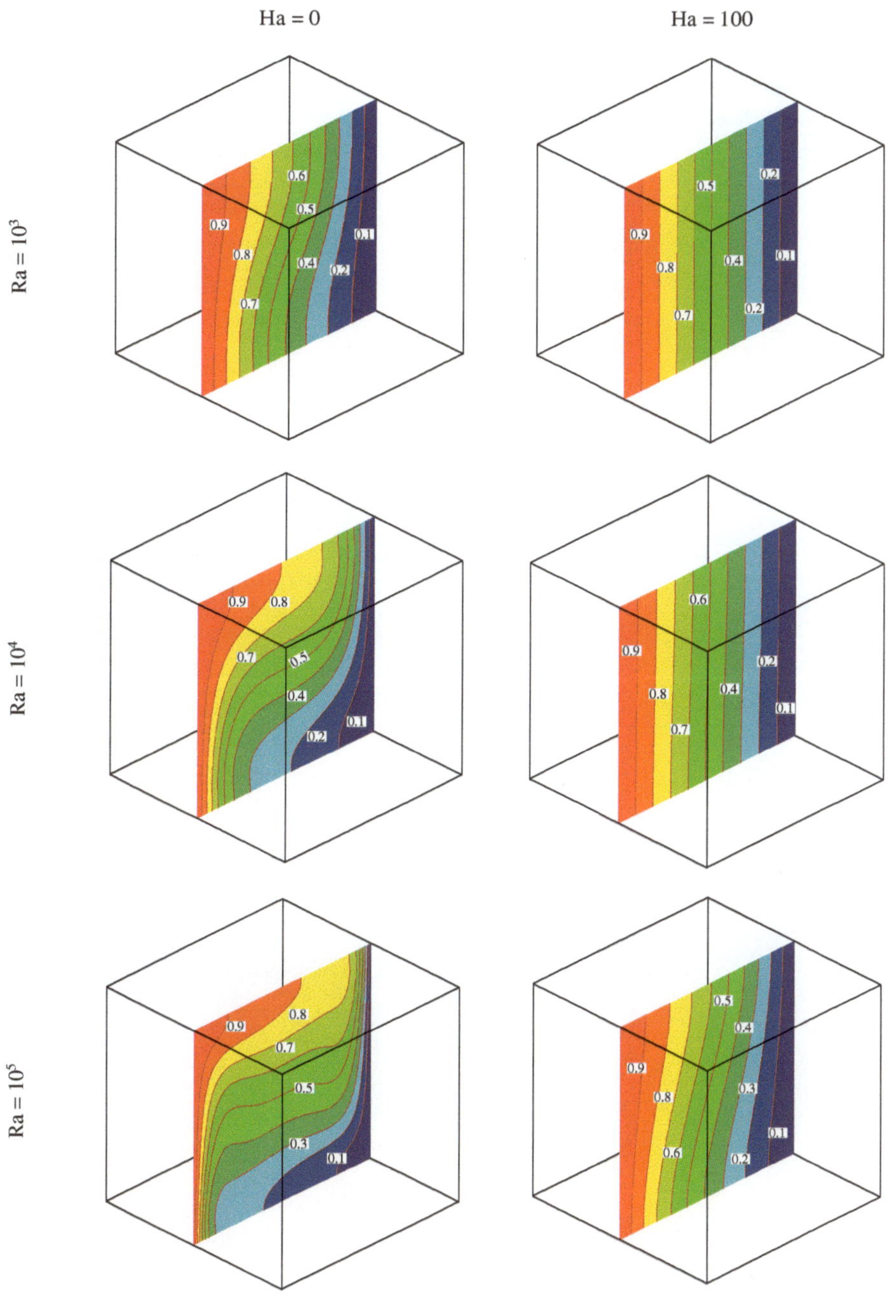

FIGURE 4.48 Effects of the Rayleigh Number and Hartmann Number on the Isotherm at $Y = y/L = 0.5$ When $\phi = 0.04$

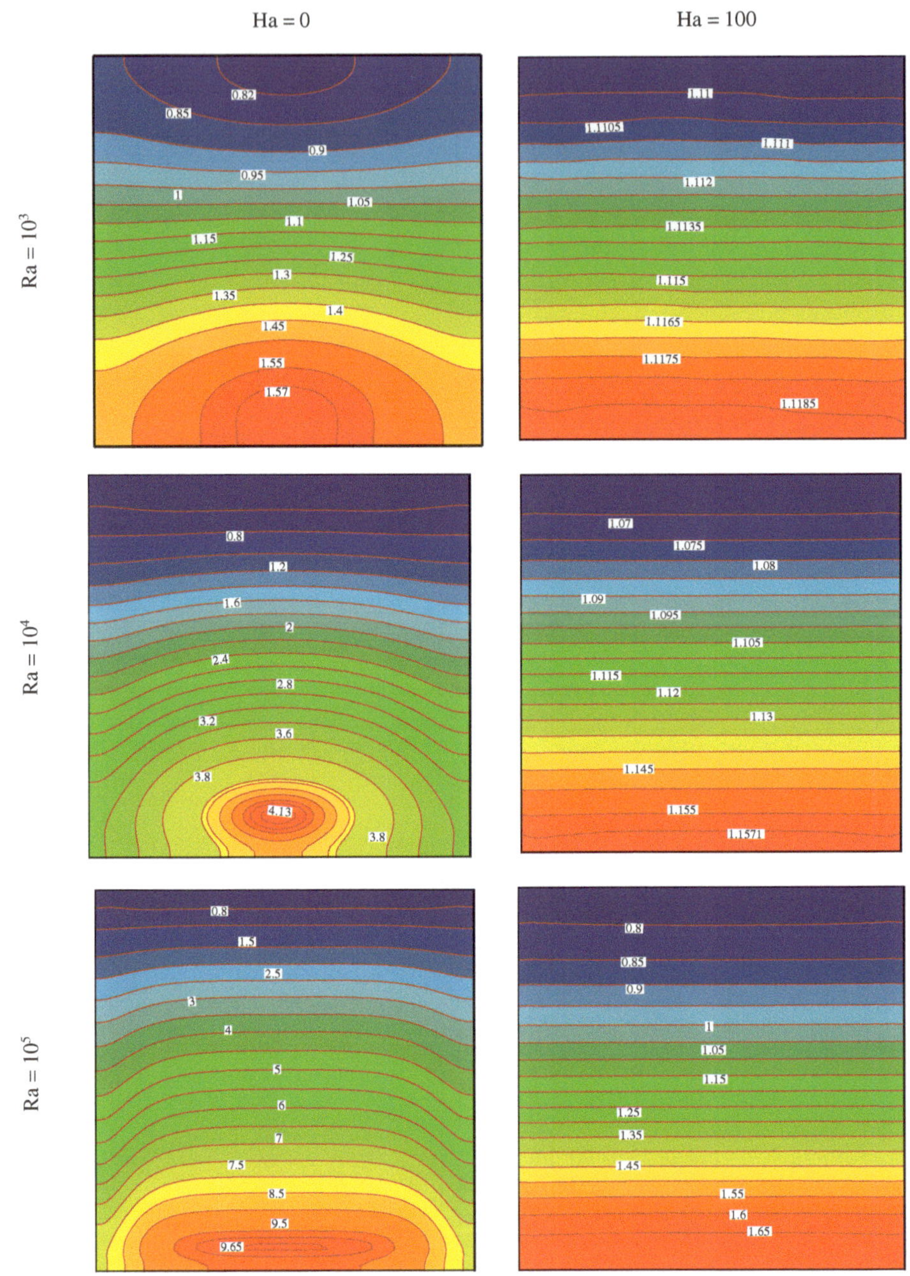

FIGURE 4.49 Effects of the Rayleigh Number and Hartmann Number on the Local Nusselt Number on the Hot Wall When $\phi = 0.04$

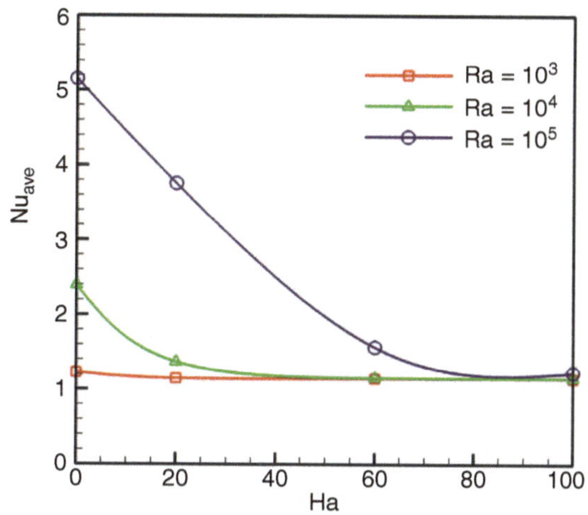

FIGURE 4.50 Effects of the Rayleigh Number and Hartmann Number on the Average Nusselt When $\phi = 0.04$

observed for the Hartmann number. The Nusselt number is an increasing function of the Rayleigh number, but it is a decreasing function of the Hartmann number.

Fig. 4.51 shows the effects of the Rayleigh number and Hartmann number on heat transfer enhancement. An increasing Hartmann number leads to an increase in the heat transfer enhancement caused by domination of the conduction mechanism. The effect of nanoparticles is more noticeable at low Rayleigh numbers than at high Rayleigh numbers. This remark can be clarified by noting that at low Rayleigh numbers,

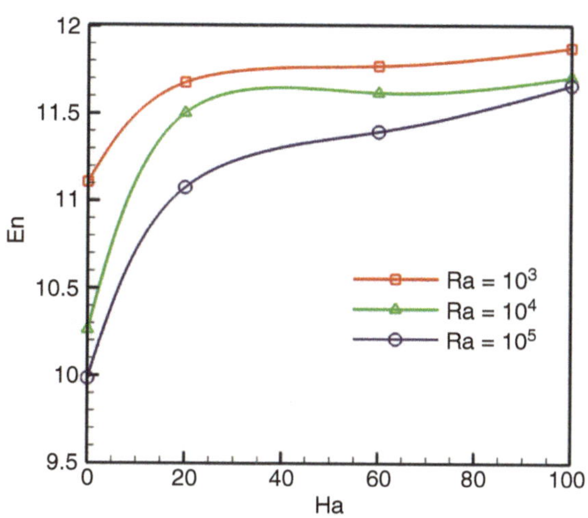

FIGURE 4.51 Effects of the Rayleigh Number and Hartmann Number on the Heat Transfer Enhancement

the heat transfer is dominant by conduction. Therefore, the addition of high thermal conductivity nanoparticles will increase the conduction and then make the enhancement more effective.

REFERENCES

[1] Chen S, Doolen G. Lattice Boltzmann method for fluid flows. Annu Rev Fluid Mech 1998;30:329–64.

[2] Mohamad AA. Lattice Boltzmann method. London: Springer-Verlag; 2011.

[3] Mohamad AA, Kuzmin A. A critical evaluation of force term in lattice Boltzmann method, natural convection problem. Int J Heat Mass Transfer 2010;53:990–6.

[4] Mohamad AA, El-Ganaoui M, Bennacer R. Lattice Boltzmann simulation of natural convection in an open ended cavity. Int J Thermal Sciences 2009;48:1870–5.

[5] Sheikholeslami M, Gorji-Bandpy M, Ganji DD. Lattice Boltzmann method for MHD natural convection heat transfer using nanofluid. Powder Technol 2014;254:82–93.

[6] Sheikholeslami M, Gorji-Bandpy M, Ganji DD. MHD free convection in an eccentric semi-annulus filled with nanofluid. J Taiwan Inst Chem Eng 2014;45:1204–16.

[7] Sheikholeslami M, Gorji-Bandpy M, Seyyedi SM, Ganji DD, Houman B, Rokni. Application of LBM in simulation of natural convection in a nanofluid filled square cavity with curve boundaries. Powder Technol 2013;247:87–94.

[8] Sheikholeslami M, Gorji-Bandpy M, Ganji DD. Natural convection in a nanofluid filled concentric annulus between an outer square cylinder and an inner elliptic cylinder. Trans B Mech Eng 2013;20(4):1–13.

[9] Sheikholeslami M, Gorji-Bandpy M, Domairry G. Free convection of nanofluid filled enclosure using lattice Boltzmann method (LBM). Appl Math Mech Engl Ed 2013;34(7):1–15.

[10] Sheikholeslami M, Gorji-Bandpy M. Free convection of ferrofluid in a cavity heated from below in the presence of an external magnetic field. Powder Technol 2014;256:490–8.

[11] Sheikholeslami M, Ellahi R. Three dimensional mesoscopic simulation of magnetic field effect on natural convection of nanofluid. Int J Heat Mass Transfer 2015;89:799–808.

[12] Ashorynejad HR, Mohamad AA, Sheikholeslami M. Magnetic field effects on natural convection flow of a nanofluid in a horizontal cylindrical annulus using lattice Boltzmann method. Int J Therm Sci 2013;64:240–50.

[13] Sheikholeslami M, Gorji-Bandpy M, Ganji DD. Numerical investigation of MHD effects on Al2O3-water nanofluid flow and heat transfer in a semi-annulus enclosure using LBM. Energy 2013;60:501–10.

[14] Sheikholeslami M, Gorji-Bandpay M, Ganji DD. Investigation of nanofluid flow and heat transfer in presence of magnetic field using KKL model. Arab J Sci Eng 2014;39(6):5007–16.

[15] Sheikholeslami M, Bandpy MG, Ellahi R, Zeeshan A. Simulation of MHD CuO–water nanofluid flow and convective heat transfer considering Lorentz forces. J Magn Magn Mater 2014;369:69–80.

[16] Sheikholeslami M, Ellahi R. Simulation of ferrofluid flow for magnetic drug targeting using Lattice Boltzmann method. Z Naturforsch A 2015;70(2):115–24.

[17] Sheikholeslami M, Ganji DD. Entropy generation of nanofluid in presence of magnetic field using Lattice Boltzmann Method. Physica A 2015;417:273–86.

[18] Frish U, Hasslacher B, Pomeau Y. Lattice gas automata for the Navier-Stokes equation. Phys Rev Lett 1986;56:1505–8.
[19] Chopard B, Luthi P, Masselot A. Cellular automata and lattice Boltzmann techniques. Geneva: University of Geneva; 1998.
[20] Succi S. The lattice Boltzmann equation for fluid dynamics and beyond. Oxford: Oxford University Press; 2001.
[21] Bhatnagar PL, Gross EP, Krook M. A Model for collision processes in gases. I. Small amplitude processes in charged and neutral one-component systems. Phys Rev 1954;94:511–25.
[22] Sokup MC, Thorne DT. Lattice Boltzmann modeling. Berlin, Heidelberg: Springer-Verlag; 2006.
[23] Yu D, Mei R, Luo L, Shyy W. Viscous flow computations with the method of lattice Boltzmann equation. Prog Aerosp Sci 2003;39:329–67.
[24] He X, Luo L. Theory of the lattice Boltzmann method: from the Boltzmann equation to the lattice Boltzmann equation. Phys Rev E 1997;56:6811–7.
[25] Yan YY, Zu YQ. Numerical simulation of heat transfer and fluid flow past a rotating isothermal cylinder – a LBM approach. Int J Heat and Mass Transfer 2008;51:2519–36.
[26] Kao PH, Yang RJ. Simulating oscillatory flows in Rayleigh Benard convection using the lattice Boltzmann method. Int J Heat and Mass Transfer 2007;50:3315–28.
[27] Barrios G, Rechtman R, Rojas J, Tovar R. The lattice Boltzmann equation for natural convection in a two-dimensional cavity with a partially heated wall. J Fluid Mech 2005;522:91–100.
[28] Dellar PJ. Lattice kinetic schemes for magnetohydrodynamics. J Comp Phys 2002;179:95–126.
[29] Kao PH, Yang RJ. Simulating oscillatory flows in Rayleigh–Benard convection using the lattice Boltzmann method. Int J Heat Mass Transfer 2007;50:3315–28.
[30] Khanafer K, Vafai K, Lightstone M. Buoyancy-driven heat transfer enhancement in a two-dimensional enclosure utilizing nanofluids. Int J Heat Mass Transfer 2003;46:3639–53.
[31] Davis GDV. Natural convection of air in a square cavity, a bench mark solution. Int J Numer Methods Fluids 1983;3:249–64.
[32] Rudraiah N, Barron RM, Venkatachalappa M, Subbaraya CK. Effect of a magnetic field on free convection in a rectangular enclosure. Int J Eng Sci 1995;33(8):1075–84.
[33] Sheikholeslami M, Gorji-Bandpay M, Ganji DD. Magnetic field effects on natural convection around a horizontal circular cylinder inside a square enclosure filled with nanofluid. Int Commun Heat Mass Transfer 2012;39:978–86.
[34] Sheikholeslami M, Gorji-Bandpy M, Vajravelu K. Lattice Boltzmann simulation of magnetohydrodynamic natural convection heat transfer of Al2O3-water nanofluid in a horizontal cylindrical enclosure with an inner triangular cylinder. Int J Heat Mass Transfer 2015;80:16–25.
[35] Sheikholeslami M, Bandpy MG, Ashorynejad HR. Lattice Boltzmann method for simulation of magnetic field effect on hydrothermal behavior of nanofluid in a cubic cavity. Physica A 2015;432:58–70.

APPENDIX

Sample codes for new numerical and semianalytical methods

A1. CVFEM CODE FOR 2D PURE DIFFUSION

```
Implicit none
!!!!!!!!!!!!Pure Diffusion!!!!!!!
Integer rows,n,k1,k2,nstep,itert
integer cols,ii,iter,loopcount
integer i,j,k,n_seg,iters
parameter (cols=51,rows=51,n=rows*cols,n_seg=4)
integer n_s(1:n),n_b(1:n_seg),B(1:n_seg,1:max(rows,cols)),S(1:n,1:12)
real*8 dy,dx,x(1:n),y(1:n)
real*8 sf(1:n),t(1:n)
real*8 apt(1:n),at(1:n,12)
real*8 aps(1:n),apv(1:n),as(1:n,12),av(1:n,12)
real*8 V(1:n),xL(1:3),yL(1:3),RHS,BC(1:n),BB(1:n),QC(1:n),QB(1:n)
real*8 diff(1:3),Vele,Nx(1:3),Ny(1:3),diff_f,vxf,vyf,delx,dely,qf
nstep=500
dx=1./real(rows-1)
dy=1./real(cols-1)
do k=1 , rows
do j=1 , cols
i=j+(k-1)*cols
y(i)=real(k-1)/real(rows-1)
x(i)=real(j-1)/real(cols-1)
```

```
enddo
enddo
!support
do k=1,rows
do j=1,cols
i=j+(k-1)*cols
n_s(i)=0 !Nodes in support
do ii=1,12
S(i,ii)=0.0 !Global node number of ii_th node in support i
enddo
enddo
enddo
!boundary
n_b(1)=cols
n_b(2)=rows
n_b(3)=cols
n_b(4)=rows
do k=1,n_seg
do j=1,n_b(k)
B(k,j)=0.0 !Global node number of j_th node on i_th boundary
enddo
enddo
!SUPPORT DATA ----------
! Support do internal points
do k=2,rows-1
do j=2,cols-1
i=j+(k-1)*cols
n_s(i)=6
S(i,1)=i+1
S(i,2)=i+1+cols
S(i,3)=i+cols
S(i,4)=i-1
```

```
S(i,5)=i-1-cols
S(i,6)=i-cols
S(i,7)=i+1
enddo
enddo
do j=2,cols-1
i=j
n_s(i)=4
S(i,1)=i+1
S(i,2)=i+1+cols
S(i,3)=i+cols
S(i,4)=i-1
S(i,5)=0 ! last point zero: a boundary node flag
enddo
do k=2,rows-1
i=cols+(k-1)*cols
n_s(i)=4
S(i,1)=i+cols
S(i,2)=i-1
S(i,3)=i-1-cols
S(i,4)=i-cols
S(i,5)=0
enddo
do j=2,cols-1
i=j+(rows-1)*cols
n_s(i)=4
S(i,1)=i-1
S(i,2)=i-1-cols
S(i,3)=i-cols
S(i,4)=i+1
S(i,5)=0
enddo
```

```
do k=2,rows-1
i=1+(k-1)*cols
n_s(i)=4
S(i,1)=i-cols
S(i,2)=i+1
S(i,3)=i+1+cols
S(i,4)=i+cols
S(i,5)=0
enddo
!Corners
! x=0, y = 0
i=1
n_s(i)=3
S(i,1)=i+1
S(i,2)=i+1+cols
S(i,3)=i+cols
S(i,4)=0
! y = 0, x = 1
i=cols
n_s(i)=2
S(i,1)=i+cols
S(i,2)=i-1
S(i,3)=0
! x =1, y = 1
i=rows*cols
n_s(i)=3
S(i,1)=i-1
S(i,2)=i-1-cols
S(i,3)=i-cols
S(i,4)=0
! x=0, y = 1
i=1+(rows-1)*cols
```

```
n_s(i)=2
S(i,1)=i-cols
S(i,2)=i+1
S(i,3)=0
!END SUPPORT DATA-----------
! BOUNDARY DATA------------
!NOTE counter-clockwise numbering and boundary segment index
! y=0 boundary (segment 1)
do i=1,n_b(1)
B(1,i)=i
enddo
! x=1 boundary (segment 2)
do i=1,n_b(2)
B(2,i)=i*cols
enddo
! y =1 boundary (segment 3)
do i=1,n_b(3)
B(3,i)=rows*cols+1-i
enddo
! x=0 boundary (segment 4)
do i=1,n_b(4)
B(4,i)=1+rows*cols-i*cols
enddo
!========================================
!=============initial conditions=======!
!========================================
do i=1,n
t(i)=0.0
enddo
do i=1,n
apt(i)=0.0
```

```
do j=1,n_s(i)+1
at(i,j)=0.0
enddo
enddo
do iter=1,nstep
print*,iter
!===============================
!==SOLVING Diffusion===========
!===============================
do i=1,n !loop on nodes
apt(i)=0.0
V(i)=0.0
do j=1,n_s(i)+1
at(i,j)=0.0
enddo
loopcount=n_s(i)
if (S(i,n_s(i)+1)==0) then
loopcount=n_s(i)-1
endif
do j=1,loopcount
k1=S(i,j)
k2=S(i,j+1)
xL(1) = x(i)
xL(2) = x(k1)
xL(3) = x(k2)
yL(1) = y(i)
yL(2) = y(k1)
yL(3) = y(k2)
diff(1)=1.0
diff(2)=1.0
diff(3)=1.0
Vele=(xL(2)*yL(3)-xL(3)*yL(2)-xL(1)*yL(3)+xL(1)*yL(2)+yL(1)*xL(3)-yL(1)*xL(2))/2.
```

```
V(i)=V(i)+Vele/3.
Nx(1)= (yL(2)-yL(3))/(2.*Vele)
Nx(2)= (yL(3)-yL(1))/(2.*Vele)
Nx(3)= (yL(1)-yL(2))/(2.*Vele)
Ny(1)=-(xL(2)-xL(3))/(2.*Vele)
Ny(2)=-(xL(3)-xL(1))/(2.*Vele)
Ny(3)=-(xL(1)-xL(2))/(2.*Vele)
! Face 1–values of diffusivity and velocity
diff_f=(5.*diff(1)+5.*diff(2)+2.*diff(3))/12.
! Face 1–face area normal components
delx= ((xL(1)+xL(2)+xL(3))/3.)-((xL(1)+xL(2))/2.)
dely= ((yL(1)+yL(2)+yL(3))/3.)-((yL(1)+yL(2))/2.)
! Face 1–Diffusion Coefficients
apt(i) = apt(i) + diff_f*(- Nx(1) * dely + Ny(1) * delx)
at(i,j) = at(i,j) + diff_f*( Nx(2) * dely – Ny(2) * delx)
at(i,j+1) = at(i,j+1) + diff_f*( Nx(3) * dely – Ny(3) * delx)
! Face 2–values of diffusivity and velocity
diff_f=(5.*diff(1)+2.*diff(2)+5*diff(3))/12.
! Face 2–face area normal components
delx= ((xL(1)+xL(3))/2.)-((xL(1)+xL(2)+xL(3))/3.)
dely= ((yL(1)+yL(3))/2.)-((yL(1)+yL(2)+yL(3))/3.)
! Face 2–Diffusion Coefficients
apt(i) = apt(i) + diff_f*(- Nx(1) * dely + Ny(1) * delx)
at(i,j) = at(i,j) + diff_f*( Nx(2) * dely – Ny(2) * delx)
at(i,j+1) = at(i,j+1) + diff_f*( Nx(3) * dely – Ny(3) * delx)
enddo !end loop on nodes/elements in support
at(i,1)=at(i,1)+at(i,n_s(i)+1)
at(i,n_s(i)+1)=0
end do !end loop on nodes in domain
!===============SET BOUNDARY
do i=1,n
BC(i)=0.0
```

```
BB(i)=0.0
enddo
do k=1,n_b(1)
BC(B(1,k))=0.0
enddo
do k=1,n_b(2)
BC(B(2,k))=1.e18
enddo
do k=1,n_b(3)
BC(B(3,k))=1.e18
enddo
do k=1,n_b(4)
BC(B(4,k))=1.e18
BB(B(4,k))=1.e18
enddo
!------------------------
do itert=1,20
do i= 1,n
RHS=BB(i)+QB(i)
do j=1,n_s(i)
RHS=RHS+at(i,j)*t(S(i,j))
enddo
t(i)=RHS/(apt(i)+BC(i)+QC(i))
enddo
enddo
enddo
end
```

A2. CVFEM CODE FOR 2D ADVECTION–DIFFUSION

```
implicit none

!!!!!!!!!!!!Advection-diffusion!!!!!!!!

integer rows,n,k1,k2,nstep,itert

integer cols,ii,iter,loopcount

integer i,j,k,n_seg

parameter (cols=51,rows=51,n=rows*cols,n_seg=4)

integer n_s(1:n),n_b(1:n_seg),B(1:n_seg,1:max(rows,cols)),S(1:n,1:12)

real*8 dy,dx,x(1:n),y(1:n),t(1:n)

real*8 ulid,apt(1:n),at(1:n,12)

real*8 V(1:n),xL(1:3),yL(1:3),RHS,BC(1:n),BB(1:n),QC(1:n),QB(1:n)

real*8 diff(1:3),Vele,Nx(1:3),Ny(1:3),diff_f,vxf,vyf,delx,dely,qf

nstep=500

dx=1./real(rows-1)

dy=1./real(cols-1)

do k=1 , rows

do j=1 , cols

i=j+(k-1)*cols

y(i)=real(k-1)/real(rows-1)

x(i)=real(j-1)/real(cols-1)

enddo

enddo

!support

do k=1,rows

do j=1,cols

i=j+(k-1)*cols

n_s(i)=0 !Nodes in support

do ii=1,12
```

```
S(i,ii)=0.0 !Global node number of ii_th node in support i
enddo
enddo
enddo
!boundary
n_b(1)=cols
n_b(2)=rows
n_b(3)=cols
n_b(4)=rows
do k=1,n_seg
do j=1,n_b(k)
B(k,j)=0.0 !Global node number of j_th node on i_th boundary
enddo
enddo
!SUPPORT DATA ----------
! Support do internal points
do k=2,rows-1
do j=2,cols-1
i=j+(k-1)*cols
n_s(i)=6
S(i,1)=i+1 !assumed morphology do support see Fig. A1
S(i,2)=i+1+cols
S(i,3)=i+cols
S(i,4)=i-1
S(i,5)=i-1-cols
S(i,6)=i-cols
S(i,7)=i+1 !last node is same as first node at internal nodes
enddo
enddo
do j=2,cols-1
i=j
n_s(i)=4
```

```
S(i,1)=i+1
S(i,2)=i+1+cols
S(i,3)=i+cols
S(i,4)=i-1
S(i,5)=0  ! last point zero: a boundary node flag
enddo
do k=2,rows-1
i=cols+(k-1)*cols
n_s(i)=4
S(i,1)=i+cols
S(i,2)=i-1
S(i,3)=i-1-cols
S(i,4)=i-cols
S(i,5)=0
enddo
do j=2,cols-1
i=j+(rows-1)*cols
n_s(i)=4
S(i,1)=i-1
S(i,2)=i-1-cols
S(i,3)=i-cols
S(i,4)=i+1
S(i,5)=0
enddo
do k=2,rows-1
i=1+(k-1)*cols
n_s(i)=4
S(i,1)=i-cols
S(i,2)=i+1
S(i,3)=i+1+cols
S(i,4)=i+cols
S(i,5)=0
```

```
enddo
!Corners
i=1
n_s(i)=3
S(i,1)=i+1
S(i,2)=i+1+cols
S(i,3)=i+cols
S(i,4)=0
i=cols
n_s(i)=2
S(i,1)=i+cols
S(i,2)=i-1
S(i,3)=0
i=rows*cols
n_s(i)=3
S(i,1)=i-1
S(i,2)=i-1-cols
S(i,3)=i-cols
S(i,4)=0
i=1+(rows-1)*cols
n_s(i)=2
S(i,1)=i-cols
S(i,2)=i+1
S(i,3)=0
!END SUPPORT DATA-----------
! BOUNDARY DATA------------
!NOTE counter-clockwise numbering and boundary segment index
do i=1,n_b(1)
B(1,i)=i
enddo
do i=1,n_b(2)
B(2,i)=i*cols
```

```
enddo
do i=1,n_b(3)
B(3,i)=rows*cols+1-i
enddo
do i=1,n_b(4)
B(4,i)=1+rows*cols-i*cols
enddo
!=======================================
!=============initial conditions=======!
!=======================================
do i=1,n
t(i)=0.0
enddo
do i=1,n
apt(i)=0.0
do j=1,n_s(i)+1
at(i,j)=0.0
enddo
enddo
do iter=1,nstep
print*,iter
!=============Calculate temperature==================
do i=1,n
apt(i)=0.0
V(i)=0.0
do j=1,n_s(i)+1
at(i,j)=0.0
enddo
loopcount=n_s(i)
if (S(i,n_s(i)+1)==0) then
loopcount=n_s(i)-1
endif
```

```
do j=1,loopcount
k1=S(i,j) !The numbers of the nodes in the j_th elemnt are
k2=S(i,j+1) ! i, k1=S(i,J), and k2=S(i,j+1)
xL(1) = x(i) !xL=local x-coordinate of a node within a element
xL(2) = x(k1)
xL(3) = x(k2)
yL(1) = y(i) !yL=local y-coordinate of a node within a element
yL(2) = y(k1)
yL(3) = y(k2)
! nodal diffusivity (kappa)
diff(1)=0.01 !!!!!
diff(2)=0.01
diff(3)=0.01
! Vele= element-volume
Vele=(xL(2)*yL(3)-xL(3)*yL(2)-xL(1)*yL(3)+xL(1)*yL(2)+yL(1)*xL(3)-yL(1)*xL(2))/2.
V(i)=V(i)+Vele/3 !Contribution to control volume
Nx(1)= (yL(2)-yL(3))/(2.*Vele)
Nx(2)= (yL(3)-yL(1))/(2.*Vele)
Nx(3)= (yL(1)-yL(2))/(2.*Vele)
Ny(1)=-(xL(2)-xL(3))/(2.*Vele)
Ny(2)=-(xL(3)-xL(1))/(2.*Vele)
Ny(3)=-(xL(1)-xL(2))/(2.*Vele)
! Face 1--values of diffusivity and velocity
diff_f=(5.*diff(1)+5.*diff(2)+2.*diff(3))/12.
vxf=0.1
vyf=0.2
! Face 1--face area normal components
delx= ((xL(1)+xL(2)+xL(3))/3.)-((xL(1)+xL(2))/2.)
dely= ((yL(1)+yL(2)+yL(3))/3.)-((yL(1)+yL(2))/2.)
qf=vxf*dely-vyf*delx !Face 1 volume flux
! Face 1--Diffusion Coefficients
apt(i) = apt(i) + diff_f*(- Nx(1) * dely + Ny(1) * delx)
```

```
at(i,j) = at(i,j) + diff_f*( Nx(2) * dely - Ny(2) * delx)
at(i,j+1) = at(i,j+1) + diff_f*( Nx(3) * dely - Ny(3) * delx)
! Face 1--Advection Coefficients
apt(i)= apt(i) +max(qf,0.)
at(i,j)=at(i,j) +max(-qf,0.)
! Face 2--values of diffusivity and velocity
diff_f=(5*diff(1)+2*diff(2)+5*diff(3))/12.
vxf= 0.1
vyf= 0.2
! Face 2--face area normal components
delx= ((xL(1)+xL(3))/2)-((xL(1)+xL(2)+xL(3))/3.)
dely= ((yL(1)+yL(3))/2)-((yL(1)+yL(2)+yL(3))/3.)
qf=vxf*dely-vyf*delx ! Face 1 volume flux
! Face 2--Diffusion Coefficients
apt(i) = apt(i) + diff_f*(- Nx(1) * dely + Ny(1) * delx)
at(i,j) = at(i,j) + diff_f*( Nx(2) * dely - Ny(2) * delx)
at(i,j+1) = at(i,j+1) + diff_f*( Nx(3) * dely - Ny(3) * delx)
! Face 2--Advection Coefficients
apt(i)= apt(i) + max(qf,0.)
at(i,j+1) = at(i,j+1) + max(-qf,0.)
end do !end loop on nodes/elements in support
at(i,1)=at(i,1)+at(i,n_s(i)+1)
at(i,n_s(i)+1)=0
end do !end loop on nodes in domain
!!!!!!!!!!!!!!!!!!!!!!!!!!!!!!!!!!!!!!!!!!!!!!!!!!!!
!SETTING BOUNDARY==========
!!!!!!!!!!!!!!!!!!!!!!!!!!!!!!!!!!!!!!!!!!!!!!!!!!!!
do i=1,n
BC(i)=0.0
BB(i)=0.0
```

```
enddo
do k=1,n_b(1)
BC(B(1,k))=1.e18
BB(B(1,k))=0.0*1.e18
enddo
do k=1,n_b(2)
BC(B(2,k))=1.e18
BB(B(2,k))=0.0*1.e18
end do
do k=1,n_b(3)
BC(B(3,k))=1.e18
BB(B(3,k))=0.0*1.e18
end do
do k=1,n_b(4)
BC(B(4,k))=1.e18
BB(B(4,k))=1*1.e18
end do
do itert=1,1
do i= 1,n
RHS=BB(i)+QB(i)
do j=1,n_s(i)
RHS=RHS+at(i,j)*t(S(i,j))
enddo
t(i)=RHS/(apt(i)+BC(i)+QC(i))
end do
end do
end do
end
```

A3. MAPLE CODE FOR HPM

```
> restart:
  with(PDEtools):
> L:=7:
> H:=0;alpha:=5*3.14/180;RE:=50;
```

$$H := 0$$

$$\alpha := 0.08722222222$$

$$RE := 50 \tag{1}$$

```
> f(x):=sum((p^i)*f[i](x),i=0..L);
```

$$f(x) := f_0(x) + p\,f_1(x) + p^2 f_2(x) + p^3 f_3(x) + p^4 f_4(x) + p^5 f_5(x) + p^6 f_6(x) + p^7 f_7(x) \tag{2}$$

```
> HO1 := (1-p)*(diff(f(x), x, x, x))+p*(diff(f(x), x, x, x)+2*
  alpha*RE*(diff(f(x), x))*f(x)+(4-H)*alpha^2*(diff(f(x), x)));
  expand(%);
  collect(%, p);
  HO1 := %;
  declare(f(x), prime = x);
```

$$HO1 := (1-p)\left(\frac{d^3}{dx^3} f_0(x) + p\left(\frac{d^3}{dx^3} f_1(x)\right) + p^2\left(\frac{d^3}{dx^3} f_2(x)\right) + p^3\left(\frac{d^3}{dx^3} f_3(x)\right) + p^4\left(\frac{d^3}{dx^3} f_4(x)\right) + p^5\left(\frac{d^3}{dx^3} f_5(x)\right) + p^6\left(\frac{d^3}{dx^3} f_6(x)\right) + p^7\left(\frac{d^3}{dx^3} f_7(x)\right)\right) + p\left(\frac{d^3}{dx^3} f_0(x) + p\left(\frac{d^3}{dx^3} f_1(x)\right) + p^2\left(\frac{d^3}{dx^3} f_2(x)\right) + p^3\left(\frac{d^3}{dx^3} f_3(x)\right) + p^4\left(\frac{d^3}{dx^3} f_4(x)\right) + p^5\left(\frac{d^3}{dx^3} f_5(x)\right) + p^6\left(\frac{d^3}{dx^3} f_6(x)\right) + p^7\left(\frac{d^3}{dx^3} f_7(x)\right) + 8.722222222\left(\frac{d}{dx} f_0(x) + p\left(\frac{d}{dx} f_1(x)\right) + p^2\left(\frac{d}{dx} f_2(x)\right) + p^3\left(\frac{d}{dx} f_3(x)\right) + p^4\left(\frac{d}{dx} f_4(x)\right) + p^5\left(\frac{d}{dx} f_5(x)\right) + p^6\left(\frac{d}{dx} f_6(x)\right) + p^7\left(\frac{d}{dx} f_7(x)\right)\right)\left(f_0(x) + p\,f_1(x) + p^2 f_2(x) + p^3 f_3(x) + p^4 f_4(x) + p^5 f_5(x) + p^6 f_6(x) + p^7 f_7(x)\right) + 0.03043086420\left(\frac{d}{dx} f_0(x)\right) + 0.03043086420\,p\left(\frac{d}{dx} f_1(x)\right) + 0.03043086420\,p^2\left(\frac{d}{dx} f_2(x)\right) + 0.03043086420\,p^3\left(\frac{d}{dx} f_3(x)\right) + 0.03043086420\,p^4\left(\frac{d}{dx} f_4(x)\right) + 0.03043086420\,p^5\left(\frac{d}{dx} f_5(x)\right) + 0.03043086420\,p^6\left(\frac{d}{dx} f_6(x)\right) + 0.03043086420\,p^7\left(\frac{d}{dx} f_7(x)\right)\right)$$

$$8.722222222\,p^5\left(\frac{d}{dx} f_4(x)\right) f_0(x) + 0.03043086420\,p\left(\frac{d}{dx} f_0(x)\right) + 8.722222222\,p^3\left(\frac{d}{dx} f_1(x)\right) f_1(x) + 8.722222222\,p^6\left(\frac{d}{dx} f_4(x)\right) f_1(x) + 8.722222222\,p^{11}\left(\frac{d}{dx} f_3(x)\right) f_7(x) + 8.722222222\,p^6\left(\frac{d}{dx} f_1(x)\right) f_4(x) + 8.722222222\,p^{13}\left(\frac{d}{dx} f_6(x)\right) f_6(x) + 8.722222222\,p^5\left(\frac{d}{dx} f_3(x)\right) f_1(x)$$

$$+ 8.722222222\, p^{6} \left(\frac{d}{dx} f_{2}(x) \right) f_{3}(x) + 8.722222222 \left(\frac{d}{dx} f_{0}(x) \right) p^{8} f_{7}(x)$$
$$+ 8.722222222\, p^{9} \left(\frac{d}{dx} f_{7}(x) \right) f_{1}(x) + 8.722222222\, p^{11} \left(\frac{d}{dx} f_{5}(x) \right) f_{5}(x)$$
$$+ 8.722222222\, p^{10} \left(\frac{d}{dx} f_{5}(x) \right) f_{4}(x) + 8.722222222\, p^{8} \left(\frac{d}{dx} f_{7}(x) \right) f_{0}(x)$$
$$+ 8.722222222\, p^{8} \left(\frac{d}{dx} f_{1}(x) \right) f_{6}(x) + 8.722222222\, p^{5} \left(\frac{d}{dx} f_{2}(x) \right) f_{2}(x)$$
$$+ 8.722222222\, p^{9} \left(\frac{d}{dx} f_{5}(x) \right) f_{3}(x) + 8.722222222\, p^{7} \left(\frac{d}{dx} f_{1}(x) \right) f_{5}(x)$$
$$+ 8.722222222\, p^{7} \left(\frac{d}{dx} f_{3}(x) \right) f_{3}(x) + 8.722222222\, p^{12} \left(\frac{d}{dx} f_{5}(x) \right) f_{6}(x)$$
$$+ 8.722222222 \left(\frac{d}{dx} f_{0}(x) \right) p^{4} f_{3}(x) + 8.722222222\, p^{4} \left(\frac{d}{dx} f_{1}(x) \right) f_{2}(x)$$
$$+ 8.722222222 \left(\frac{d}{dx} f_{0}(x) \right) p^{3} f_{2}(x) + 8.722222222\, p^{7} \left(\frac{d}{dx} f_{2}(x) \right) f_{4}(x)$$
$$+ 8.722222222\, p^{10} \left(\frac{d}{dx} f_{4}(x) \right) f_{5}(x) + 8.722222222 \left(\frac{d}{dx} f_{0}(x) \right) p^{7} f_{6}(x)$$
$$+ 8.722222222\, p^{14} \left(\frac{d}{dx} f_{6}(x) \right) f_{7}(x) + 8.722222222\, p^{3} \left(\frac{d}{dx} f_{2}(x) \right) f_{0}(x)$$
$$+ 8.722222222\, p^{7} \left(\frac{d}{dx} f_{4}(x) \right) f_{2}(x) + 8.722222222\, p^{12} \left(\frac{d}{dx} f_{4}(x) \right) f_{7}(x)$$
$$+ 8.722222222\, p^{9} \left(\frac{d}{dx} f_{4}(x) \right) f_{4}(x) + 8.722222222\, p^{8} \left(\frac{d}{dx} f_{3}(x) \right) f_{4}(x)$$
$$+ 8.722222222\, p^{6} \left(\frac{d}{dx} f_{5}(x) \right) f_{0}(x) + 8.722222222\, p^{9} \left(\frac{d}{dx} f_{2}(x) \right) f_{6}(x)$$
$$+ 8.722222222\, p^{11} \left(\frac{d}{dx} f_{4}(x) \right) f_{6}(x) + 8.722222222\, p^{9} \left(\frac{d}{dx} f_{3}(x) \right) f_{5}(x)$$
$$+ 8.722222222\, p^{9} \left(\frac{d}{dx} f_{1}(x) \right) f_{7}(x) + 8.722222222\, p^{11} \left(\frac{d}{dx} f_{6}(x) \right) f_{4}(x)$$
$$+ 8.722222222\, p^{10} \left(\frac{d}{dx} f_{2}(x) \right) f_{7}(x) + 8.722222222\, p^{9} \left(\frac{d}{dx} f_{6}(x) \right) f_{2}(x)$$
$$+ 8.722222222\, p^{10} \left(\frac{d}{dx} f_{3}(x) \right) f_{6}(x) + 8.722222222\, p^{6} \left(\frac{d}{dx} f_{3}(x) \right) f_{2}(x)$$
$$+ 8.722222222\, p^{2} \left(\frac{d}{dx} f_{1}(x) \right) f_{0}(x) + 8.722222222 \left(\frac{d}{dx} f_{0}(x) \right) p^{6} f_{5}(x)$$
$$+ 8.722222222\, p^{10} \left(\frac{d}{dx} f_{6}(x) \right) f_{3}(x) + 8.722222222\, p^{5} \left(\frac{d}{dx} f_{1}(x) \right) f_{3}(x)$$
$$+ 8.722222222\, p^{8} \left(\frac{d}{dx} f_{6}(x) \right) f_{1}(x) + 8.722222222 \left(\frac{d}{dx} f_{0}(x) \right) p^{2} f_{1}(x)$$
$$+ 8.722222222\, p^{8} \left(\frac{d}{dx} f_{5}(x) \right) f_{2}(x) + 8.722222222\, p^{7} \left(\frac{d}{dx} f_{6}(x) \right) f_{0}(x)$$
$$+ 8.722222222 \left(\frac{d}{dx} f_{0}(x) \right) p^{5} f_{4}(x) + 8.722222222\, p^{4} \left(\frac{d}{dx} f_{3}(x) \right) f_{0}(x)$$
$$+ 8.722222222\, p^{8} \left(\frac{d}{dx} f_{4}(x) \right) f_{3}(x) + 8.722222222\, p^{10} \left(\frac{d}{dx} f_{7}(x) \right) f_{2}(x)$$

$$+ 8.722222222\, p^{11} \left(\frac{d}{dx} f_7(x) \right) f_3(x) + 8.722222222\, p^{12} \left(\frac{d}{dx} f_7(x) \right) f_4(x)$$

$$+ 8.722222222\, p^{13} \left(\frac{d}{dx} f_7(x) \right) f_5(x) + 8.722222222\, p^{14} \left(\frac{d}{dx} f_7(x) \right) f_6(x)$$

$$+ 8.722222222\, p^{15} \left(\frac{d}{dx} f_7(x) \right) f_7(x) + 8.722222222\, p^{7} \left(\frac{d}{dx} f_5(x) \right) f_1(x)$$

$$+ 8.722222222\, p \left(\frac{d}{dx} f_0(x) \right) f_0(x) + 8.722222222\, p^{12} \left(\frac{d}{dx} f_6(x) \right) f_5(x)$$

$$+ 8.722222222\, p^{13} \left(\frac{d}{dx} f_5(x) \right) f_7(x) + 8.722222222\, p^{8} \left(\frac{d}{dx} f_2(x) \right) f_5(x)$$

$$+ p \left(\frac{d^3}{dx^3} f_1(x) \right) + p^2 \left(\frac{d^3}{dx^3} f_2(x) \right) + p^3 \left(\frac{d^3}{dx^3} f_3(x) \right) + p^4 \left(\frac{d^3}{dx^3} f_4(x) \right)$$

$$+ p^5 \left(\frac{d^3}{dx^3} f_5(x) \right) + p^6 \left(\frac{d^3}{dx^3} f_6(x) \right) + p^7 \left(\frac{d^3}{dx^3} f_7(x) \right)$$

$$+ 0.03043086420\, p^{8} \left(\frac{d}{dx} f_7(x) \right) + 0.03043086420\, p^{7} \left(\frac{d}{dx} f_6(x) \right)$$

$$+ 0.03043086420\, p^{6} \left(\frac{d}{dx} f_5(x) \right) + 0.03043086420\, p^{2} \left(\frac{d}{dx} f_1(x) \right)$$

$$+ 0.03043086420\, p^{3} \left(\frac{d}{dx} f_2(x) \right) + 0.03043086420\, p^{4} \left(\frac{d}{dx} f_3(x) \right)$$

$$+ 0.03043086420\, p^{5} \left(\frac{d}{dx} f_4(x) \right) + \frac{d^3}{dx^3} f_0(x) + 8.722222222\, p^{4} \left(\frac{d}{dx} f_2(x) \right) f_1(x)$$

$$8.722222222\, p^{15} \left(\frac{d}{dx} f_7(x) \right) f_7(x) + \Bigg(8.722222222 \left(\frac{d}{dx} f_7(x) \right) f_6(x)$$

$$+ 8.722222222 \left(\frac{d}{dx} f_6(x) \right) f_7(x) \Bigg) p^{14} + \Bigg(8.722222222 \left(\frac{d}{dx} f_5(x) \right) f_7(x)$$

$$+ 8.722222222 \left(\frac{d}{dx} f_6(x) \right) f_6(x) + 8.722222222 \left(\frac{d}{dx} f_7(x) \right) f_5(x) \Bigg) p^{13}$$

$$+ \Bigg(8.722222222 \left(\frac{d}{dx} f_5(x) \right) f_6(x) + 8.722222222 \left(\frac{d}{dx} f_7(x) \right) f_4(x)$$

$$+ 8.722222222 \left(\frac{d}{dx} f_4(x) \right) f_7(x) + 8.722222222 \left(\frac{d}{dx} f_6(x) \right) f_5(x) \Bigg) p^{12}$$

$$+ \Bigg(8.722222222 \left(\frac{d}{dx} f_7(x) \right) f_3(x) + 8.722222222 \left(\frac{d}{dx} f_4(x) \right) f_6(x)$$

$$+ 8.722222222 \left(\frac{d}{dx} f_3(x) \right) f_7(x) + 8.722222222 \left(\frac{d}{dx} f_6(x) \right) f_4(x)$$

$$+ 8.722222222 \left(\frac{d}{dx} f_5(x) \right) f_5(x) \Bigg) p^{11} + \Bigg(8.722222222 \left(\frac{d}{dx} f_6(x) \right) f_3(x)$$

$$+ 8.722222222 \left(\frac{d}{dx} f_5(x) \right) f_4(x) + 8.722222222 \left(\frac{d}{dx} f_4(x) \right) f_5(x)$$

$$+ 8.722222222 \left(\frac{d}{dx} f_3(x) \right) f_6(x) + 8.722222222 \left(\frac{d}{dx} f_2(x) \right) f_7(x)$$

$$+ 8.722222222 \left(\frac{d}{dx} f_7(x) \right) f_2(x) \Bigg) p^{10} + \Bigg(8.722222222 \left(\frac{d}{dx} f_2(x) \right) f_6(x)$$

$$+ 8.722222222 \left(\frac{d}{dx} f_6(x)\right) f_2(x) + 8.722222222 \left(\frac{d}{dx} f_3(x)\right) f_5(x)$$

$$+ 8.722222222 \left(\frac{d}{dx} f_1(x)\right) f_7(x) + 8.722222222 \left(\frac{d}{dx} f_4(x)\right) f_4(x)$$

$$+ 8.722222222 \left(\frac{d}{dx} f_5(x)\right) f_3(x) + 8.722222222 \left(\frac{d}{dx} f_7(x)\right) f_1(x) \Big) p^9$$

$$+ \Big(0.03043086420 \left(\frac{d}{dx} f_7(x)\right) + 8.722222222 \left(\frac{d}{dx} f_1(x)\right) f_6(x)$$

$$+ 8.722222222 \left(\frac{d}{dx} f_7(x)\right) f_0(x) + 8.722222222 \left(\frac{d}{dx} f_3(x)\right) f_4(x)$$

$$+ 8.722222222 \left(\frac{d}{dx} f_4(x)\right) f_3(x) + 8.722222222 \left(\frac{d}{dx} f_6(x)\right) f_1(x)$$

$$+ 8.722222222 \left(\frac{d}{dx} f_0(x)\right) f_7(x) + 8.722222222 \left(\frac{d}{dx} f_5(x)\right) f_2(x)$$

$$+ 8.722222222 \left(\frac{d}{dx} f_2(x)\right) f_5(x) \Big) p^8 + \Big(8.722222222 \left(\frac{d}{dx} f_0(x)\right) f_6(x) + \frac{d^3}{dx^3} f_7(x)$$

$$+ 8.722222222 \left(\frac{d}{dx} f_6(x)\right) f_0(x) + 0.03043086420 \left(\frac{d}{dx} f_6(x)\right)$$

$$+ 8.722222222 \left(\frac{d}{dx} f_1(x)\right) f_5(x) + 8.722222222 \left(\frac{d}{dx} f_2(x)\right) f_4(x)$$

$$+ 8.722222222 \left(\frac{d}{dx} f_3(x)\right) f_3(x) + 8.722222222 \left(\frac{d}{dx} f_4(x)\right) f_2(x)$$

$$+ 8.722222222 \left(\frac{d}{dx} f_5(x)\right) f_1(x) \Big) p^7 + \Big(8.722222222 \left(\frac{d}{dx} f_3(x)\right) f_2(x)$$

$$+ 8.722222222 \left(\frac{d}{dx} f_2(x)\right) f_3(x) + \frac{d^3}{dx^3} f_6(x) + 0.03043086420 \left(\frac{d}{dx} f_5(x)\right)$$

$$+ 8.722222222 \left(\frac{d}{dx} f_4(x)\right) f_1(x) + 8.722222222 \left(\frac{d}{dx} f_1(x)\right) f_4(x)$$

$$+ 8.722222222 \left(\frac{d}{dx} f_5(x)\right) f_0(x) + 8.722222222 \left(\frac{d}{dx} f_0(x)\right) f_5(x) \Big) p^6 + \Big(\frac{d^3}{dx^3} f_5(x)$$

$$+ 0.03043086420 \left(\frac{d}{dx} f_4(x)\right) + 8.722222222 \left(\frac{d}{dx} f_4(x)\right) f_0(x)$$

$$+ 8.722222222 \left(\frac{d}{dx} f_3(x)\right) f_1(x) + 8.722222222 \left(\frac{d}{dx} f_1(x)\right) f_3(x)$$

$$+ 8.722222222 \left(\frac{d}{dx} f_2(x)\right) f_2(x) + 8.722222222 \left(\frac{d}{dx} f_0(x)\right) f_4(x) \Big) p^5 + \Big(\frac{d^3}{dx^3} f_4(x)$$

$$+ 8.722222222 \left(\frac{d}{dx} f_0(x)\right) f_3(x) + 0.03043086420 \left(\frac{d}{dx} f_3(x)\right)$$

$$+ 8.722222222 \left(\frac{d}{dx} f_3(x)\right) f_0(x) + 8.722222222 \left(\frac{d}{dx} f_2(x)\right) f_1(x)$$

$$+ 8.722222222 \left(\frac{d}{dx} f_1(x)\right) f_2(x) \Big) p^4 + \Big(\frac{d^3}{dx^3} f_3(x) + 8.722222222 \left(\frac{d}{dx} f_2(x)\right) f_0(x)$$

$$+ 8.722222222 \left(\frac{d}{dx} f_0(x)\right) f_2(x) + 8.722222222 \left(\frac{d}{dx} f_1(x)\right) f_1(x)$$

$$+ 0.03043086420 \left(\frac{d}{dx} f_2(x) \right) \Big) p^3 + \Big(8.722222222 \left(\frac{d}{dx} f_0(x) \right) f_1(x) + \frac{d^3}{dx^3} f_2(x)$$

$$+ 8.722222222 \left(\frac{d}{dx} f_1(x) \right) f_0(x) + 0.03043086420 \left(\frac{d}{dx} f_1(x) \right) \Big) p^2 + \Big(\frac{d^3}{dx^3} f_1(x)$$

$$+ 8.722222222 \left(\frac{d}{dx} f_0(x) \right) f_0(x) + 0.03043086420 \left(\frac{d}{dx} f_0(x) \right) \Big) p + \frac{d^3}{dx^3} f_0(x)$$

$$HO1 := 8.722222222\, p^{15} \left(\frac{d}{dx} f_7(x) \right) f_7(x) + \Big(8.722222222 \left(\frac{d}{dx} f_7(x) \right) f_6(x)$$

$$+ 8.722222222 \left(\frac{d}{dx} f_6(x) \right) f_7(x) \Big) p^{14} + \Big(8.722222222 \left(\frac{d}{dx} f_5(x) \right) f_7(x)$$

$$+ 8.722222222 \left(\frac{d}{dx} f_6(x) \right) f_6(x) + 8.722222222 \left(\frac{d}{dx} f_7(x) \right) f_5(x) \Big) p^{13}$$

$$+ \Big(8.722222222 \left(\frac{d}{dx} f_5(x) \right) f_6(x) + 8.722222222 \left(\frac{d}{dx} f_7(x) \right) f_4(x)$$

$$+ 8.722222222 \left(\frac{d}{dx} f_4(x) \right) f_7(x) + 8.722222222 \left(\frac{d}{dx} f_6(x) \right) f_5(x) \Big) p^{12}$$

$$+ \Big(8.722222222 \left(\frac{d}{dx} f_7(x) \right) f_3(x) + 8.722222222 \left(\frac{d}{dx} f_4(x) \right) f_6(x)$$

$$+ 8.722222222 \left(\frac{d}{dx} f_3(x) \right) f_7(x) + 8.722222222 \left(\frac{d}{dx} f_6(x) \right) f_4(x)$$

$$+ 8.722222222 \left(\frac{d}{dx} f_5(x) \right) f_5(x) \Big) p^{11} + \Big(8.722222222 \left(\frac{d}{dx} f_6(x) \right) f_3(x)$$

$$+ 8.722222222 \left(\frac{d}{dx} f_5(x) \right) f_4(x) + 8.722222222 \left(\frac{d}{dx} f_4(x) \right) f_5(x)$$

$$+ 8.722222222 \left(\frac{d}{dx} f_3(x) \right) f_6(x) + 8.722222222 \left(\frac{d}{dx} f_2(x) \right) f_7(x)$$

$$+ 8.722222222 \left(\frac{d}{dx} f_7(x) \right) f_2(x) \Big) p^{10} + \Big(8.722222222 \left(\frac{d}{dx} f_2(x) \right) f_6(x)$$

$$+ 8.722222222 \left(\frac{d}{dx} f_6(x) \right) f_2(x) + 8.722222222 \left(\frac{d}{dx} f_3(x) \right) f_5(x)$$

$$+ 8.722222222 \left(\frac{d}{dx} f_1(x) \right) f_7(x) + 8.722222222 \left(\frac{d}{dx} f_4(x) \right) f_4(x)$$

$$+ 8.722222222 \left(\frac{d}{dx} f_5(x) \right) f_3(x) + 8.722222222 \left(\frac{d}{dx} f_7(x) \right) f_1(x) \Big) p^9$$

$$+ \Big(0.03043086420 \left(\frac{d}{dx} f_7(x) \right) + 8.722222222 \left(\frac{d}{dx} f_1(x) \right) f_6(x)$$

$$+ 8.722222222 \left(\frac{d}{dx} f_7(x) \right) f_0(x) + 8.722222222 \left(\frac{d}{dx} f_3(x) \right) f_4(x)$$

$$+ 8.722222222 \left(\frac{d}{dx} f_4(x) \right) f_3(x) + 8.722222222 \left(\frac{d}{dx} f_6(x) \right) f_1(x)$$

$$+ 8.722222222 \left(\frac{d}{dx} f_0(x) \right) f_7(x) + 8.722222222 \left(\frac{d}{dx} f_5(x) \right) f_2(x)$$

$$+ 8.722222222 \left(\frac{d}{dx} f_2(x) \right) f_5(x) \Big) p^8 + \Big(8.722222222 \left(\frac{d}{dx} f_0(x) \right) f_6(x) + \frac{d^3}{dx^3} f_7(x)$$

$$+ 8.722222222 \left(\frac{d}{dx} f_6(x) \right) f_0(x) + 0.03043086420 \left(\frac{d}{dx} f_6(x) \right)$$

$$+ 8.722222222 \left(\frac{d}{dx} f_1(x) \right) f_5(x) + 8.722222222 \left(\frac{d}{dx} f_2(x) \right) f_4(x)$$

$$+ 8.722222222 \left(\frac{d}{dx} f_3(x) \right) f_3(x) + 8.722222222 \left(\frac{d}{dx} f_4(x) \right) f_2(x)$$

$$+8.722222222\left(\frac{d}{dx}f_5(x)\right)f_1(x)\Bigg)p^7+\Bigg(8.722222222\left(\frac{d}{dx}f_3(x)\right)f_2(x)$$

$$+8.722222222\left(\frac{d}{dx}f_2(x)\right)f_3(x)+\frac{d^3}{dx^3}f_6(x)+0.03043086420\left(\frac{d}{dx}f_5(x)\right)$$

$$+8.722222222\left(\frac{d}{dx}f_4(x)\right)f_1(x)+8.722222222\left(\frac{d}{dx}f_1(x)\right)f_4(x)$$

$$+8.722222222\left(\frac{d}{dx}f_5(x)\right)f_0(x)+8.722222222\left(\frac{d}{dx}f_0(x)\right)f_5(x)\Bigg)p^6+\Bigg(\frac{d^3}{dx^3}f_5(x)$$

$$+0.03043086420\left(\frac{d}{dx}f_4(x)\right)+8.722222222\left(\frac{d}{dx}f_4(x)\right)f_0(x)$$

$$+8.722222222\left(\frac{d}{dx}f_3(x)\right)f_1(x)+8.722222222\left(\frac{d}{dx}f_1(x)\right)f_3(x)$$

$$+8.722222222\left(\frac{d}{dx}f_2(x)\right)f_2(x)+8.722222222\left(\frac{d}{dx}f_0(x)\right)f_4(x)\Bigg)p^5+\Bigg(\frac{d^3}{dx^3}f_4(x)$$

$$+8.722222222\left(\frac{d}{dx}f_0(x)\right)f_3(x)+0.03043086420\left(\frac{d}{dx}f_3(x)\right)$$

$$+8.722222222\left(\frac{d}{dx}f_3(x)\right)f_0(x)+8.722222222\left(\frac{d}{dx}f_2(x)\right)f_1(x)$$

$$+8.722222222\left(\frac{d}{dx}f_1(x)\right)f_2(x)\Bigg)p^4+\Bigg(\frac{d^3}{dx^3}f_3(x)+8.722222222\left(\frac{d}{dx}f_2(x)\right)f_0(x)$$

$$+8.722222222\left(\frac{d}{dx}f_0(x)\right)f_2(x)+8.722222222\left(\frac{d}{dx}f_1(x)\right)f_1(x)$$

$$+0.03043086420\left(\frac{d}{dx}f_2(x)\right)\Bigg)p^3+\Bigg(8.722222222\left(\frac{d}{dx}f_0(x)\right)f_1(x)+\frac{d^3}{dx^3}f_2(x)$$

$$+8.722222222\left(\frac{d}{dx}f_1(x)\right)f_0(x)+0.03043086420\left(\frac{d}{dx}f_1(x)\right)\Bigg)p^2+\Bigg(\frac{d^3}{dx^3}f_1(x)$$

$$+8.722222222\left(\frac{d}{dx}f_0(x)\right)f_0(x)+0.03043086420\left(\frac{d}{dx}f_0(x)\right)\Bigg)p+\frac{d^3}{dx^3}f_0(x)$$

derivatives with respect to x of functions of one variable will now be displayed with ' **(3)**

```
> for i from 0 to L+1 do equa[1][i] := coeff(HO1, p, i) = 0 end
  do;
```

$$equa_{1_0} := f_0''' = 0$$

$$equa_{1_1} := f_1''' + 8.722222222\, f_0''\, f_0(x) + 0.03043086420\, f_0'' = 0$$

$$equa_{1_2} := 8.722222222\, f_0''\, f_1(x) + f_2''' + 8.722222222\, f_1''\, f_0(x) + 0.03043086420\, f_1'' = 0$$

$$equa_{1_3} := f_3''' + 8.722222222\, f_2''\, f_0(x) + 8.722222222\, f_0''\, f_2(x) + 8.722222222\, f_1''\, f_1(x)$$
$$+ 0.03043086420\, f_2'' = 0$$

$$equa_{1_4} := f_4''' + 8.722222222\, f_0''\, f_3(x) + 0.03043086420\, f_3'' + 8.722222222\, f_3''\, f_0(x)$$
$$+ 8.722222222\, f_2''\, f_1(x) + 8.722222222\, f_1''\, f_2(x) = 0$$

$$equa_{1_5} := f_5''' + 0.03043086420\, f_4'' + 8.722222222\, f_4''\, f_0(x) + 8.722222222\, f_3''\, f_1(x)$$
$$+ 8.722222222\, f_1''\, f_3(x) + 8.722222222\, f_2''\, f_2(x) + 8.722222222\, f_0''\, f_4(x) = 0$$

$$equa_{1_6} := 8.722222222\, f_3''\, f_2(x) + 8.722222222\, f_2''\, f_3(x) + f_6''' + 0.03043086420\, f_5'' + 8.722222222\, f_4''\, f_1(x) + 8.722222222\, f_1''\, f_4(x) + 8.722222222\, f_5''\, f_0(x) + 8.722222222\, f_0''\, f_5(x) = 0$$

$$equa_{1_7} := 8.722222222\, f_0''\, f_6(x) + f_7''' + 8.722222222\, f_6''\, f_0(x) + 0.03043086420\, f_6'' + 8.722222222\, f_1''\, f_5(x) + 8.722222222\, f_2''\, f_4(x) + 8.722222222\, f_3''\, f_3(x) + 8.722222222\, f_4''\, f_2(x) + 8.722222222\, f_5''\, f_1(x) = 0$$

$$equa_{1_8} := 0.03043086420\, f_7'' + 8.722222222\, f_1''\, f_6(x) + 8.722222222\, f_7''\, f_0(x) + 8.722222222\, f_3''\, f_4(x) + 8.722222222\, f_4''\, f_3(x) + 8.722222222\, f_6''\, f_1(x) + 8.722222222\, f_0''\, f_7(x) + 8.722222222\, f_5''\, f_2(x) + 8.722222222\, f_2''\, f_5(x) = 0 \quad \textbf{(4)}$$

```
> con[1][0]:=f[0](0) = 1, (D(f[0]))(0) = 0,f[0](1) = 0:
  for j from 1 to L do:
  con[1][j]:=f[j](0) = 0, (D(f[j]))(0) = 0,f[j](1) = 0:
  end do:

  for i from 0 to L do;
  dsolve({equa[1][i],con[1][i]},f[i](x)):
  f[i](x):=rhs(%):
  f[i](x):=evalf(%);
  end do:

  for u from 0 to L-1 do:
  f[u](_z1):=subs(x=_z1,f[u](x));
  f[u+1](x):=value(simplify(f[u+1](x)));
  f[u+1](x):=simplify(%);
  end do:

  f(x):=evalf(simplify(sum(f[n](x),n=0..L)));
```

$$f(x) := 1.290392648\, x^4 - 0.07075724583\, x^{14} - 0.8316430126\, x^6 - 0.01180590830\, x^{18} + 0.4859809463\, x^8 - 0.2703818500\, x^{10} + 0.03130321695\, x^{16} + 0.000001159639883\, x^{28} + 0.003611710909\, x^{20} - 0.0008522399964\, x^{22} - 3.495198750\, 10^{-8}\, x^{30} - 0.00001693928892\, x^{26} + 1. - 1.769189719\, x^2 + 0.0001464245398\, x^{24} + 0.1432108435\, x^{12} \quad \textbf{(5)}$$

A4. MAPLE CODE FOR GOHAM

```
> restart:
  with(PDEtools):
> k:=2:L:=2:
> H:=0;alpha:=5*3.14/180;RE:=50;
```

$$H := 0$$

$$\alpha := 0.08722222222$$

$$RE := 50 \quad (1)$$

```
> f(x):=sum((p^i)*f[i](x),i=0..L);
  A:=sum((p^i)*c[i],i=1..k);
```

$$f(x) := f_0(x) + p\, f_1(x) + p^2 f_2(x)$$

$$A := p\, c_1 + p^2 c_2 \quad (2)$$

```
> HO1 := (1-p)*(diff(f(x), x, x, x))+A*(diff(f(x), x, x, x)+2*
  alpha*RE*(diff(f(x), x))*f(x)+(4-H)*alpha^2*(diff(f(x), x)));
  expand(%);
  collect(%, p);
  HO1 := %;
  declare(f(x), prime = x);
```

$$HO1 := (1-p)\left(\frac{d^3}{dx^3} f_0(x) + p\left(\frac{d^3}{dx^3} f_1(x)\right) + p^2\left(\frac{d^3}{dx^3} f_2(x)\right)\right) + \left(p\, c_1 + p^2 c_2\right)\left(\frac{d^3}{dx^3} f_0(x) + p\left(\frac{d^3}{dx^3} f_1(x)\right) + p^2\left(\frac{d^3}{dx^3} f_2(x)\right) + 8.722222222\left(\frac{d}{dx} f_0(x) + p\left(\frac{d}{dx} f_1(x)\right) + p^2\left(\frac{d}{dx} f_2(x)\right)\right)\left(f_0(x) + p\, f_1(x) + p^2 f_2(x)\right) + 0.03043086420\left(\frac{d}{dx} f_0(x)\right) + 0.03043086420\, p\left(\frac{d}{dx} f_1(x)\right) + 0.03043086420\, p^2\left(\frac{d}{dx} f_2(x)\right)\right)$$

$$8.722222222\, p\, c_1\left(\frac{d}{dx} f_0(x)\right) f_0(x) + 8.722222222\, p^2 c_1\left(\frac{d}{dx} f_0(x)\right) f_1(x)$$
$$+ 8.722222222\, p^3 c_1\left(\frac{d}{dx} f_0(x)\right) f_2(x) + 8.722222222\, p^2 c_1\left(\frac{d}{dx} f_1(x)\right) f_0(x)$$
$$+ 8.722222222\, p^3 c_1\left(\frac{d}{dx} f_1(x)\right) f_1(x) + 8.722222222\, p^4 c_1\left(\frac{d}{dx} f_1(x)\right) f_2(x)$$
$$+ 8.722222222\, p^3 c_1\left(\frac{d}{dx} f_2(x)\right) f_0(x) + 8.722222222\, p^4 c_1\left(\frac{d}{dx} f_2(x)\right) f_1(x)$$
$$+ 8.722222222\, p^5 c_1\left(\frac{d}{dx} f_2(x)\right) f_2(x) + 8.722222222\, p^2 c_2\left(\frac{d}{dx} f_0(x)\right) f_0(x)$$
$$+ 8.722222222\, p^3 c_2\left(\frac{d}{dx} f_0(x)\right) f_1(x) + 8.722222222\, p^4 c_2\left(\frac{d}{dx} f_0(x)\right) f_2(x)$$
$$+ 8.722222222\, p^3 c_2\left(\frac{d}{dx} f_1(x)\right) f_0(x) + 8.722222222\, p^4 c_2\left(\frac{d}{dx} f_1(x)\right) f_1(x)$$

$$+ 8.722222222\, p^5 c_2 \left(\frac{d}{dx} f_1(x)\right) f_2(x) + 8.722222222\, p^4 c_2 \left(\frac{d}{dx} f_2(x)\right) f_0(x)$$

$$+ 8.722222222\, p^5 c_2 \left(\frac{d}{dx} f_2(x)\right) f_1(x) + 8.722222222\, p^6 c_2 \left(\frac{d}{dx} f_2(x)\right) f_2(x)$$

$$+ p^4 c_2 \left(\frac{d^3}{dx^3} f_2(x)\right) + p^2 \left(\frac{d^3}{dx^3} f_2(x)\right) + p \left(\frac{d^3}{dx^3} f_1(x)\right) + \frac{d^3}{dx^3} f_0(x)$$

$$+ p^2 c_2 \left(\frac{d^3}{dx^3} f_0(x)\right) + p\, c_1 \left(\frac{d^3}{dx^3} f_0(x)\right) + p^3 c_1 \left(\frac{d^3}{dx^3} f_2(x)\right) + p^2 c_1 \left(\frac{d^3}{dx^3} f_1(x)\right)$$

$$+ p^3 c_2 \left(\frac{d^3}{dx^3} f_1(x)\right) + 0.03043086420\, p\, c_1 \left(\frac{d}{dx} f_0(x)\right)$$

$$+ 0.03043086420\, p^2 c_1 \left(\frac{d}{dx} f_1(x)\right) + 0.03043086420\, p^3 c_1 \left(\frac{d}{dx} f_2(x)\right)$$

$$+ 0.03043086420\, p^2 c_2 \left(\frac{d}{dx} f_0(x)\right) + 0.03043086420\, p^3 c_2 \left(\frac{d}{dx} f_1(x)\right)$$

$$+ 0.03043086420\, p^4 c_2 \left(\frac{d}{dx} f_2(x)\right) - p \left(\frac{d^3}{dx^3} f_0(x)\right) - p^2 \left(\frac{d^3}{dx^3} f_1(x)\right)$$

$$- p^3 \left(\frac{d^3}{dx^3} f_2(x)\right)$$

$$8.722222222\, p^6 c_2 \left(\frac{d}{dx} f_2(x)\right) f_2(x) + \Big(8.722222222\, c_1 \left(\frac{d}{dx} f_2(x)\right) f_2(x)$$

$$+ 8.722222222\, c_2 \left(\frac{d}{dx} f_2(x)\right) f_1(x) + 8.722222222\, c_2 \left(\frac{d}{dx} f_1(x)\right) f_2(x)\Big) p^5$$

$$+ \Big(8.722222222\, c_2 \left(\frac{d}{dx} f_2(x)\right) f_0(x) + 8.722222222\, c_1 \left(\frac{d}{dx} f_2(x)\right) f_1(x)$$

$$+ c_2 \left(\frac{d^3}{dx^3} f_2(x)\right) + 8.722222222\, c_2 \left(\frac{d}{dx} f_0(x)\right) f_2(x)$$

$$+ 8.722222222\, c_1 \left(\frac{d}{dx} f_1(x)\right) f_2(x) + 0.03043086420\, c_2 \left(\frac{d}{dx} f_2(x)\right)$$

$$+ 8.722222222\, c_2 \left(\frac{d}{dx} f_1(x)\right) f_1(x)\Big) p^4 + \Big(8.722222222\, c_1 \left(\frac{d}{dx} f_2(x)\right) f_0(x)$$

$$+ c_2 \left(\frac{d^3}{dx^3} f_1(x)\right) + 8.722222222\, c_2 \left(\frac{d}{dx} f_0(x)\right) f_1(x)$$

$$+ 8.722222222\, c_2 \left(\frac{d}{dx} f_1(x)\right) f_0(x) + 8.722222222\, c_1 \left(\frac{d}{dx} f_0(x)\right) f_2(x)$$

$$- 1. \left(\frac{d^3}{dx^3} f_2(x)\right) + 0.03043086420\, c_2 \left(\frac{d}{dx} f_1(x)\right) + 0.03043086420\, c_1 \left(\frac{d}{dx} f_2(x)\right)$$

$$+ c_1 \left(\frac{d^3}{dx^3} f_2(x)\right) + 8.722222222\, c_1 \left(\frac{d}{dx} f_1(x)\right) f_1(x)\Big) p^3 + \Big(c_1 \left(\frac{d^3}{dx^3} f_1(x)\right)$$

$$+ c_2 \left(\frac{d^3}{dx^3} f_0(x)\right) - 1. \left(\frac{d^3}{dx^3} f_1(x)\right) + 8.722222222\, c_1 \left(\frac{d}{dx} f_0(x)\right) f_1(x)$$

$$+0.03043086420\, c_1 \left(\frac{d}{dx} f_1(x)\right) + 8.722222222\, c_1 \left(\frac{d}{dx} f_1(x)\right) f_0(x)$$
$$+8.722222222\, c_2 \left(\frac{d}{dx} f_0(x)\right) f_0(x) + \frac{d^3}{dx^3} f_2(x) + 0.03043086420\, c_2 \left(\frac{d}{dx} f_0(x)\right)\Bigg) p^2$$
$$+\left(c_1 \left(\frac{d^3}{dx^3} f_0(x)\right) + \frac{d^3}{dx^3} f_1(x) - 1.\left(\frac{d^3}{dx^3} f_0(x)\right) + 0.03043086420\, c_1 \left(\frac{d}{dx} f_0(x)\right)\right.$$
$$\left.+8.722222222\, c_1 \left(\frac{d}{dx} f_0(x)\right) f_0(x)\right) p + \frac{d^3}{dx^3} f_0(x)$$

$$HO1 := 8.722222222\, p^6 c_2 \left(\frac{d}{dx} f_2(x)\right) f_2(x) + \Bigg(8.722222222\, c_1 \left(\frac{d}{dx} f_2(x)\right) f_2(x)$$
$$+8.722222222\, c_2 \left(\frac{d}{dx} f_2(x)\right) f_1(x) + 8.722222222\, c_2 \left(\frac{d}{dx} f_1(x)\right) f_2(x)\Bigg) p^5$$
$$+\Bigg(8.722222222\, c_2 \left(\frac{d}{dx} f_2(x)\right) f_0(x) + 8.722222222\, c_1 \left(\frac{d}{dx} f_2(x)\right) f_1(x)$$
$$+c_2 \left(\frac{d^3}{dx^3} f_2(x)\right) + 8.722222222\, c_2 \left(\frac{d}{dx} f_0(x)\right) f_2(x)$$
$$+8.722222222\, c_1 \left(\frac{d}{dx} f_1(x)\right) f_2(x) + 0.03043086420\, c_2 \left(\frac{d}{dx} f_2(x)\right)$$
$$+8.722222222\, c_2 \left(\frac{d}{dx} f_1(x)\right) f_1(x)\Bigg) p^4 + \Bigg(8.722222222\, c_1 \left(\frac{d}{dx} f_2(x)\right) f_0(x)$$
$$+c_2 \left(\frac{d^3}{dx^3} f_1(x)\right) + 8.722222222\, c_2 \left(\frac{d}{dx} f_0(x)\right) f_1(x)$$
$$+8.722222222\, c_2 \left(\frac{d}{dx} f_1(x)\right) f_0(x) + 8.722222222\, c_1 \left(\frac{d}{dx} f_0(x)\right) f_2(x)$$
$$-1.\left(\frac{d^3}{dx^3} f_2(x)\right) + 0.03043086420\, c_2 \left(\frac{d}{dx} f_1(x)\right) + 0.03043086420\, c_1 \left(\frac{d}{dx} f_2(x)\right)$$
$$+c_1 \left(\frac{d^3}{dx^3} f_2(x)\right) + 8.722222222\, c_1 \left(\frac{d}{dx} f_1(x)\right) f_1(x)\Bigg) p^3 + \Bigg(c_1 \left(\frac{d^3}{dx^3} f_1(x)\right)$$
$$+c_2 \left(\frac{d^3}{dx^3} f_0(x)\right) - 1.\left(\frac{d^3}{dx^3} f_1(x)\right) + 8.722222222\, c_1 \left(\frac{d}{dx} f_0(x)\right) f_1(x)$$
$$+0.03043086420\, c_1 \left(\frac{d}{dx} f_1(x)\right) + 8.722222222\, c_1 \left(\frac{d}{dx} f_1(x)\right) f_0(x)$$
$$+8.722222222\, c_2 \left(\frac{d}{dx} f_0(x)\right) f_0(x) + \frac{d^3}{dx^3} f_2(x) + 0.03043086420\, c_2 \left(\frac{d}{dx} f_0(x)\right)\Bigg) p^2$$
$$+\left(c_1 \left(\frac{d^3}{dx^3} f_0(x)\right) + \frac{d^3}{dx^3} f_1(x) - 1.\left(\frac{d^3}{dx^3} f_0(x)\right) + 0.03043086420\, c_1 \left(\frac{d}{dx} f_0(x)\right)\right.$$
$$\left.+8.722222222\, c_1 \left(\frac{d}{dx} f_0(x)\right) f_0(x)\right) p + \frac{d^3}{dx^3} f_0(x)$$

derivatives with respect to x of functions of one variable will now be displayed with ' **(3)**

```
> for i from 0 to L+1 do equa[1][i] := coeff(HO1, p, i) = 0 end
  do;
```

$$equa_{1_0} := f_0''' = 0$$

$$equa_{1_1} := c_1\, f_0''' + f_1''' - 1.\, f_0''' + 0.03043086420\, c_1\, f_0'' + 8.722222222\, c_1\, f_0''\, f_0(x) = 0$$

$$equa_{1_2} := c_1\, f_1''' + c_2\, f_0''' - 1.\, f_1''' + 8.722222222\, c_1\, f_0''\, f_1(x) + 0.03043086420\, c_1\, f_1'' + 8.722222222\, c_1\, f_1''\, f_0(x) + 8.722222222\, c_2\, f_0''\, f_0(x) + f_2''' + 0.03043086420\, c_2\, f_0'' = 0$$

$$equa_{1_3} := 8.722222222\, c_1\, f_2''\, f_0(x) + c_2\, f_1''' + 8.722222222\, c_2\, f_0''\, f_1(x) + 8.722222222\, c_2\, f_1''\, f_0(x) + 8.722222222\, c_1\, f_0''\, f_2(x) - 1.\, f_2''' + 0.03043086420\, c_2\, f_1'' + 0.03043086420\, c_1\, f_2'' + c_1\, f_2''' + 8.722222222\, c_1\, f_1''\, f_1(x) = 0 \quad \mathbf{(4)}$$

```
> con[1][0]:=f[0](0) = 1, (D(f[0]))(0) = 0,f[0](1) = 0:
  for j from 1 to L do:
  con[1][j]:=f[j](0) = 0, (D(f[j]))(0) = 0,f[j](1) = 0:
  end do:

  for i from 0 to L do;
  dsolve({equa[1][i],con[1][i]},f[i](x)):
  f[i](x):=rhs(%):
  f[i](x):=evalf(%);
  end do:
  for u from 0 to L-1 do:
  f[u](_z1):=subs(x=_z1,f[u](x));
  f[u+1](x):=value(simplify(f[u+1](x)));
  f[u+1](x):=simplify(%);
  end do:

  f(x):=evalf(simplify(sum(f[n](x),n=0..L)));
```

$$f(x) := -1.\, x^2 + 1. - 0.2907407407\, c_1\, x^6 + 1.458775515\, c_1\, x^4 - 1.168034774\, c_1\, x^2 - 0.01408836305\, c_1^2\, x^{10} + 0.1363260451\, c_1^2\, x^8 - 0.2372298773\, x^6\, c_1^2 - 0.1453703703\, x^6\, c_2 - 0.3034126257\, x^4\, c_1^2 + 0.7293877571\, x^4\, c_2 + 0.4184048209\, x^2\, c_1^2 - 0.5840173865\, x^2\, c_2 \quad \mathbf{(5)}$$

```
> R:=diff(f(x), x, x, x)+2*alpha*RE*(diff(f(x), x))*f(x)+(4-H)*
  alpha^2*(diff(f(x), x));
  R2:=R^2:
  j:=int(R2,x=0.1..0.5):


  e1:=diff(j,c[1])=0:
  e2:=diff(j,c[2])=0:
  q:={e1,e2}:
  q:=evalf(solve(q));
```

$$R := -34.71132168\, c_1\, x^3 + 34.93952374\, c_1\, x - 10.11043324\, c_1^2\, x^7 + 45.76223650\, c_1^2\, x^5 - 28.50451771\, c_1^2\, x^3 - 17.35566084\, c_2\, x^3 - 7.256438177\, c_1^2\, x + 17.46976186\, c_2\, x + 8.722222222\, \big(-2.\, x - 1.744444444\, c_1\, x^5 + 5.835102060\, c_1\, x^3 - 2.336069548\, c_1\, x - 0.1408836305\, c_1^2\, x^9 + 1.090608361\, c_1^2\, x^7 - 1.423379264\, c_1^2\, x^5 - 0.8722222218\, x^5\, c_2$$

$$- 1.213650503\, c_1^2 x^3 + 2.917551028\, c_2 x^3 + 0.8368096418\, c_1^2 x - 1.168034773\, c_2 x \big) \big(-1.\, x^2 + 1. - 0.2907407407\, c_1 x^6 + 1.458775515\, c_1 x^4 - 1.168034774\, c_1 x^2 - 0.01408836305\, c_1^2 x^{10} + 0.1363260451\, c_1^2 x^8 - 0.2372298773\, x^6 c_1^2 - 0.1453703703\, x^6 c_2 - 0.3034126257\, x^4 c_1^2 + 0.7293877571\, x^4 c_2 + 0.4184048209\, x^2 c_1^2 - 0.5840173865\, x^2 c_2 \big) - 0.06086172840\, x - 0.05308495198\, c_1 x^5 - 0.004287210628\, c_1^2 x^9 - 0.02654247598\, x^5 c_2$$

$$q := \{c_1 = 1.196705929, c_2 = -0.09681806333\}, \{c_1 = 4.598568575, c_2 = -14.43011732\}, \{c_1 = 5.162837978, c_2 = 0.7197425540\}, \{c_1 = 43.29519980 + 49.87903556\, I, c_2 = -590.7904959 + 3317.545520\, I\}, \{c_1 = 8.714059057 + 19.94893112\, I, c_2 = -333.8966638 + 259.4485736\, I\}, \{c_1 = 14.15469894 + 46.23787729\, I, c_2 = -1570.575613 + 1024.541612\, I\}, \{c_1 = -14.15469909 + 46.23787754\, I, c_2 = -1513.956832 - 1209.493140\, I\}, \{c_1 = -8.714059124 + 19.94893115\, I, c_2 = -299.0404275 - 339.2443009\, I\}, \{c_1 = -43.29519974 + 49.87903582\, I, c_2 = -417.6097220 - 3517.061676\, I\}, \{c_1 = -4.075657643\; 10^{-10}, c_2 = -12.60598898\}, \{c_1 = -1.921121623\; 10^{-8}, c_2 = 1.330102691\}, \{c_1 = -2.294461235\; 10^{-8}, c_2 = -25.14396918\}, \{c_1 = -1.196705923, c_2 = 4.690005635\}, \{c_1 = -4.598568587, c_2 = 3.964157131\}, \{c_1 = -5.162837994, c_2 = 21.37109466\}, \{c_1 = -43.29519974 - 49.87903582\, I, c_2 = -417.6097220 + 3517.061676\, I\}, \{c_1 = -8.714059124 - 19.94893115\, I, c_2 = -299.0404275 + 339.2443009\, I\}, \{c_1 = -14.15469909 - 46.23787754\, I, c_2 = -1513.956832 + 1209.493140\, I\}, \{c_1 = 14.15469894 - 46.23787729\, I, c_2 = -1570.575613 - 1024.541612\, I\}, \{c_1 = 8.714059057 - 19.94893112\, I, c_2 = -333.8966638 - 259.4485736\, I\}, \{c_1 = 43.29519980 - 49.87903556\, I, c_2 = -590.7904959 - 3317.545520\, I\} \quad (6)$$

```
> f(x):=subs(c[1] = -1.196705923, c[2] = 4.690005635,f(x));
```

$$f(x) := -1.742051038\, x^2 + 1. - 0.6735947985\, x^6 + 1.240588634\, x^4 - 0.02017601610\, x^{10} + 0.1952332198\, x^8 \quad (7)$$

A5. MAPLE CODE FOR DTM

```
> restart;
> F[0]:=1;
  F[1]:=0;
  F[2]:=beta;
```

$$F_0 := 1$$
$$F_1 := 0$$
$$F_2 := \beta \qquad (1)$$

```
> n:=20;
  for k from 0 to n do
  F[k+3]:=solve((k+1)*(k+2)*(k+3)*F[k+3]+2*alpha*RE*sum((k-m+1)*F
  [k-m+1]*F[m],m=0..k)+(4-H)*(alpha)^2*(k+1)*F[k+1],F[k+3]);  end
  do;
```

$$n := 20$$
$$F_3 := 0$$
$$F_4 := -\frac{1}{6}\alpha RE\beta - \frac{1}{3}\alpha^2\beta + \frac{1}{12}\alpha^2\beta H$$
$$F_5 := 0$$
$$F_6 := \frac{1}{90}\alpha^2 RE^2\beta + \frac{2}{45}\alpha^3 RE\beta - \frac{1}{90}\alpha^3 RE\beta H - \frac{1}{30}\alpha RE\beta^2 + \frac{2}{45}\alpha^4\beta - \frac{1}{45}\alpha^4\beta H + \frac{1}{360}\alpha^4\beta H^2$$
$$F_7 := 0$$
$$F_8 := -\frac{1}{2520}\alpha^3 RE^3\beta - \frac{1}{420}\alpha^4 RE^2\beta + \frac{1}{1680}\alpha^4 RE^2\beta H + \frac{1}{140}\alpha^2 RE^2\beta^2 - \frac{1}{210}\alpha^5 RE\beta + \frac{1}{420}\alpha^5 RE\beta H - \frac{1}{3360}\alpha^5 RE\beta H^2 + \frac{1}{70}\alpha^3 RE\beta^2 - \frac{1}{280}\alpha^3 RE\beta^2 H - \frac{1}{315}\alpha^6\beta + \frac{1}{420}\alpha^6\beta H - \frac{1}{1680}\alpha^6\beta H^2 + \frac{1}{20160}\alpha^6\beta H^3$$
$$F_9 := 0$$
$$F_{10} := \frac{1}{113400}\alpha^4 RE^4\beta + \frac{4}{14175}\alpha^7 RE\beta + \frac{1}{700}\alpha^5 RE\beta^2 H - \frac{1}{56700}\alpha^5 RE^3\beta H - \frac{1}{9450}\alpha^6 RE^2\beta H + \frac{1}{75600}\alpha^6 RE^2\beta H^2 + \frac{1}{1400}\alpha^4 RE^2\beta^2 H - \frac{1}{4725}\alpha^7 RE\beta H + \frac{1}{18900}\alpha^7 RE\beta H^2 - \frac{1}{226800}\alpha^7 RE\beta H^3 - \frac{1}{5600}\alpha^5 RE\beta^2 H^2 + \frac{1}{1350}\alpha^2 RE^2\beta^3 + \frac{2}{14175}\alpha^8\beta + \frac{1}{14175}\alpha^5 RE^3\beta - \frac{1}{1400}\alpha^3 RE^3\beta^2 + \frac{1}{4725}\alpha^6 RE^2\beta - \frac{1}{350}\alpha^4 RE^2\beta^2 - \frac{1}{350}\alpha^5 RE\beta^2 - \frac{2}{14175}\alpha^8\beta H + \frac{1}{18900}\alpha^8\beta H^2 - \frac{1}{113400}\alpha^8\beta H^3 + \frac{1}{1814400}\alpha^8\beta H^4$$
$$F_{11} := 0$$
$$F_{12} := \frac{1}{1496880}\alpha^{10}\beta H^3 + \frac{1}{187110}\alpha^{10}\beta H + \frac{2}{7425}\alpha^5 RE^3\beta^2 - \frac{1}{187110}\alpha^7 RE^3\beta + \frac{1}{2993760}\alpha^6 RE^4\beta H - \frac{2}{7425}\alpha^6 RE^2\beta^2 H + \frac{127}{1247400}\alpha^4 RE^2\beta^3 H$$

$$
\begin{aligned}
&- \frac{1}{2993760}\alpha^7 RE^3\beta H^2 + \frac{1}{124740}\alpha^8 RE^2\beta H + \frac{8}{22275}\alpha^7 RE\beta^2 + \frac{4}{7425}\alpha^6 RE^2\beta^2 \\
&- \frac{1}{374220}\alpha^{10}\beta H^2 - \frac{1}{498960}\alpha^8 RE^2\beta H^2 + \frac{1}{5987520}\alpha^8 RE^2\beta H^3 \\
&- \frac{1}{14850}\alpha^5 RE^3\beta^2 H - \frac{1}{11975040}\alpha^{10}\beta H^4 + \frac{1}{374220}\alpha^7 RE^3\beta H - \frac{2}{467775}\alpha^{10}\beta \\
&+ \frac{1}{22275}\alpha^4 RE^4\beta^2 - \frac{1}{748440}\alpha^6 RE^4\beta - \frac{1}{93555}\alpha^9 RE\beta - \frac{127}{623700}\alpha^3 RE^3\beta^3 \\
&- \frac{1}{7484400}\alpha^5 RE^5\beta - \frac{1}{93555}\alpha^8 RE^2\beta - \frac{127}{311850}\alpha^4 RE^2\beta^3 + \frac{1}{239500800}\alpha^{10}\beta H^5 \\
&+ \frac{1}{29700}\alpha^6 RE^2\beta^2 H^2 + \frac{1}{93555}\alpha^9 RE\beta H - \frac{1}{249480}\alpha^9 RE\beta H^2 \\
&+ \frac{1}{1496880}\alpha^9 RE\beta H^3 - \frac{1}{23950080}\alpha^9 RE\beta H^4 - \frac{2}{7425}\alpha^7 RE\beta^2 H \\
&+ \frac{1}{14850}\alpha^7 RE\beta^2 H^2 - \frac{1}{178200}\alpha^7 RE\beta^2 H^3
\end{aligned}
$$

$$F_{13} := 0$$

$$
\begin{aligned}
F_{14} := &- \frac{1}{70200}\alpha^3 RE^3\beta^4 + \frac{1}{11351340}\alpha^8 RE^4\beta + \frac{1546}{14189175}\alpha^5 RE^3\beta^3 \\
&+ \frac{1546}{14189175}\alpha^6 RE^2\beta^3 + \frac{1}{56756700}\alpha^7 RE^5\beta + \frac{4}{42567525}\alpha^{12}\beta \\
&- \frac{151}{9459450}\alpha^6 RE^4\beta^2 - \frac{151}{3153150}\alpha^7 RE^3\beta^2 - \frac{302}{4729725}\alpha^8 RE^2\beta^2 \\
&- \frac{151}{75675600}\alpha^5 RE^5\beta^2 + \frac{1}{681080400}\alpha^6 RE^6\beta + \frac{4}{14189175}\alpha^{11} RE\beta \\
&+ \frac{773}{28378350}\alpha^4 RE^4\beta^3 + \frac{2}{8513505}\alpha^9 RE^3\beta + \frac{1}{2837835}\alpha^{10} RE^2\beta \\
&- \frac{151}{4729725}\alpha^9 RE\beta^2 - \frac{1}{22702680}\alpha^{11} RE\beta H^3 - \frac{1}{2837835}\alpha^{11} RE\beta H \\
&- \frac{1}{227026800}\alpha^7 RE^5\beta H + \frac{151}{6306300}\alpha^7 RE^3\beta^2 H - \frac{773}{28378350}\alpha^5 RE^3\beta^3 H \\
&+ \frac{1}{181621440}\alpha^8 RE^4\beta H^2 - \frac{1}{5675670}\alpha^9 RE^3\beta H + \frac{1}{5675670}\alpha^{11} RE\beta H^2 \\
&+ \frac{1}{22702680}\alpha^9 RE^3\beta H^2 - \frac{1}{272432160}\alpha^9 RE^3\beta H^3 + \frac{151}{37837800}\alpha^6 RE^4\beta^2 H \\
&+ \frac{1}{181621440}\alpha^{11} RE\beta H^4 - \frac{1}{22702680}\alpha^8 RE^4\beta H - \frac{1}{3632428800}\alpha^{11} RE\beta H^5 \\
&- \frac{151}{50450400}\alpha^7 RE^3\beta^2 H^2 - \frac{1}{2837835}\alpha^{10} RE^2\beta H + \frac{1}{7567560}\alpha^{10} RE^2\beta H^2 \\
&- \frac{1}{45405360}\alpha^{10} RE^2\beta H^3 + \frac{1}{726485760}\alpha^{10} RE^2\beta H^4 + \frac{151}{3153150}\alpha^8 RE^2\beta^2 H \\
&- \frac{151}{12612600}\alpha^8 RE^2\beta^2 H^2 + \frac{151}{151351200}\alpha^8 RE^2\beta^2 H^3 - \frac{773}{14189175}\alpha^6 RE^2\beta^3 H \\
&+ \frac{773}{113513400}\alpha^6 RE^2\beta^3 H^2 + \frac{151}{4729725}\alpha^9 RE\beta^2 H - \frac{151}{12612600}\alpha^9 RE\beta^2 H^2 \\
&+ \frac{151}{75675600}\alpha^9 RE\beta^2 H^3 - \frac{151}{1210809600}\alpha^9 RE\beta^2 H^4 + \frac{1}{43589145600}\alpha^{12}\beta H^6 \\
&- \frac{1}{1816214400}\alpha^{12}\beta H^5 + \frac{1}{181621440}\alpha^{12}\beta H^4 + \frac{1}{11351340}\alpha^{12}\beta H^2 \\
&- \frac{2}{14189175}\alpha^{12}\beta H - \frac{1}{34054020}\alpha^{12}\beta H^3
\end{aligned}
$$

$$F_{15} := 0$$

$$
\begin{aligned}
F_{16} := {} & \frac{29}{3003000}\,\alpha^{5}\,RE^{3}\,\beta^{4} + \frac{1}{31135104000}\,\alpha^{14}\,\beta\,H^{5} - \frac{8053}{425675250}\,\alpha^{8}\,RE^{2}\,\beta^{3} \\
& - \frac{8053}{3405402000}\,\alpha^{5}\,RE^{5}\,\beta^{3} + \frac{101}{37837800}\,\alpha^{8}\,RE^{4}\,\beta^{2} - \frac{1}{291891600}\,\alpha^{10}\,RE^{4}\,\beta \\
& + \frac{1}{10461394944000}\,\alpha^{14}\,\beta\,H^{7} + \frac{1}{364864500}\,\alpha^{14}\,\beta\,H + \frac{101}{151351200}\,\alpha^{7}\,RE^{5}\,\beta^{2} \\
& + \frac{29}{6006000}\,\alpha^{4}\,RE^{4}\,\beta^{4} + \frac{101}{18918900}\,\alpha^{10}\,RE^{2}\,\beta^{2} - \frac{8053}{567567000}\,\alpha^{6}\,RE^{4}\,\beta^{3} \\
& - \frac{1}{4670265600}\,\alpha^{14}\,\beta\,H^{4} - \frac{1}{121621500}\,\alpha^{12}\,RE^{2}\,\beta + \frac{101}{47297250}\,\alpha^{11}\,RE\,\beta^{2} \\
& - \frac{1}{182432250}\,\alpha^{13}\,RE\,\beta + \frac{1}{1167566400}\,\alpha^{14}\,\beta\,H^{3} - \frac{8053}{283783500}\,\alpha^{7}\,RE^{3}\,\beta^{3} \\
& - \frac{1}{5837832000}\,\alpha^{8}\,RE^{6}\,\beta - \frac{1}{972972000}\,\alpha^{9}\,RE^{5}\,\beta - \frac{1}{81729648000}\,\alpha^{7}\,RE^{7}\,\beta \\
& + \frac{101}{18918900}\,\alpha^{9}\,RE^{3}\,\beta^{2} - \frac{1}{373621248000}\,\alpha^{14}\,\beta\,H^{6} + \frac{101}{1513512000}\,\alpha^{6}\,RE^{6}\,\beta^{2} \\
& - \frac{1}{145945800}\,\alpha^{11}\,RE^{3}\,\beta - \frac{1}{486486000}\,\alpha^{14}\,\beta\,H^{2} - \frac{101}{48432384000}\,\alpha^{11}\,RE\,\beta^{2}\,H^{5} \\
& + \frac{101}{75675600}\,\alpha^{11}\,RE\,\beta^{2}\,H^{2} - \frac{101}{37837800}\,\alpha^{11}\,RE\,\beta^{2}\,H + \frac{1}{778377600}\,\alpha^{12}\,RE^{2}\,\beta\,H^{3} \\
& + \frac{1}{97297200}\,\alpha^{12}\,RE^{2}\,\beta\,H + \frac{1}{23351328000}\,\alpha^{8}\,RE^{6}\,\beta\,H - \frac{101}{75675600}\,\alpha^{8}\,RE^{4}\,\beta^{2}\,H \\
& + \frac{8053}{2270268000}\,\alpha^{6}\,RE^{4}\,\beta^{3}\,H - \frac{1}{15567552000}\,\alpha^{9}\,RE^{5}\,\beta\,H^{2} + \frac{1}{389188800}\,\alpha^{10}\,RE^{4}\,\beta\,H \\
& - \frac{1}{194594400}\,\alpha^{12}\,RE^{2}\,\beta\,H^{2} - \frac{1}{1556755200}\,\alpha^{10}\,RE^{4}\,\beta\,H^{2} + \frac{1}{18681062400}\,\alpha^{10}\,RE^{4}\,\beta\,H^{3} \\
& - \frac{101}{605404800}\,\alpha^{7}\,RE^{5}\,\beta^{2}\,H - \frac{1}{6227020800}\,\alpha^{12}\,RE^{2}\,\beta\,H^{4} + \frac{1}{1945944000}\,\alpha^{9}\,RE^{5}\,\beta\,H \\
& + \frac{1}{124540416000}\,\alpha^{12}\,RE^{2}\,\beta\,H^{5} + \frac{101}{605404800}\,\alpha^{8}\,RE^{4}\,\beta^{2}\,H^{2} + \frac{1}{145945800}\,\alpha^{11}\,RE^{3}\,\beta\,H \\
& - \frac{1}{389188800}\,\alpha^{11}\,RE^{3}\,\beta\,H^{2} + \frac{1}{2335132800}\,\alpha^{11}\,RE^{3}\,\beta\,H^{3} - \frac{1}{37362124800}\,\alpha^{11}\,RE^{3}\,\beta\,H^{4} \\
& - \frac{101}{25225200}\,\alpha^{9}\,RE^{3}\,\beta^{2}\,H + \frac{101}{100900800}\,\alpha^{9}\,RE^{3}\,\beta^{2}\,H^{2} - \frac{101}{1210809600}\,\alpha^{9}\,RE^{3}\,\beta^{2}\,H^{3} \\
& + \frac{8053}{567567000}\,\alpha^{7}\,RE^{3}\,\beta^{3}\,H - \frac{8053}{4540536000}\,\alpha^{7}\,RE^{3}\,\beta^{3}\,H^{2} - \frac{101}{18918900}\,\alpha^{10}\,RE^{2}\,\beta^{2}\,H \\
& + \frac{101}{50450400}\,\alpha^{10}\,RE^{2}\,\beta^{2}\,H^{2} - \frac{101}{302702400}\,\alpha^{10}\,RE^{2}\,\beta^{2}\,H^{3} + \frac{101}{4843238400}\,\alpha^{10}\,RE^{2}\,\beta^{2}\,H^{4} \\
& - \frac{101}{302702400}\,\alpha^{11}\,RE\,\beta^{2}\,H^{3} - \frac{1}{747242496000}\,\alpha^{13}\,RE\,\beta\,H^{6} \\
& + \frac{1}{31135104000}\,\alpha^{13}\,RE\,\beta\,H^{5} - \frac{1}{3113510400}\,\alpha^{13}\,RE\,\beta\,H^{4} - \frac{1}{194594400}\,\alpha^{13}\,RE\,\beta\,H^{2} \\
& + \frac{1}{121621500}\,\alpha^{13}\,RE\,\beta\,H + \frac{1}{583783200}\,\alpha^{13}\,RE\,\beta\,H^{3} - \frac{29}{12012000}\,\alpha^{5}\,RE^{3}\,\beta^{4}\,H \\
& + \frac{8053}{567567000}\,\alpha^{8}\,RE^{2}\,\beta^{3}\,H - \frac{8053}{2270268000}\,\alpha^{8}\,RE^{2}\,\beta^{3}\,H^{2} + \frac{8053}{27243216000}\,\alpha^{8}\,RE^{2}\,\beta^{3}\,H^{3} \\
& + \frac{101}{2421619200}\,\alpha^{11}\,RE\,\beta^{2}\,H^{4} - \frac{1}{638512875}\,\alpha^{14}\,\beta
\end{aligned}
$$

$$F_{17} := 0$$

$$
\begin{aligned}
F_{18} := &- \frac{7279}{4168212048000}\,\alpha^{7}\,RE^{7}\,\beta^{2} - \frac{7279}{21709437750}\,\alpha^{12}\,RE^{2}\,\beta^{2} \\
&+ \frac{1}{3201186852864000}\,\alpha^{16}\,\beta\,H^{8} + \frac{156017}{65128313250}\,\alpha^{10}\,RE^{2}\,\beta^{3} - \frac{3463}{1072071000}\,\alpha^{7}\,RE^{3}\,\beta^{4} \\
&+ \frac{156017}{173675502000}\,\alpha^{10}\,RE^{2}\,\beta^{3}\,H^{2} - \frac{156017}{1042053012000}\,\alpha^{10}\,RE^{2}\,\beta^{3}\,H^{3} \\
&- \frac{3463}{17153136000}\,\alpha^{7}\,RE^{3}\,\beta^{4}\,H^{2} + \frac{156017}{16672848192000}\,\alpha^{10}\,RE^{2}\,\beta^{3}\,H^{4} \\
&+ \frac{7279}{22230464256000}\,\alpha^{12}\,RE^{2}\,\beta^{2}\,H^{5} - \frac{7279}{34735100400}\,\alpha^{12}\,RE^{2}\,\beta^{2}\,H^{2} \\
&+ \frac{7279}{17367550200}\,\alpha^{12}\,RE^{2}\,\beta^{2}\,H - \frac{1}{44659414800}\,\alpha^{13}\,RE^{3}\,\beta\,H^{3} \\
&- \frac{1}{5582426850}\,\alpha^{13}\,RE^{3}\,\beta\,H - \frac{1}{3126159036000}\,\alpha^{9}\,RE^{7}\,\beta\,H \\
&+ \frac{7279}{138940401600}\,\alpha^{9}\,RE^{5}\,\beta^{2}\,H - \frac{156017}{521026506000}\,\alpha^{7}\,RE^{5}\,\beta^{3}\,H \\
&+ \frac{1}{1786376592000}\,\alpha^{10}\,RE^{6}\,\beta\,H^{2} - \frac{1}{37216179000}\,\alpha^{11}\,RE^{5}\,\beta\,H \\
&+ \frac{1}{11164853700}\,\alpha^{13}\,RE^{3}\,\beta\,H^{2} + \frac{1}{148864716000}\,\alpha^{11}\,RE^{5}\,\beta\,H^{2} \\
&- \frac{1}{1786376592000}\,\alpha^{11}\,RE^{5}\,\beta\,H^{3} + \frac{7279}{1389404016000}\,\alpha^{8}\,RE^{6}\,\beta^{2}\,H \\
&+ \frac{1}{357275318400}\,\alpha^{13}\,RE^{3}\,\beta\,H^{4} + \frac{1}{7145506368000}\,\alpha^{15}\,RE\,\beta\,H^{6} \\
&+ \frac{1}{9304044750}\,\alpha^{15}\,RE\,\beta\,H^{2} + \frac{3463}{2144142000}\,\alpha^{7}\,RE^{3}\,\beta^{4}\,H - \frac{156017}{65128313250}\,\alpha^{10}\,RE^{2}\,\beta^{3}\,H \\
&+ \frac{1}{781539759000}\,\alpha^{9}\,RE^{7}\,\beta + \frac{1}{178637659200}\,\alpha^{16}\,\beta\,H^{4} + \frac{1}{27912134250}\,\alpha^{11}\,RE^{5}\,\beta \\
&+ \frac{2}{13956067125}\,\alpha^{13}\,RE^{3}\,\beta + \frac{1}{11164853700}\,\alpha^{12}\,RE^{4}\,\beta - \frac{7279}{26051325300}\,\alpha^{10}\,RE^{4}\,\beta^{2} \\
&- \frac{4}{97692469875}\,\alpha^{16}\,\beta\,H - \frac{7279}{65128313250}\,\alpha^{13}\,RE\,\beta^{2} + \frac{156017}{130256626500}\,\alpha^{7}\,RE^{5}\,\beta^{3} \\
&+ \frac{1}{111648537000}\,\alpha^{10}\,RE^{6}\,\beta - \frac{7279}{69470200800}\,\alpha^{9}\,RE^{5}\,\beta^{2} + \frac{41}{161109000}\,\alpha^{4}\,RE^{4}\,\beta^{5} \\
&- \frac{7279}{347351004000}\,\alpha^{8}\,RE^{6}\,\beta^{2} + \frac{8}{97692469875}\,\alpha^{15}\,RE\,\beta - \frac{7279}{17367550200}\,\alpha^{11}\,RE^{3}\,\beta^{2} \\
&- \frac{1}{893188296000}\,\alpha^{16}\,\beta\,H^{5} - \frac{1}{55824268500}\,\alpha^{16}\,\beta\,H^{3} + \frac{156017}{1042053012000}\,\alpha^{6}\,RE^{6}\,\beta^{3} \\
&+ \frac{156017}{43418875500}\,\alpha^{8}\,RE^{4}\,\beta^{3} + \frac{156017}{32564156625}\,\alpha^{9}\,RE^{3}\,\beta^{3} - \frac{3463}{4288284000}\,\alpha^{5}\,RE^{5}\,\beta^{4} \\
&+ \frac{2}{13956067125}\,\alpha^{14}\,RE^{2}\,\beta - \frac{3463}{1072071000}\,\alpha^{6}\,RE^{4}\,\beta^{4} + \frac{1}{7145506368000}\,\alpha^{16}\,\beta\,H^{6} \\
&- \frac{1}{223297074000}\,\alpha^{10}\,RE^{6}\,\beta\,H - \frac{1}{7145506368000}\,\alpha^{13}\,RE^{3}\,\beta\,H^{5} \\
&- \frac{7279}{1111523212800}\,\alpha^{9}\,RE^{5}\,\beta^{2}\,H^{2} - \frac{1}{11164853700}\,\alpha^{12}\,RE^{4}\,\beta\,H \\
&+ \frac{1}{29772943200}\,\alpha^{12}\,RE^{4}\,\beta\,H^{2} - \frac{1}{178637659200}\,\alpha^{12}\,RE^{4}\,\beta\,H^{3} \\
&+ \frac{1}{2858202547200}\,\alpha^{12}\,RE^{4}\,\beta\,H^{4} + \frac{7279}{34735100400}\,\alpha^{10}\,RE^{4}\,\beta^{2}\,H
\end{aligned}
$$

$$
\begin{aligned}
&- \frac{7279}{138940401600} \alpha^{10} RE^4 \beta^2 H^2 + \frac{7279}{1667284819200} \alpha^{10} RE^4 \beta^2 H^3 \\
&- \frac{156017}{86837751000} \alpha^8 RE^4 \beta^3 H + \frac{156017}{694702008000} \alpha^8 RE^4 \beta^3 H^2 \\
&+ \frac{7279}{17367550200} \alpha^{11} RE^3 \beta^2 H - \frac{7279}{46313467200} \alpha^{11} RE^3 \beta^2 H^2 \\
&+ \frac{7279}{277880803200} \alpha^{11} RE^3 \beta^2 H^3 - \frac{7279}{4446092851200} \alpha^{11} RE^3 \beta^2 H^4 \\
&+ \frac{7279}{138940401600} \alpha^{12} RE^2 \beta^2 H^3 + \frac{1}{28582025472000} \alpha^{14} RE^2 \beta H^6 \\
&- \frac{1}{1190917728000} \alpha^{14} RE^2 \beta H^5 + \frac{1}{119091772800} \alpha^{14} RE^2 \beta H^4 \\
&+ \frac{1}{7443235800} \alpha^{14} RE^2 \beta H^2 - \frac{1}{4652022375} \alpha^{14} RE^2 \beta H - \frac{1}{22329707400} \alpha^{14} RE^2 \beta H^3 \\
&+ \frac{3463}{4288284000} \alpha^6 RE^4 \beta^4 H - \frac{156017}{43418875500} \alpha^9 RE^3 \beta^3 H \\
&+ \frac{156017}{173675502000} \alpha^9 RE^3 \beta^3 H^2 - \frac{156017}{2084106024000} \alpha^9 RE^3 \beta^3 H^3 \\
&- \frac{7279}{1111523212800} \alpha^{12} RE^2 \beta^2 H^4 + \frac{1}{12504636144000} \alpha^8 RE^8 \beta \\
&+ \frac{1}{27912134250} \alpha^{16} \beta H^2 - \frac{1}{595458864000} \alpha^{15} RE \beta H^5 \\
&- \frac{7279}{1111523212800} \alpha^{13} RE \beta^2 H^4 - \frac{1}{200074178304000} \alpha^{15} RE \beta H^7 \\
&- \frac{2}{13956067125} \alpha^{15} RE \beta H + \frac{1}{89318829600} \alpha^{15} RE \beta H^4 + \frac{7279}{43418875500} \alpha^{13} RE \beta^2 H \\
&- \frac{7279}{69470200800} \alpha^{13} RE \beta^2 H^2 - \frac{1}{22329707400} \alpha^{15} RE \beta H^3 \\
&+ \frac{7279}{11115232128000} \alpha^{13} RE \beta^2 H^5 + \frac{7279}{208410602400} \alpha^{13} RE \beta^2 H^3 \\
&- \frac{7279}{266765571072000} \alpha^{13} RE \beta^2 H^6 + \frac{2}{97692469875} \alpha^{16} \beta \\
&- \frac{1}{100037089152000} \alpha^{16} \beta H^7
\end{aligned}
$$

$$F_{19} := 0$$

$$
\begin{aligned}
F_{20} := &\frac{266309}{249987465000} \alpha^8 RE^4 \beta^4 + \frac{266309}{499974930000} \alpha^7 RE^5 \beta^4 \\
&- \frac{1}{337903056691200000} \alpha^{18} \beta H^8 - \frac{2}{1031198293125} \alpha^{16} RE^2 \beta \\
&- \frac{1}{1178512335000} \alpha^{13} RE^5 \beta + \frac{809}{49104680625} \alpha^{14} RE^2 \beta^2 - \frac{1}{589256167500} \alpha^{14} RE^4 \beta \\
&- \frac{1}{226274368320000} \alpha^{18} \beta H^6 + \frac{809}{21998896920000} \alpha^8 RE^8 \beta^2 \\
&- \frac{1457537}{197990072280000} \alpha^7 RE^7 \beta^3 + \frac{809}{78567489000} \alpha^{11} RE^5 \beta^2 \\
&- \frac{1}{131993381520000} \alpha^{10} RE^8 \beta + \frac{1}{1216451004088320000} \alpha^{18} \beta H^9 \\
&- \frac{1457537}{19799007228000} \alpha^8 RE^6 \beta^3 - \frac{1}{3535537005000} \alpha^{12} RE^6 \beta \\
&+ \frac{1}{37712394720000} \alpha^{18} \beta H^5 - \frac{34621}{336717810000} \alpha^5 RE^5 \beta^5 - \frac{34621}{168358905000} \alpha^6 RE^4 \beta^5
\end{aligned}
$$

$$- \frac{1}{1031198293125} \alpha^{17} RE \beta - \frac{1}{2062396586250} \alpha^{18} \beta H^2 - \frac{1457537}{6187189758750} \alpha^{12} RE^2 \beta^3$$
$$- \frac{1}{441942125625} \alpha^{15} RE^3 \beta - \frac{1}{2375880867360000} \alpha^{9} RE^9 \beta$$
$$+ \frac{1618}{343732764375} \alpha^{15} RE \beta^2 - \frac{1457537}{2474875903500} \alpha^{10} RE^4 \beta^3$$
$$+ \frac{1}{2111894104320000} \alpha^{18} \beta H^7 - \frac{1457537}{4949751807000} \alpha^{9} RE^5 \beta^3$$
$$+ \frac{1}{3535537005000} \alpha^{18} \beta H^3 + \frac{809}{32736453750} \alpha^{13} RE^3 \beta^2 - \frac{1}{9428098680000} \alpha^{18} \beta H^4$$
$$+ \frac{809}{1571349780000} \alpha^{9} RE^7 \beta^2 + \frac{1}{2062396586250} \alpha^{18} \beta H + \frac{809}{39283744500} \alpha^{12} RE^4 \beta^2$$
$$+ \frac{266309}{2999849580000} \alpha^{6} RE^6 \beta^4 - \frac{1}{16499172690000} \alpha^{11} RE^7 \beta$$
$$+ \frac{809}{261891630000} \alpha^{10} RE^6 \beta^2 + \frac{266309}{374981197500} \alpha^{9} RE^3 \beta^4 - \frac{1457537}{2474875903500} \alpha^{11} RE^3 \beta^3$$
$$- \frac{1}{67580611338240000} \alpha^{17} RE \beta H^8 + \frac{809}{130945815000} \alpha^{15} RE \beta^2 H^2$$
$$- \frac{809}{2815858805760000} \alpha^{15} RE \beta^2 H^7 - \frac{1457537}{6599669076000} \alpha^{11} RE^3 \beta^3 H^2$$
$$+ \frac{1457537}{39598014456000} \alpha^{11} RE^3 \beta^3 H^3 + \frac{266309}{3999799440000} \alpha^{8} RE^4 \beta^4 H^2$$
$$- \frac{1457537}{633568231296000} \alpha^{11} RE^3 \beta^3 H^4 - \frac{809}{33522128640000} \alpha^{13} RE^3 \beta^2 H^5$$
$$+ \frac{809}{52378326000} \alpha^{13} RE^3 \beta^2 H^2 - \frac{809}{26189163000} \alpha^{13} RE^3 \beta^2 H$$
$$+ \frac{1}{3771239472000} \alpha^{14} RE^4 \beta H^3 + \frac{1}{471404934000} \alpha^{14} RE^4 \beta H$$
$$+ \frac{1}{527973526080000} \alpha^{10} RE^8 \beta H - \frac{809}{523783260000} \alpha^{10} RE^6 \beta^2 H$$
$$+ \frac{1457537}{79196028912000} \alpha^{8} RE^6 \beta^3 H - \frac{1}{263986763040000} \alpha^{11} RE^7 \beta H^2$$
$$+ \frac{1}{4714049340000} \alpha^{12} RE^6 \beta H - \frac{1}{942809868000} \alpha^{14} RE^4 \beta H^2$$
$$- \frac{1}{18856197360000} \alpha^{12} RE^6 \beta H^2 + \frac{1}{226274368320000} \alpha^{12} RE^6 \beta H^3$$
$$- \frac{809}{6285399120000} \alpha^{9} RE^7 \beta^2 H - \frac{1}{30169915776000} \alpha^{14} RE^4 \beta H^4$$
$$- \frac{1}{301699157760000} \alpha^{16} RE^2 \beta H^6 - \frac{1}{392837445000} \alpha^{16} RE^2 \beta H^2$$
$$- \frac{266309}{499974930000} \alpha^{8} RE^4 \beta^4 H + \frac{1457537}{2474875903500} \alpha^{11} RE^3 \beta^3 H$$
$$- \frac{1}{3771239472000} \alpha^{17} RE \beta H^4 + \frac{2}{1031198293125} \alpha^{17} RE \beta H$$
$$- \frac{809}{98209361250} \alpha^{15} RE \beta^2 H - \frac{809}{314269956000} \alpha^{15} RE \beta^2 H^3$$
$$- \frac{809}{8380532160000} \alpha^{15} RE \beta^2 H^5 + \frac{1}{18856197360000} \alpha^{17} RE \beta H^5$$
$$+ \frac{1}{1178512335000} \alpha^{17} RE \beta H^3 - \frac{1}{150849578880000} \alpha^{17} RE \beta H^6$$

$$
\begin{aligned}
&+ \frac{1}{32998345380000}\,\alpha^{11}\,RE^{7}\,\beta\,H + \frac{1}{603398315520000}\,\alpha^{14}\,RE^{4}\,\beta\,H^{5} \\
&+ \frac{809}{4190266080000}\,\alpha^{10}\,RE^{6}\,\beta^{2}\,H^{2} + \frac{1}{1178512335000}\,\alpha^{13}\,RE^{5}\,\beta\,H \\
&- \frac{1}{3142699560000}\,\alpha^{13}\,RE^{5}\,\beta\,H^{2} + \frac{1}{18856197360000}\,\alpha^{13}\,RE^{5}\,\beta\,H^{3} \\
&- \frac{1}{301699157760000}\,\alpha^{13}\,RE^{5}\,\beta\,H^{4} - \frac{809}{104756652000}\,\alpha^{11}\,RE^{5}\,\beta^{2}\,H \\
&+ \frac{809}{419026608000}\,\alpha^{11}\,RE^{5}\,\beta^{2}\,H^{2} - \frac{809}{5028319296000}\,\alpha^{11}\,RE^{5}\,\beta^{2}\,H^{3} \\
&+ \frac{1457537}{9899503614000}\,\alpha^{9}\,RE^{5}\,\beta^{3}\,H - \frac{1457537}{79196028912000}\,\alpha^{9}\,RE^{5}\,\beta^{3}\,H^{2} \\
&- \frac{809}{39283744500}\,\alpha^{12}\,RE^{4}\,\beta^{2}\,H + \frac{809}{104756652000}\,\alpha^{12}\,RE^{4}\,\beta^{2}\,H^{2} \\
&- \frac{809}{628539912000}\,\alpha^{12}\,RE^{4}\,\beta^{2}\,H^{3} + \frac{809}{10056638592000}\,\alpha^{12}\,RE^{4}\,\beta^{2}\,H^{4} \\
&- \frac{809}{209513304000}\,\alpha^{13}\,RE^{3}\,\beta^{2}\,H^{3} - \frac{1}{1810194946560000}\,\alpha^{15}\,RE^{3}\,\beta\,H^{6} \\
&+ \frac{1}{75424789440000}\,\alpha^{15}\,RE^{3}\,\beta\,H^{5} - \frac{1}{7542478944000}\,\alpha^{15}\,RE^{3}\,\beta\,H^{4} \\
&- \frac{1}{471404934000}\,\alpha^{15}\,RE^{3}\,\beta\,H^{2} + \frac{1}{294628083750}\,\alpha^{15}\,RE^{3}\,\beta\,H \\
&+ \frac{1}{1414214802000}\,\alpha^{15}\,RE^{3}\,\beta\,H^{3} - \frac{266309}{1999899720000}\,\alpha^{7}\,RE^{5}\,\beta^{4}\,H \\
&+ \frac{1457537}{3299834538000}\,\alpha^{10}\,RE^{4}\,\beta^{3}\,H - \frac{1457537}{13199338152000}\,\alpha^{10}\,RE^{4}\,\beta^{3}\,H^{2} \\
&+ \frac{1457537}{158392057824000}\,\alpha^{10}\,RE^{4}\,\beta^{3}\,H^{3} + \frac{809}{1676106432000}\,\alpha^{13}\,RE^{3}\,\beta^{2}\,H^{4} \\
&+ \frac{809}{100566385920000}\,\alpha^{15}\,RE\,\beta^{2}\,H^{6} - \frac{1}{589256167500}\,\alpha^{17}\,RE\,\beta\,H^{2} \\
&+ \frac{809}{1257079824000}\,\alpha^{15}\,RE\,\beta^{2}\,H^{4} + \frac{1}{25141596480000}\,\alpha^{16}\,RE^{2}\,\beta\,H^{5} \\
&+ \frac{809}{838053216000}\,\alpha^{14}\,RE^{2}\,\beta^{2}\,H^{4} + \frac{1}{8447576417280000}\,\alpha^{16}\,RE^{2}\,\beta\,H^{7} \\
&+ \frac{1}{294628083750}\,\alpha^{16}\,RE^{2}\,\beta\,H - \frac{1}{3771239472000}\,\alpha^{16}\,RE^{2}\,\beta\,H^{4} \\
&- \frac{809}{32736453750}\,\alpha^{14}\,RE^{2}\,\beta^{2}\,H + \frac{809}{52378326000}\,\alpha^{14}\,RE^{2}\,\beta^{2}\,H^{2} \\
&+ \frac{1}{942809868000}\,\alpha^{16}\,RE^{2}\,\beta\,H^{3} - \frac{809}{8380532160000}\,\alpha^{14}\,RE^{2}\,\beta^{2}\,H^{5} \\
&- \frac{809}{157134978000}\,\alpha^{14}\,RE^{2}\,\beta^{2}\,H^{3} + \frac{809}{201132771840000}\,\alpha^{14}\,RE^{2}\,\beta^{2}\,H^{6} \\
&- \frac{1457537}{316784115648000}\,\alpha^{12}\,RE^{2}\,\beta^{3}\,H^{4} + \frac{1457537}{39598014456000}\,\alpha^{12}\,RE^{2}\,\beta^{3}\,H^{3}
\end{aligned}
$$

$$- \frac{266309}{499974930000} \alpha^9 RE^3 \beta^4 H + \frac{266309}{1999899720000} \alpha^9 RE^3 \beta^4 H^2$$
$$+ \frac{1457537}{4949751807000} \alpha^{12} RE^2 \beta^3 H - \frac{1457537}{9899503614000} \alpha^{12} RE^2 \beta^3 H^2$$
$$+ \frac{1}{2111894104320000} \alpha^{17} RE \beta H^7 - \frac{266309}{23998796640000} \alpha^9 RE^3 \beta^4 H^3$$
$$+ \frac{1457537}{6335682312960000} \alpha^{12} RE^2 \beta^3 H^5 + \frac{34621}{673435620000} \alpha^6 RE^4 \beta^5 H$$
$$- \frac{2}{9280784638125} \alpha^{18} \beta$$

$$F_{21} := 0$$

$$F_{22} := \frac{1}{3049047113112000} \alpha^{12} RE^8 \beta - \frac{863}{756212081625} \alpha^{15} RE^3 \beta^2$$
$$+ \frac{891188}{47641361142375} \alpha^{14} RE^2 \beta^3 - \frac{3452}{5293484571375} \alpha^{16} RE^2 \beta^2$$
$$- \frac{863}{12099393306000} \alpha^{11} RE^7 \beta^2 - \frac{863}{3024848326500} \alpha^{12} RE^6 \beta^2$$
$$+ \frac{222797}{3176090742825} \alpha^{12} RE^4 \beta^3 + \frac{1}{653367238524000} \alpha^{20} \beta H^4$$
$$+ \frac{222797}{12704362971300} \alpha^{10} RE^6 \beta^3 - \frac{863}{756212081625} \alpha^{14} RE^4 \beta^2$$
$$- \frac{863}{84695753142000} \alpha^{10} RE^8 \beta^2 + \frac{1}{40835452407750} \alpha^{16} RE^4 \beta$$
$$+ \frac{222797}{63521814856500} \alpha^9 RE^7 \beta^3 + \frac{222797}{762261778278000} \alpha^8 RE^8 \beta^3$$
$$- \frac{863}{5293484571375} \alpha^{17} RE \beta^2 - \frac{14650351}{2032698075408000} \alpha^7 RE^7 \beta^4$$
$$- \frac{1}{2177890795080000} \alpha^{20} \beta H^5 - \frac{1}{14050009097220096000} \alpha^{20} \beta H^9$$
$$- \frac{14650351}{84695753142000} \alpha^9 RE^5 \beta^4 + \frac{1}{10453875816384000} \alpha^{20} \beta H^6$$
$$- \frac{14650351}{63521814856500} \alpha^{10} RE^4 \beta^4 + \frac{891188}{15880453714125} \alpha^{13} RE^3 \beta^3$$
$$- \frac{14650351}{127043629713000} \alpha^{11} RE^3 \beta^4 - \frac{7}{1611090000} \alpha^5 RE^5 \beta^6$$
$$+ \frac{445594}{9528272228475} \alpha^{11} RE^5 \beta^3 + \frac{4}{428772250281375} \alpha^{19} RE \beta$$
$$+ \frac{1}{571696333708500} \alpha^{13} RE^7 \beta + \frac{2246207}{27223634938500} \alpha^8 RE^4 \beta^5$$
$$- \frac{863}{1355132050272000} \alpha^9 RE^9 \beta^2 - \frac{14650351}{254087259426000} \alpha^8 RE^6 \beta^4$$
$$+ \frac{1}{780556060956672000} \alpha^{20} \beta H^8 + \frac{2246207}{108894539754000} \alpha^6 RE^6 \beta^5$$

$$
\begin{aligned}
&+ \frac{2246207}{27223634938500}\,\alpha^{7} RE^{5} \beta^{5} + \frac{1}{190565444569500}\,\alpha^{20} \beta H^{2} \\
&+ \frac{1}{163341809631000}\,\alpha^{14} RE^{6} \beta - \frac{1}{73177130714688000}\,\alpha^{20} \beta H^{7} \\
&- \frac{2}{428772250281375}\,\alpha^{20} \beta H + \frac{1}{562000363888803840000}\,\alpha^{20} \beta H^{10} \\
&- \frac{1}{285848166854250}\,\alpha^{20} \beta H^{3} + \frac{1}{68059087346250}\,\alpha^{15} RE^{5} \beta \\
&+ \frac{4}{142924083427125}\,\alpha^{17} RE^{3} \beta - \frac{863}{1209939330600}\,\alpha^{13} RE^{5} \beta^{2} \\
&+ \frac{1}{27441424018008000}\,\alpha^{11} RE^{9} \beta + \frac{863}{19359029289600}\,\alpha^{13} RE^{5} \beta^{2} H^{3} \\
&- \frac{1}{435578159016000}\,\alpha^{15} RE^{5} \beta H^{3} - \frac{863}{3226504881600}\,\alpha^{13} RE^{5} \beta^{2} H^{2} \\
&- \frac{222797}{254087259426000}\,\alpha^{9} RE^{7} \beta^{3} H + \frac{1}{548828480360160000}\,\alpha^{10} RE^{10} \beta \\
&- \frac{1}{54447269877000}\,\alpha^{15} RE^{5} \beta H - \frac{1}{109765696072032000}\,\alpha^{11} RE^{9} \beta H \\
&+ \frac{863}{24198786612000}\,\alpha^{11} RE^{7} \beta^{2} H + \frac{1}{47641361142375}\,\alpha^{18} RE^{2} \beta \\
&+ \frac{1}{780556060956672000}\,\alpha^{19} RE \beta H^{8} + \frac{863}{10841056402176000}\,\alpha^{17} RE \beta^{2} H^{7} \\
&+ \frac{1}{5226937908192000}\,\alpha^{19} RE \beta H^{6} - \frac{1}{2810001819444019200 0}\,\alpha^{19} RE \beta H^{9} \\
&- \frac{863}{19359029289600}\,\alpha^{17} RE \beta^{2} H^{4} + \frac{863}{96795146448000}\,\alpha^{17} RE \beta^{2} H^{5} \\
&- \frac{1}{871156318032000}\,\alpha^{19} RE \beta H^{5} + \frac{1}{47641361142375}\,\alpha^{19} RE \beta H^{2} \\
&- \frac{863}{346913804869632000}\,\alpha^{17} RE \beta^{2} H^{8} + \frac{1726}{5293484571375}\,\alpha^{17} RE \beta^{2} H \\
&- \frac{1}{48784753809792000}\,\alpha^{19} RE \beta H^{7} - \frac{1}{81670904815500}\,\alpha^{19} RE \beta H^{3} \\
&+ \frac{1}{217789079508000}\,\alpha^{19} RE \beta H^{4} - \frac{863}{774361171584000}\,\alpha^{17} RE \beta^{2} H^{6} \\
&- \frac{1}{47641361142375}\,\alpha^{19} RE \beta H - \frac{863}{3024848326500}\,\alpha^{17} RE \beta^{2} H^{2} \\
&+ \frac{863}{6049696653000}\,\alpha^{17} RE \beta^{2} H^{3} + \frac{1}{312222424382668800 0}\,\alpha^{18} RE^{2} \beta H^{8} \\
&- \frac{863}{1008282775500}\,\alpha^{16} RE^{2} \beta^{2} H^{2} + \frac{863}{21682112804352000}\,\alpha^{16} RE^{2} \beta^{2} H^{7} \\
&+ \frac{222797}{8469575314200}\,\alpha^{12} RE^{4} \beta^{3} H^{2} - \frac{222797}{50817451885200}\,\alpha^{12} RE^{4} \beta^{3} H^{3} \\
&- \frac{14650351}{1355132050272000}\,\alpha^{9} RE^{5} \beta^{4} H^{2} + \frac{222797}{813079230163200}\,\alpha^{12} RE^{4} \beta^{3} H^{4}
\end{aligned}
$$

$$
\begin{aligned}
&+\frac{863}{774361171584000}\,\alpha^{14}\,RE^{4}\,\beta^{2}\,H^{5}-\frac{863}{1209939330600}\,\alpha^{14}\,RE^{4}\,\beta^{2}\,H^{2}\\
&+\frac{863}{604969665300}\,\alpha^{14}\,RE^{4}\,\beta^{2}\,H+\frac{1}{48784753809792000}\,\alpha^{12}\,RE^{8}\,\beta\,H^{2}\\
&-\frac{1}{762261778278000}\,\alpha^{13}\,RE^{7}\,\beta\,H+\frac{1}{108894539754000}\,\alpha^{15}\,RE^{5}\,\beta\,H^{2}\\
&+\frac{1}{3049047113112000}\,\alpha^{13}\,RE^{7}\,\beta\,H^{2}-\frac{1}{36588565357344000}\,\alpha^{13}\,RE^{7}\,\beta\,H^{3}\\
&+\frac{863}{338783012568000}\,\alpha^{10}\,RE^{8}\,\beta^{2}\,H+\frac{1}{3484625272128000}\,\alpha^{15}\,RE^{5}\,\beta\,H^{4}\\
&+\frac{1}{20907751632768000}\,\alpha^{17}\,RE^{3}\,\beta\,H^{6}+\frac{1}{27223634938500}\,\alpha^{17}\,RE^{3}\,\beta\,H^{2}\\
&+\frac{14650351}{169391506284000}\,\alpha^{9}\,RE^{5}\,\beta^{4}\,H-\frac{222797}{3176090742825}\,\alpha^{12}\,RE^{4}\,\beta^{3}\,H\\
&+\frac{1}{174231263606400}\,\alpha^{18}\,RE^{2}\,\beta\,H^{4}-\frac{2}{47641361142375}\,\alpha^{18}\,RE^{2}\,\beta\,H\\
&+\frac{863}{756212081625}\,\alpha^{16}\,RE^{2}\,\beta^{2}\,H+\frac{863}{2419878661200}\,\alpha^{16}\,RE^{2}\,\beta^{2}\,H^{3}\\
&+\frac{863}{64530097632000}\,\alpha^{16}\,RE^{2}\,\beta^{2}\,H^{5}-\frac{1}{871156318032000}\,\alpha^{18}\,RE^{2}\,\beta\,H^{5}\\
&-\frac{1}{54447269877000}\,\alpha^{18}\,RE^{2}\,\beta\,H^{3}+\frac{1}{6969250544256000}\,\alpha^{18}\,RE^{2}\,\beta\,H^{6}\\
&-\frac{1}{6098094226224000}\,\alpha^{12}\,RE^{8}\,\beta\,H-\frac{1}{69692505442560000}\,\alpha^{15}\,RE^{5}\,\beta\,H^{5}\\
&-\frac{863}{193590292896000}\,\alpha^{11}\,RE^{7}\,\beta^{2}\,H^{2}-\frac{1}{163341809631000}\,\alpha^{14}\,RE^{6}\,\beta\,H\\
&+\frac{1}{435578159016000}\,\alpha^{14}\,RE^{6}\,\beta\,H^{2}-\frac{1}{2613468954096000}\,\alpha^{14}\,RE^{6}\,\beta\,H^{3}\\
&+\frac{1}{41815503265536000}\,\alpha^{14}\,RE^{6}\,\beta\,H^{4}+\frac{863}{4033131102000}\,\alpha^{12}\,RE^{6}\,\beta^{2}\,H\\
&-\frac{863}{16132524408000}\,\alpha^{12}\,RE^{6}\,\beta^{2}\,H^{2}+\frac{863}{193590292896000}\,\alpha^{12}\,RE^{6}\,\beta^{2}\,H^{3}\\
&-\frac{222797}{25408725942600}\,\alpha^{10}\,RE^{6}\,\beta^{3}\,H+\frac{222797}{203269807540800}\,\alpha^{10}\,RE^{6}\,\beta^{3}\,H^{2}\\
&+\frac{863}{1209939330600}\,\alpha^{13}\,RE^{5}\,\beta^{2}\,H-\frac{863}{309744468633600}\,\alpha^{13}\,RE^{5}\,\beta^{2}\,H^{4}\\
&+\frac{863}{4839757322400}\,\alpha^{14}\,RE^{4}\,\beta^{2}\,H^{3}+\frac{1}{167262013062144000}\,\alpha^{16}\,RE^{4}\,\beta\,H^{6}\\
&-\frac{1}{6969250544256000}\,\alpha^{16}\,RE^{4}\,\beta\,H^{5}+\frac{1}{696925054425600}\,\alpha^{16}\,RE^{4}\,\beta\,H^{4}\\
&+\frac{1}{43557815901600}\,\alpha^{16}\,RE^{4}\,\beta\,H^{2}-\frac{1}{27223634938500}\,\alpha^{16}\,RE^{4}\,\beta\,H\\
&-\frac{1}{130673447704800}\,\alpha^{16}\,RE^{4}\,\beta\,H^{3}+\frac{14650351}{1016349037704000}\,\alpha^{8}\,RE^{6}\,\beta^{4}\,H
\end{aligned}
$$

$$- \frac{222797}{6352181485650} \alpha^{11} RE^5 \beta^3 H + \frac{222797}{25408725942600} \alpha^{11} RE^5 \beta^3 H^2$$
$$- \frac{222797}{304904711311200} \alpha^{11} RE^5 \beta^3 H^3 - \frac{863}{38718058579200} \alpha^{14} RE^4 \beta^2 H^4$$
$$- \frac{863}{774361171584000} \alpha^{16} RE^2 \beta^2 H^6 + \frac{1}{27223634938500} \alpha^{18} RE^2 \beta H^2$$
$$- \frac{863}{9679514644800} \alpha^{16} RE^2 \beta^2 H^4 - \frac{1}{1742312636064000} \alpha^{17} RE^3 \beta H^5$$
$$- \frac{863}{12906019526400} \alpha^{15} RE^3 \beta^2 H^4 - \frac{1}{585417045717504000} \alpha^{17} RE^3 \beta H^7$$
$$- \frac{1}{20417726203875} \alpha^{17} RE^3 \beta H + \frac{1}{261346895409600} \alpha^{17} RE^3 \beta H^4$$
$$+ \frac{863}{504141387750} \alpha^{15} RE^3 \beta^2 H - \frac{863}{806626220400} \alpha^{15} RE^3 \beta^2 H^2$$
$$- \frac{1}{65336723852400} \alpha^{17} RE^3 \beta H^3 + \frac{863}{129060195264000} \alpha^{15} RE^3 \beta^2 H^5$$
$$+ \frac{863}{2419878661200} \alpha^{15} RE^3 \beta^2 H^3 - \frac{863}{3097444686336000} \alpha^{15} RE^3 \beta^2 H^6$$
$$+ \frac{222797}{203269807540800} \alpha^{13} RE^3 \beta^3 H^4 - \frac{222797}{25408725942600} \alpha^{13} RE^3 \beta^3 H^3$$
$$+ \frac{14650351}{84695753142000} \alpha^{10} RE^4 \beta^4 H - \frac{14650351}{338783012568000} \alpha^{10} RE^4 \beta^4 H^2$$
$$- \frac{222797}{3176090742825} \alpha^{13} RE^3 \beta^3 H + \frac{222797}{6352181485650} \alpha^{13} RE^3 \beta^3 H^2$$
$$- \frac{1}{97569507619584000} \alpha^{18} RE^2 \beta H^7 + \frac{14650351}{4065396150816000} \alpha^{10} RE^4 \beta^4 H^3$$
$$- \frac{222797}{4065396150816000} \alpha^{13} RE^3 \beta^3 H^5 - \frac{2246207}{108894539754000} \alpha^7 RE^5 \beta^5 H$$
$$- \frac{14650351}{338783012568000} \alpha^{11} RE^3 \beta^4 H^2 + \frac{14650351}{2032698075408000} \alpha^{11} RE^3 \beta^4 H^3$$
$$+ \frac{2246207}{435578159016000} \alpha^8 RE^4 \beta^5 H^2 - \frac{14650351}{32523169206528000} \alpha^{11} RE^3 \beta^4 H^4$$
$$- \frac{2246207}{54447269877000} \alpha^8 RE^4 \beta^5 H + \frac{14650351}{127043629713000} \alpha^{11} RE^3 \beta^4 H$$
$$+ \frac{222797}{203269807540800} \alpha^{14} RE^2 \beta^3 H^4 - \frac{445594}{15880453714125} \alpha^{14} RE^2 \beta^3 H$$
$$+ \frac{222797}{12704362971300} \alpha^{14} RE^2 \beta^3 H^2 - \frac{222797}{2032698075408000} \alpha^{14} RE^2 \beta^3 H^5$$
$$- \frac{222797}{38113088913900} \alpha^{14} RE^2 \beta^3 H^3 + \frac{222797}{48784753809792000} \alpha^{14} RE^2 \beta^3 H^6$$
$$+ \frac{4}{2143861251406875} \alpha^{20} \beta$$

$$F_{23} := 0$$

```
> U:=sum(F[r]*t^r,r=0..n+3):

  p:=subs(H=0,alpha=5*3.14/180,RE=50,U):
  e:=subs(t=1,p)=0:
  j:={e}:
  j:=solve(j):
  sj:=evalf(j);
```

$$sj := \{\beta = 10.87898850\}, \{\beta = 5.893976044 + 7.129389126\,I\}, \{\beta = -5.183783796 + 7.863444546\,I\}, \{\beta = -1.766812734\}, \{\beta = -5.183783796 - 7.863444546\,I\}, \{\beta = 5.893976044 - 7.129389126\,I\} \tag{3}$$

```
> p:=subs(beta = -1.766812734,p);
```

$$p := 1 - 1.651976178\,t^2 + 1.204931200\,t^4 - 0.7482642437\,t^6 + 0.4269827593\,t^8 - 0.2316735377\,t^{10} + 0.1215464199\,t^{12} - 0.06225961446\,t^{14} + 0.03132423392\,t^{16} - 0.01554180730\,t^{18} + 0.007625768132\,t^{20} - 0.003707733575\,t^{22} \tag{4}$$

A6. MAPLE CODE FOR ADM

```
> restart:
> PDEtools[declare](f(x),prime=x):
```

f(x) will now be displayed as f

derivatives with respect to x of functions of one variable will now be displayed with ' **(1)**

```
> H:=0;alpha:=5*3.14/180;RE:=50;
```

$$H := 0$$

$$\alpha := 0.08722222222$$

$$RE := 50 \tag{2}$$

```
> A[m]:=-2*alpha*RE*sum(diff(f[m-n](x),x)*(f[n](x)),n=0..m)-(4-H)*
  (alpha)^2*(diff(f[m](x),x));
```

$$A_m := -8.722222222\left(\sum_{n=0}^{m} f'''_{m-n}\,f_n\right) - 0.03043086420\,f'''_m \tag{3}$$

```
> for i from 0 to 6 do A[i]:= expand(subs(m = i, A[m])) end do;
```

$$A_0 := -8.722222222\,f'''_0\,f_0 - 0.03043086420\,f'''_0$$

$$A_1 := -8.722222222\,f'''_1\,f_0 - 8.722222222\,f'''_0\,f_1 - 0.03043086420\,f'''_1$$

$$A_2 := -8.722222222\,f'''_2\,f_0 - 8.722222222\,f'''_1\,f_1 - 8.722222222\,f'''_0\,f_2 - 0.03043086420\,f'''_2$$

$$A_3 := -8.722222222\,f'''_3\,f_0 - 8.722222222\,f'''_2\,f_1 - 8.722222222\,f'''_1\,f_2 - 8.722222222\,f'''_0\,f_3 - 0.03043086420\,f'''_3$$

$$A_4 := -8.722222222\,f'''_4\,f_0 - 8.722222222\,f'''_3\,f_1 - 8.722222222\,f'''_2\,f_2 - 8.722222222\,f'''_1\,f_3 - 8.722222222\,f'''_0\,f_4 - 0.03043086420\,f'''_4$$

$$A_5 := -8.722222222\,f'''_5\,f_0 - 8.722222222\,f'''_4\,f_1 - 8.722222222\,f'''_3\,f_2 - 8.722222222\,f'''_2\,f_3 - 8.722222222\,f'''_1\,f_4 - 8.722222222\,f'''_0\,f_5 - 0.03043086420\,f'''_5$$

$$A_6 := -8.722222222\, f_6''' f_0 - 8.722222222\, f_5''' f_1 - 8.722222222\, f_4''' f_2 - 8.722222222\, f_3''' f_3 - 8.722222222\, f_2''' f_4 - 8.722222222\, f_1''' f_5 - 8.722222222\, f_0''' f_6 - 0.03043086420\, f_6''' \quad (4)$$

```
> f[0](x):=a*x^2/2+1;
```

$$f_0 := \frac{1}{2}\, a\, x^2 + 1 \quad (5)$$

```
> for m from 0 by 1 to 5 do;
  f[m+1](x):=collect(simplify(int(int(int(A[m],x),x),x)),x):end
  do;
```

$$f_1 := -0.03634259259\, a^2 x^6 - 0.3646938787\, a\, x^4$$

$$f_2 := 0.001761045381\, a^3 x^{10} + 0.03408151129\, a^2 x^8 + 0.1064013001\, a\, x^6$$

$$f_3 := -0.00007384725559\, a^4 x^{14} - 0.002118566073\, a^3 x^{12} - 0.01491518225\, a^2 x^{10} - 0.01663024406\, a\, x^8$$

$$f_4 := 0.000002876753716\, a^5 x^{18} + 0.0001095453285\, a^4 x^{16} + 0.001240269629\, a^3 x^{14} + 0.004102500165\, a^2 x^{12} + 0.001617319522\, a\, x^{10}$$

$$f_5 := -0.0004712189947\, a^3 x^{16} - 0.0007979893838\, a^2 x^{14} - 0.0001072411872\, a\, x^{12} - 0.00008017885756\, a^4 x^{18} - 0.000005086533018\, a^5 x^{20} - 1.070986563\; 10^{-7}\, a^6 x^{22}$$

$$f_6 := 2.200899762\; 10^{-7}\, a^6 x^{24} + 0.00003857331794\, a^4 x^{20} + 0.0001305645587\, a^3 x^{18} + 0.0001167941967\, a^2 x^{16} + 0.000005157389604\, a\, x^{14} + 0.000004465825411\, a^5 x^{22} + 3.866579693\; 10^{-9}\, a^7 x^{26} \quad (6)$$

```
> f(x):=sum(f[r](x),r=0..5):
> e1:=subs(x=1,f(x))=0;
  s:={e1}:
  s:=evalf(solve(s,{a}));
```

$$e1 := 0.2265872556\, a + 1 - 0.01387175276\, a^2 + 0.000411529942\, a^3 - 0.0000444807847\, a^4 - 0.000002209779302\, a^5 - 1.070986563\; 10^{-7}\, a^6 = 0$$

$$s := \{a = 12.88012130\}, \{a = 3.801204743 + 16.02081577\, I\}, \{a = -18.78808143 + 20.06115405\, I\}, \{a = -3.539486758\}, \{a = -18.78808143 - 20.06115405\, I\}, \{a = 3.801204743 - 16.02081577\, I\} \quad (7)$$

```
> f(x):=subs({a = -3.539486758},f(x));
```

$$f := -1.769743379\, x^2 + 1 - 0.8319047756\, x^6 + 1.290829154\, x^4 - 0.2706706656\, x^{10} + 0.4858345607\, x^8 - 0.07658425177\, x^{14} + 0.1457182312\, x^{12} - 0.01418216798\, x^{18} + 0.03808819527\, x^{16} + 0.002825682275\, x^{20} - 0.0002105841952\, x^{22} \quad (8)$$

A7. MAPLE CODE FOR HAM

```
> restart;
> N:=4;alpha:=5*3.14/180;r:=10;Ha:=5;H:=1;
```

$$N := 4$$
$$\alpha := 0.08722222222$$
$$r := 10$$
$$Ha := 5$$
$$H := 1 \qquad (1)$$

```
> dsolve(diff(f(x),x,x,x));
  Rf:=diff(f[m-1](x),x,x,x)+2*alpha*r*sum(f[m-1-n](x)*diff(f[n]
  (x),x),n=0..m-1)+(4-Ha)*(alpha)^2*diff(f[m-1](x),x);
  dsolve(diff(f[m](x),x,x,x)-CHI[m]*(diff(f[m-1](x),x,x,x))=h*H*
  Rf,f[m](x));
```

$$f(x) = \frac{1}{2}\,\_C1\,x^2 + \_C2\,x + \_C3$$

$$Rf := \frac{d^3}{dx^3} f_{m-1}(x) + 1.744444444 \left(\sum_{n=0}^{m-1} f_{m-1-n}(x) \left(\frac{d}{dx} f_n(x) \right) \right) - 0.007607716049 \left(\frac{d}{dx} f_{m-1}(x) \right)$$

$$f_m(x) = \int \left(\int \left(\int \left(X_m \left(\frac{d^3}{dx^3} f_{m-1}(x) \right) + h \left(\frac{d^3}{dx^3} f_{m-1}(x) \right) + \frac{436111111}{250000000}\, h \left(\sum_{n=0}^{m-1} f_{m-1-n}(x) \left(\frac{d}{dx} f_n(x) \right) \right) - \frac{7607716049}{1000000000000}\, h \left(\frac{d}{dx} f_{m-1}(x) \right) \right) dx\, dx + \_C1\, x \right) dx + \_C2\, x + \_C3 \qquad (2)$$

```
> f[0](x):=1-x^2;
  for m from 1 by 1 to N do
  CHI[m]:=`if` (m>1,1,0);
  f[m](x)  := int(int(int(CHI[m]*(diff(f[m-1](x), x, x, x))+h*H*
  (diff(f[m-1](x), x, x, x))+2*h*H*alpha*r*(sum(f[m-1-n](x)*(diff
  (f[n](x), x)), n = 0 .. m-1))+4*h*H*alpha^2*(diff(f[m-1](x), x))
  -h*H*alpha^2*(diff(f[m-1](x), x))*Ha, x), x)+_C1*x, x)+_C2*x+
  _C3;
  s1:=evalf(subs(x=0,f[m](x)))=0;
  s2:=evalf(subs(x=0,diff(f[m](x),x)))=0;
  s3:=evalf(subs(x=1,f[m](x)))=0;
  s:={s1,s2,s3}:
  f[m](x):=simplify(subs(solve(s,{_C1,_C2,_C3}),f[m](x)));
  end do;
  f(x):=sum(f[l](x),l=0..N);
```

$$f_0(x) := 1 - x^2$$
$$X_1 := 0$$

$$f_1(x) := -3.488888888\,h\,(-0.008333333333\,x^6 + 0.04166666668\,x^4) + 0.0006339763375\,x^4\,h + 0.5000000000\,\_C1\,x^2 + \_C2\,x + \_C3$$

$$s1 := \_C3 = 0$$

$$s2 := \_C2 = 0$$

$$s3 := -0.1156623200\,h + 0.5000000000\,\_C1 + \_C2 + \_C3 = 0$$

$$s := \{\_C2 = 0, \_C3 = 0, -0.1156623200\,h + 0.5000000000\,\_C1 + \_C2 + \_C3 = 0\}$$

$$f_1(x) := 0.02907407407\,h\,x^6 - 0.1447363940\,x^4\,h + 0.1156623200\,x^2\,h$$

$$X_2 := 1$$

$$f_2(x) := 0.02907407407\,h\,x^6 - 0.1447363940\,x^4\,h + h\,(0.02907407407\,h\,x^6 - 0.1447363940\,x^4\,h) + 1.744444444\,h\,(-0.0003230452674\,h\,x^{10} + 0.003103758359\,h\,x^8 - 0.008679957133\,h\,x^6 + 0.009638526668\,x^4\,h) - 0.007607716040\,h\,(0.0005191798941\,h\,x^8 - 0.004824546467\,h\,x^6 + 0.009638526668\,x^4\,h) + 0.5000000000\,\_C1\,x^2 + \_C2\,x + \_C3$$

$$s1 := \_C3 = 0$$

$$s2 := \_C2 = 0$$

$$s3 := -0.1156623199\,h - 0.1091799223\,h^2 + 0.5000000000\,\_C1 + \_C2 + \_C3 = 0$$

$$s := \{\_C2 = 0, \_C3 = 0, -0.1156623199\,h - 0.1091799223\,h^2 + 0.5000000000\,\_C1 + \_C2 + \_C3 = 0\}$$

$$f_2(x) := 0.02907407407\,h\,x^6 - 0.1447363940\,x^4\,h + 0.01396907485\,h^2\,x^6 - 0.1279958469\,x^4\,h^2 - 0.0005635345219\,h^2\,x^{10} + 0.005410384252\,h^2\,x^8 + 0.1156623199\,x^2\,h + 0.1091799223\,x^2\,h^2$$

$$X_3 := 1$$

$$f_3(x) := 0.02907407407\,h\,x^6 - 0.1447363940\,x^4\,h + 0.01396907485\,h^2\,x^6 - 0.1279958469\,x^4\,h^2 - 0.0005635345220\,h^2\,x^{10} + 0.005410384254\,h^2\,x^8 + h\,(0.02907407407\,h\,x^6 - 0.1447363940\,x^4\,h + 0.01396907485\,h^2\,x^6 - 0.1279958469\,x^4\,h^2 - 0.0005635345220\,h^2\,x^{10} + 0.005410384254\,h^2\,x^8) + 1.744444444\,h\,(0.000005418601173\,h^2\,x^{14} - 0.00007713632887\,h^2\,x^{12} - 0.002777777778\,(0.02907407407\,h + 0.01396907485\,h^2)\,x^{10} - 0.005952380951\,(-0.1447363940\,h - 0.1279958469\,h^2)\,x^8 - 0.01666666667\,(0.1156623199\,h + 0.1091799223\,h^2)\,x^6 + 0.0001537454081\,h^2\,x^{10} - 0.0002989383414\,h^2\,x^8 + 0.0002229628712\,h^2\,x^6 + 0.001388888889\,(-0.04053137508\,h^2 - 0.1744444444\,h)\,x^{10} + 0.002976190476\,(0.7533900204\,h + 0.5957978367\,h^2)\,x^8 + 0.008333333333\,(-0.8102702158\,h - 0.7303432322\,h^2)\,x^6 + 0.04166666668\,(0.2313246398\,h + 0.2183598446\,h^2)\,x^4) - 0.007607716040\,h\,(0.0005191798941\,h\,x^8 - 0.004824546467\,h\,x^6 + 0.0002494477651\,h^2\,x^8 - 0.004266528230\,h^2\,x^6 - 0.000004269200923\,h^2\,x^{12} + 0.00006011538058\,h^2\,x^{10} + 0.009638526658\,x^4\,h + 0.009098326858\,x^4\,h^2) + 0.5000000000\,\_C1\,x^2 + \_C2\,x + \_C3$$

$$s1 := \_C3 = 0$$

$$s2 := \_C2 = 0$$

$$s3 := -0.1156623199\,h - 0.1091799223\,h^2 + h\,(-0.1156623199\,h - 0.1091799223\,h^2) + 1.744444444\,h\,(0.003638511424\,h^2 + 0.003739282621\,h)$$

$$- 0.007607716040\, h\, (0.005333160083\, h + 0.005137092573\, h^2) + 0.5000000000\, \_C1 + \_C2 + \_C3 = 0$$

$$s := \{\_C2 = 0, \_C3 = 0, -0.1156623199\, h - 0.1091799223\, h^2 + h\, (-0.1156623199\, h - 0.1091799223\, h^2) + 1.744444444\, h\, (0.003638511424\, h^2 + 0.003739282621\, h) - 0.007607716040\, h\, (0.005333160083\, h + 0.005137092573\, h^2) + 0.5000000000\, \_C1 + \_C2 + \_C3 = 0\}$$

$$f_3(x) := 0.1156623199\, x^2 h + 0.0005991479658\, h^3 x^6 - 0.1121935386\, x^4 h^3 - 0.0004616822094\, h^3 x^{10} + 0.009309325179\, h^3 x^8 + 0.02793814971\, h^2 x^6 - 0.2559916938\, x^4 h^2 - 0.001127069044\, h^2 x^{10} + 0.01082076850\, h^2 x^8 + 0.2183598446\, x^2 h^2 - 0.1447363940\, x^4 h + 0.02907407407\, h x^6 + 0.1028718228\, x^2 h^3 + 0.000009452448710\, h^3 x^{14} - 0.0001345275615\, h^3 x^{12}$$

$$X_4 := 1$$

$$\begin{aligned}
f_4(x) := {} & 0.0005991479660\, h^3 x^6 - 0.1121935386\, x^4 h^3 - 0.0004616822094\, h^3 x^{10} \\
& + 0.009309325179\, h^3 x^8 + 0.02793814972\, h^2 x^6 - 0.2559916938\, x^4 h^2 \\
& - 0.001127069044\, h^2 x^{10} + 0.01082076850\, h^2 x^8 - 0.1447363940\, x^4 h \\
& + 0.02907407407\, h x^6 + 0.000009452448707\, h^3 x^{14} - 0.0001345275615\, h^3 x^{12} \\
& + h\, (0.0005991479660\, h^3 x^6 - 0.1121935386\, x^4 h^3 - 0.0004616822094\, h^3 x^{10} \\
& + 0.009309325179\, h^3 x^8 + 0.02793814972\, h^2 x^6 - 0.2559916938\, x^4 h^2 \\
& - 0.001127069044\, h^2 x^{10} + 0.01082076850\, h^2 x^8 - 0.1447363940\, x^4 h \\
& + 0.02907407407\, h x^6 + 0.000009452448707\, h^3 x^{14} - 0.0001345275615\, h^3 x^{12}) \\
& + 1.744444444\, h\, (0.001927705333\, (0.1156623199\, h + 0.1091799223\, h^2)\, h x^6 \\
& + 0.001388888889\, (0.1744444444\, (0.1156623199\, h + 0.1091799223\, h^2)\, h \\
& - 0.5789455760\, (-0.1447363940\, h - 0.1279958469\, h^2)\, h \\
& + 0.2313246400\, (0.02907407407\, h + 0.01396907485\, h^2)\, h)\, x^{10} \\
& + 0.0009638526667\, h\, (0.2313246398\, h + 0.2183598446\, h^2)\, x^6 \\
& + 0.0004578754579\, (0.003002491356\, h^3 + 0.01127069044\, h^2)\, x^{14} \\
& + 0.0007575757576\, (-0.07909142352\, h^3 - 0.09783683844\, h^2)\, x^{12} + 0.001388888889\, (\\
& -0.1744444444\, h + 0.07087971364\, h^3 - 0.08106275030\, h^2)\, x^{10} \\
& + 0.002976190476\, (0.7533900204\, h + 0.4523690422\, h^3 + 1.191595673\, h^2)\, x^8 \\
& + 0.008333333333\, (-0.6545178000\, h^3 - 1.460686464\, h^2 - 0.8102702158\, h)\, x^6 \\
& + 0.04166666668\, (0.2313246398\, h + 0.4367196892\, h^2 + 0.2057436456\, h^3)\, x^4 \\
& + 0.000001595191156\, h^3 x^{16} - 8.443363772\, 10^{-8}\, h^3 x^{18} - 0.0009157509157\, (\\
& -0.0004616822094\, h^3 - 0.001127069044\, h^2)\, x^{14} - 0.001515151515\, (0.009309325179\, h^3 \\
& + 0.01082076850\, h^2)\, x^{12} - 0.002777777778\, (0.02907407407\, h + 0.0005991479658\, h^3 \\
& + 0.02793814971\, h^2)\, x^{10} - 0.005952380951\, (-0.1121935386\, h^3 - 0.2559916938\, h^2 \\
& - 0.1447363940\, h)\, x^8 - 0.01666666667\, (0.1156623199\, h + 0.2183598446\, h^2 \\
& + 0.1028718228\, h^3)\, x^6 + 0.0004578754579\, (-0.006916433157\, h^3 \\
& + 0.02907407407\, h\, (0.1744444444\, h + 0.08381444910\, h^2))\, x^{14} \\
& + 0.0007575757576\, (0.005006220758\, h^3 - 0.1447363940\, h\, (0.1744444444\, h \\
& + 0.08381444910\, h^2) + 0.02907407407\, h\, (-0.5789455760\, h - 0.5119833876\, h^2))\, x^{12} \\
& + 0.001388888889\, (0.1156623200\, h\, (0.1744444444\, h + 0.08381444910\, h^2)
\end{aligned}$$

$$-0.1447363940\,h\,(-0.5789455760\,h-0.5119833876\,h^2)$$
$$+0.02907407407\,h\,(0.2313246398\,h+0.2183598446\,h^2))\,)\,x^{10}$$
$$+0.002976190476\,(0.1156623200\,h\,(-0.5789455760\,h-0.5119833876\,h^2)$$
$$-0.1447363940\,h\,(0.2313246398\,h+0.2183598446\,h^2))\,x^8+0.002976190476\,($$
$$-0.5789455760\,(0.1156623199\,h+0.1091799223\,h^2)\,h+0.2313246400\,($$
$$-0.1447363940\,h-0.1279958469\,h^2)\,h)\,x^8$$
$$+0.0004578754579\,(0.1744444444\,(0.02907407407\,h+0.01396907485\,h^2)\,h$$
$$-0.003262677448\,h^3)\,x^{14}+0.0007575757576\,(0.1744444444\,(-0.1447363940\,h$$
$$-0.1279958469\,h^2)\,h-0.5789455760\,(0.02907407407\,h+0.01396907485\,h^2)\,h$$
$$+0.001251555189\,h^3)\,x^{12})-0.007607716040\,h\,(0.009638526658\,x^4\,h$$
$$+0.00001069907082\,h^3\,x^8-0.003739784620\,h^3\,x^6-0.000003497592496\,h^3\,x^{12}$$
$$+0.0001034369464\,h^3\,x^{10}+0.0004988955308\,h^2\,x^8-0.008533056460\,h^2\,x^6$$
$$-0.000008538401850\,h^2\,x^{12}+0.0001202307611\,h^2\,x^{10}+0.01819665372\,x^4\,h^2$$
$$-0.004824546467\,h\,x^6+0.0005191798941\,h\,x^8+0.008572651900\,x^4\,h^3$$
$$+3.938520294\,10^{-8}\,h^3\,x^{16}-7.391624257\,10^{-7}\,h^3\,x^{14})+0.5000000000\,\_C1\,x^2+\_C2\,x$$
$$+\_C3$$

$$s1:=\_C3=0$$
$$s2:=\_C2=0$$

$$s3:=-0.1028718228\,h^3-0.2183598446\,h^2-0.1156623199\,h+h\,(-0.1028718228\,h^3$$
$$-0.2183598446\,h^2-0.1156623199\,h)+1.744444444\,h\,(0.003739282621\,h$$
$$+0.00718059719\,h^2+0.0005734702477\,h\,(0.2313246398\,h+0.2183598446\,h^2)$$
$$+0.000446936975\,(0.1156623199\,h+0.1091799223\,h^2)\,h+0.0000165299947\,($$
$$-0.1447363940\,h-0.1279958469\,h^2)\,h-0.0000374370812\,(0.02907407407\,h$$
$$+0.01396907485\,h^2)\,h+0.00006430563278\,h\,(0.1744444444\,h+0.08381444910\,h^2)$$
$$+0.0001652361394\,h\,(-0.5789455760\,h-0.5119833876\,h^2)+0.003444106888\,h^3)$$
$$-0.007607716040\,h\,(0.005333160083\,h+0.004942805928\,h^3+0.01027418515\,h^2)$$
$$+0.5000000000\,\_C1+\_C2+\_C3=0$$

$$s:=\{\_C2=0,\_C3=0,-0.1028718228\,h^3-0.2183598446\,h^2-0.1156623199\,h+h\,($$
$$-0.1028718228\,h^3-0.2183598446\,h^2-0.1156623199\,h)$$
$$+1.744444444\,h\,(0.003739282621\,h+0.00718059719\,h^2$$
$$+0.0005734702477\,h\,(0.2313246398\,h+0.2183598446\,h^2)$$
$$+0.000446936975\,(0.1156623199\,h+0.1091799223\,h^2)\,h+0.0000165299947\,($$
$$-0.1447363940\,h-0.1279958469\,h^2)\,h-0.0000374370812\,(0.02907407407\,h$$
$$+0.01396907485\,h^2)\,h+0.00006430563278\,h\,(0.1744444444\,h+0.08381444910\,h^2)$$
$$+0.0001652361394\,h\,(-0.5789455760\,h-0.5119833876\,h^2)+0.003444106888\,h^3)$$
$$-0.007607716040\,h\,(0.005333160083\,h+0.004942805928\,h^3+0.01027418515\,h^2)$$
$$+0.5000000000\,\_C1+\_C2+\_C3=0\}$$

$$f_4(x):=0.1156623199\,x^2\,h+0.001797443923\,h^3\,x^6-0.3365806159\,x^4\,h^3$$
$$-0.001385046628\,h^3\,x^{10}+0.02792797553\,h^3\,x^8+0.04190722458\,h^2\,x^6$$
$$-0.3839875407\,x^4\,h^2-0.001690603566\,h^2\,x^{10}+0.01623115275\,h^2\,x^8$$
$$+0.3275397668\,x^2\,h^2+0.000002782422718\,h^4\,x^{16}-1.472897902\,10^{-7}\,h^4\,x^{18}$$
$$+0.09674058375\,x^2\,h^4-0.1447363940\,x^4\,h+0.02907407407\,h\,x^6+0.3086154684\,x^2\,h^3$$
$$+0.00002835734613\,h^3\,x^{14}-0.0004035826844\,h^3\,x^{12}-0.01114375831\,h^4\,x^6$$

$$-0.09730424192\,x^4 h^4 + 0.0001582785433\,h^4 x^{10} + 0.01186940545\,h^4 x^8 + 0.000008356132825\,h^4 x^{14} - 0.0003312587453\,h^4 x^{12}$$

$$f(x) := 1 + 0.4626492797\,x^2 h + 0.002396591889\,h^3 x^6 - 0.4487741545\,x^4 h^3 - 0.001846728837\,h^3 x^{10} + 0.03723730071\,h^3 x^8 + 0.08381444914\,h^2 x^6 - 0.7679750814\,x^4 h^2 - 0.003381207132\,h^2 x^{10} + 0.03246230550\,h^2 x^8 + 0.6550795337\,x^2 h^2 + 0.000002782422718\,h^4 x^{16} - 1.472897902\ 10^{-7}\,h^4 x^{18} + 0.09674058375\,x^2 h^4 - 0.5789455760\,x^4 h + 0.1162962963\,h\,x^6 - x^2 + 0.4114872912\,x^2 h^3 + 0.00003780979484\,h^3 x^{14} - 0.0005381102459\,h^3 x^{12} - 0.01114375831\,h^4 x^6 - 0.09730424192\,x^4 h^4 + 0.0001582785433\,h^4 x^{10} + 0.01186940545\,h^4 x^8 + 0.000008356132825\,h^4 x^{14} - 0.0003312587453\,h^4 x^{12} \quad \textbf{(3)}$$

```
> hh:=evalf(subs(x=1,diff(f(x),x))):
  plot(hh,h=-1.5..-0.2);
> A(x):=subs(h=-0.9,f(x));
  plot(A(x),x=0..1);
```

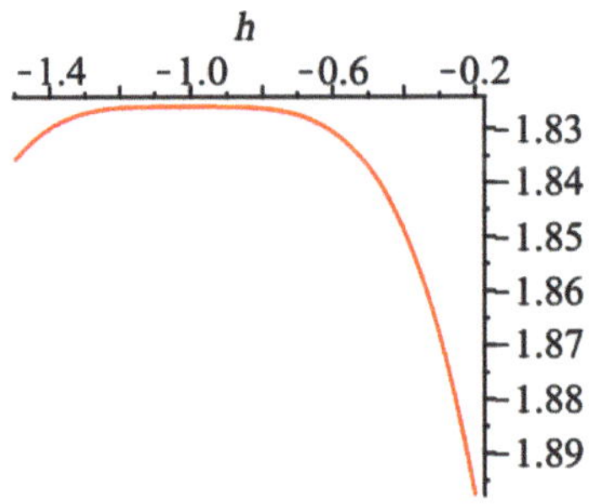

$$A(x) := 1 + 0.000001825547545\,x^{16} - 9.663683135\ 10^{-8}\,x^{18} + 0.1623062480\,x^4 + 0.00693599216\,x^8 - 0.04583549822\,x^6 + 0.0001749435065\,x^{12} - 0.00002208088169\,x^{14} - 1.122272668\,x^2 - 0.001288665903\,x^{10}$$

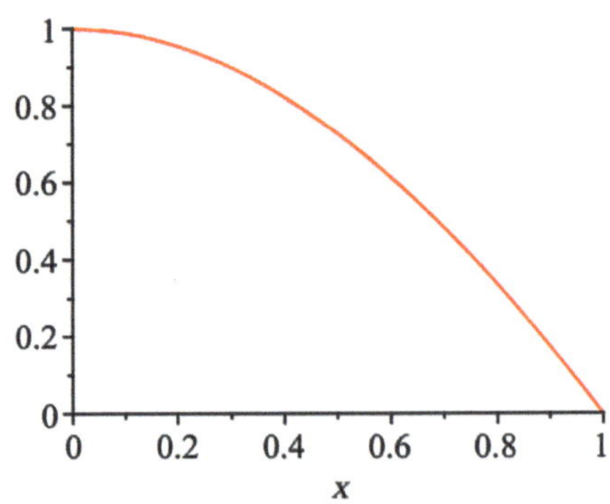

A8. FORTRAN CODE FOR LBM (LID DRIVEN CAVITY)

```
!!!!!!!!!!!!!!!!!!!!!!!!!!!LBM lid-driven cavity Isothermal!!!!!!!!!!!!!!!!!!!!!!!!!!!!!!!!!!!

parameter (n=100,m=100)

real f(0:8,0:n,0:m)

real feq,rho(0:n,0:m)

real w(0:8),cx(0:8),cy(0:8)

real u(0:n,0:m), v(0:n,0:m)

integer i

real visco, omega

open(2,file='uvfield.txt')

open(3,file='uvely.txt')

open(4,file='uvelx.txt')

uo=0.10

sumvelo=0.0

rhoo=5.00

dx=1.0

dy=dx

dt=1.0

visco=0.01

Re=uo*m/visco

print *, "Re=", Re


omega=1/(3.*visco+0.5)

mstep=50000


cx(:)=(/0.0,1.0,0.0,-1.0,0.0,1.0,-1.0,-1.0,1.0/)

cy(:)=(/0.0,0.0,1.0,0.0,-1.0,1.0,1.0,-1.0,-1.0/)

w(:)=(/4./9.,1./9.,1./9.,1./9.,1./9.,1./36.,1./36.,1./36.,1./36./)
```

```
!!!!!!!!!!!!!!!!!!!!!!Initial condition!!!!!!!!!!!!!!!!!!
do j=0,m
do i=0,n
rho(i,j)=rhoo
u(i,j)=0.0
v(i,j)=0.0
end do
end do
do i=1,n-1
u(i,m)=uo
v(i,m)=0.0
end do
!!!!!!!!!!!!!!!!!!!!!!!!!!!!!!!!!!!!!!!!!!!!!!!!!!!!!!!!!!!!
! main loop
1 do kk=1,mstep
call collesion(u,v,f,feq,rho,omega,w,cx,cy,n,m)
call streaming(f,n,m)
call sfbound(f,n,m,uo)
call rhouv(f,rho,u,v,cx,cy,n,m)
print *,kk,u(n/2,m/2),v(n/2,m/2), mstep-kk
END DO
! end of the main loop
call result(u,v,rho,uo,n,m)
stop
end
! end of the main program
subroutine collesion(u,v,f,feq,rho,omega,w,cx,cy,n,m)
real f(0:8,0:n,0:m),rho(0:n,0:m)
real w(0:8),cx(0:8),cy(0:8)
real u(0:n,0:m),v(0:n,0:m)
DO i=0,n
```

```
DO j=0,m
t1=u(i,j)*u(i,j)+v(i,j)*v(i,j)
DO k=0,8
t2=u(i,j)*cx(k)+v(i,j)*cy(k)
feq=rho(i,j)*w(k)*(1.0+3.0*t2+4.50*t2*t2-1.50*t1)
f(k,i,j)=omega*feq+(1-omega)*f(k,i,j)
END DO
END DO
END DO
return
end

subroutine streaming(f,n,m)
real f(0:8,0:n,0:m)
! streaming
do j=0,m
do i=n,1,-1 ! RIGHT TO LEFT
f(1,i,j)=f(1,i-1,j)
end do
do i=0,n-1 ! LEFT TO RIGHT
f(3,i,j)=f(3,i+1,j)
end do
end do
do j=m,1,-1 ! TOP TO BOTTOM
do i=0,n
f(2,i,j)=f(2,i,j-1)
end do
do i=n,1,-1
f(5,i,j)=f(5,i-1,j-1)
end do
do i=0,n-1
f(6,i,j)=f(6,i+1,j-1)
```

```
end do
end do
do j=0,m-1 !BOTTOM TO TOP
do i=0,n
f(4,i,j)=f(4,i,j+1)
end do
do i=0,n-1
f(7,i,j)=f(7,i+1,j+1)
end do
do i=n,1,-1
f(8,i,j)=f(8,i-1,j+1)
end do
end do
return
end

subroutine sfbound(f,n,m,uo)
real f(0:8,0:n,0:m)
do j=0,m
! bounce back on west boundary
f(1,0,j)=f(3,0,j)
f(5,0,j)=f(7,0,j)
f(8,0,j)=f(6,0,j)
! bounce back on east boundary
f(3,n,j)=f(1,n,j)
f(7,n,j)=f(5,n,j)
f(6,n,j)=f(8,n,j)
end do
! bounce back on south boundary
do i=0,n
f(2,i,0)=f(4,i,0)
f(5,i,0)=f(7,i,0)
```

```
f(6,i,0)=f(8,i,0)
end do
! moving lid, north boundary
do i=1,n-1
rhon=f(0,i,m)+f(1,i,m)+f(3,i,m)+2.*(f(2,i,m)+f(6,i,m)+f(5,i,m))
f(4,i,m)=f(2,i,m)
f(8,i,m)=f(6,i,m)+rhon*uo/6.0
f(7,i,m)=f(5,i,m)-rhon*uo/6.0
end do
return
end
subroutine rhouv(f,rho,u,v,cx,cy,n,m)
real f(0:8,0:n,0:m),rho(0:n,0:m),u(0:n,0:m),v(0:n,0:m),cx(0:8),cy(0:8)
do j=0,m
do i=0,n
ssum=0.0
do k=0,8
ssum=ssum+f(k,i,j)
end do
rho(i,j)=ssum
end do
end do
do i=1,n
rho(i,m)=f(0,i,m)+f(1,i,m)+f(3,i,m)+2.*(f(2,i,m)+f(6,i,m)+f(5,i,m))
end do
do i=1,n
do j=1,m-1
usum=0.0
vsum=0.0
do k=0,8
usum=usum+f(k,i,j)*cx(k)
vsum=vsum+f(k,i,j)*cy(k)
```

```
end do
u(i,j)=usum/rho(i,j)
v(i,j)=vsum/rho(i,j)
end do
end do
return
end
subroutine result(u,v,rho,uo,n,m)
real u(0:n,0:m),v(0:n,0:m)
real rho(-1:n,-1:m),strf(-1:n,-1:m)
! streamfunction calculations
strf(0,0)=0.
do i=0,n
rhoav=0.5*(rho(i-1,0)+rho(i,0))
if(i.ne.0) strf(i,0)=strf(i-1,0)-rhoav*0.5*(v(i-1,0)+v(i,0))
do j=1,m
rhom=0.5*(rho(i,j)+rho(i,j-1))
strf(i,j)=strf(i,j-1)+rhom*0.5*(u(i,j-1)+u(i,j))
end do
end do

write(2,*)"VARIABLES=X, Y, U, V, S"
write(2,*)"ZONE ","I=",n+1,"J=",m+1,",","F=BLOCK"
do j=0,m
write(2,*)(i,i=0,n)
end do
do j=0,m
write(2,*)(j,i=0,n)
end do
do j=0,m
write(2,*)(u(i,j),i=0,n)
end do
```

```
do j=0,m
write(2,*)(v(i,j),i=0,n)
end do
do j=0,m
write(2,*)(strf(i,j),i=0,n)
end do
do j=0,m
write(3,*)j/float(m),u(n/2,j)/uo,u(n/4,j)/uo,u(3*n/4,j)/uo
end do
do i=0,m
write(4,*) i/float(n),v(i,m/2)/uo
end do
return
end
! ============end of the program
```

A9. FORTRAN CODE FOR LBM (NATURAL CONVECTION IN A CAVITY)

```
!!!!!!!!!!!!!!!!LBM 2D Natural convection in a Differentially Heated Enclosure!!!!!!!!!!!!!!!!!!!!
parameter (n=100,m=100)
real f(0:8,0:n,0:m)
real feq(0:8,0:n,0:m),rho(0:n,0:m)
real w(0:8),cx(0:8),cy(0:8)
real u(0:n,0:m),v(0:n,0:m)
real g(0:8,0:n,0:m),geq(0:8,0:n,0:m),th(0:n,0:m)
integer i
cx(:)=(/0.0,1.0,0.0,-1.0,0.0,1.0,-1.0,-1.0,1.0/)
cy(:)=(/0.0,0.0,1.0,0.0,-1.0,1.0,1.0,-1.0,-1.0/)
w(:)=(/4./9.,1./9.,1./9.,1./9.,1./9.,1./36.,1./36.,1./36.,1./36./)
sumvelo=0.0
rhoo=6.00
```

```
dx=1.0
dy=dx
dt=1.0
tw=1.0
th=0.0
ra=1.0e3
pr=0.71
visco=0.02
alpha=visco/pr
pr=visco/alpha
gbeta=ra*visco*alpha/(float(m*m*m))
omega=1/(3.*visco+0.5)
omegat=1/(3.*alpha+0.5)
mstep=60000
!!!!!!!!!!!!!!!initial condition!!!!!!!!!!!!!!
do j=0,m
do i=0,n
rho(i,j)=rhoo
u(i,j)=0.0
v(i,j)=0.0
end do
end do
do i=0,n
u(i,m)=0.0
v(i,m)=0.0
end do
!!!!!!!!!!!!!!!!!!!!!!!!!!!!!!!!!!!!!!!!!!!!!!!!!!!!
!main loop
1 do kk=1,mstep
call collesion(u,v,f,feq,rho,omega,w,cx,cy,n,m,th,gbeta)
call streaming(f,n,m)
```

```
call bounceb(f,n,m)
call rhouv(f,rho,u,v,cx,cy,n,m)
! -----------------------------
!collesion for scalar
call collt(u,v,g,geq,th,omegat,w,cx,cy,n,m)
!streaming for scalar
call streaming(g,n,m)
call gbound(g,tw,w,n,m)
call tcalcu(g,th,n,m)
END DO
! end of the main loop
call result(u,v,rho,th,uo,n,m,ra)
stop
end
! end of the main program
subroutine collesion(u,v,f,feq,rho,omega,w,cx,cy,n,m,th,gbeta)
real f(0:8,0:n,0:m)
real feq(0:8,0:n,0:m),rho(0:n,0:m)
real w(0:8),cx(0:8),cy(0:8)
real u(0:n,0:m),v(0:n,0:m),th(0:n,0:m)
tref=0.5
do i=0,n
do j=0,m
t1=u(i,j)*u(i,j)+v(i,j)*v(i,j)
do k=0,8
t2=u(i,j)*cx(k)+v(i,j)*cy(k)
force=3.0*w(k)*gbeta*(th(i,j)-tref)*cy(k)*rho(i,j)
if(i.eq.0.or.i.eq.n) force=0.0
if(j.eq.0.or.j.eq.m) force=0.0
feq(k,i,j)=rho(i,j)*w(k)*(1.0+3.0*t2+4.5*t2*t2-1.50*t1)
f(k,i,j)=omega*feq(k,i,j)+(1.-omega)*f(k,i,j)+force
end do
```

```
end do
end do
return
end
subroutine collt(u,v,g,geq,th,omegat,w,cx,cy,n,m)
real g(0:8,0:n,0:m),geq(0:8,0:n,0:m),th(0:n,0:m)
real w(0:8),cx(0:8),cy(0:8)
real u(0:n,0:m),v(0:n,0:m)
do i=0,n
do j=0,m
do k=0,8
geq(k,i,j)=th(i,j)*w(k)*(1.0+3.0*(u(i,j)*cx(k)+v(i,j)*cy(k)))
g(k,i,j)=omegat*geq(k,i,j)+(1.0-omegat)*g(k,i,j)
end do
end do
end do
return
end
subroutine streaming(f,n,m)
real f(0:8,0:n,0:m)
! streaming
do j=0,m
do i=n,1,-1 ! RIGHT TO LEFT
f(1,i,j)=f(1,i-1,j)
end do
do i=0,n-1 ! LEFT TO RIGHT
f(3,i,j)=f(3,i+1,j)
end do
end do
do j=m,1,-1 ! TOP TO BOTTOM
do i=0,n
f(2,i,j)=f(2,i,j-1)
```

```
end do
do i=n,1,-1
f(5,i,j)=f(5,i-1,j-1)
end do
do i=0,n-1
f(6,i,j)=f(6,i+1,j-1)
end do
end do
do j=0,m-1 !BOTTOM TO TOP
do i=0,n
f(4,i,j)=f(4,i,j+1)
end do
do i=0,n-1
f(7,i,j)=f(7,i+1,j+1)
end do
do i=n,1,-1
f(8,i,j)=f(8,i-1,j+1)
end do
end do
return
end
subroutine bounceb(f,n,m)
real f(0:8,0:n,0:m)
do j=0,m
! West Boundary
f(1,0,j)=f(3,0,j)
f(5,0,j)=f(7,0,j)
f(8,0,j)=f(6,0,j)
! East Boundary
f(3,n,j)=f(1,n,j)
f(7,n,j)=f(5,n,j)
f(6,n,j)=f(8,n,j)
```

```
end do
do i=0,n
! South Boundary
f(2,i,0)=f(4,i,0)
f(5,i,0)=f(7,i,0)
f(6,i,0)=f(8,i,0)
! North Boundary
f(4,i,m)=f(2,i,m)
f(8,i,m)=f(6,i,m)
f(7,i,m)=f(5,i,m)
end do
return
end
subroutine gbound(g,tw,w,n,m)
real g(0:8,0:n,0:m)
real w(0:8),tw
tw=1.0
! Boundary Conditions
! West Boundary Condition, T=1
do j=0,m
g(1,0,j)=tw*(w(1)+w(3))-g(3,0,j)
g(5,0,j)=tw*(w(5)+w(7))-g(7,0,j)
g(8,0,j)=tw*(w(8)+w(6))-g(6,0,j)
end do
! East Boundary Condition, T=0
do j=0,m
g(6,n,j)=-g(8,n,j)
g(3,n,j)=-g(1,n,j)
g(7,n,j)=-g(5,n,j)
end do
! Top Boundary Condition, Adiabatic
do i=0,n
```

```
g(8,i,m)=g(8,i,m-1)
g(7,i,m)=g(7,i,m-1)
g(6,i,m)=g(6,i,m-1)
g(5,i,m)=g(5,i,m-1)
g(4,i,m)=g(4,i,m-1)
g(3,i,m)=g(3,i,m-1)
g(2,i,m)=g(2,i,m-1)
g(1,i,m)=g(1,i,m-1)
g(0,i,m)=g(0,i,m-1)
end do
! Bottom Boundary Condition, Adiabatic
do i=0,n
g(1,i,0)=g(1,i,1)
g(2,i,0)=g(2,i,1)
g(3,i,0)=g(3,i,1)
g(4,i,0)=g(4,i,1)
g(5,i,0)=g(5,i,1)
g(6,i,0)=g(6,i,1)
g(7,i,0)=g(7,i,1)
g(8,i,0)=g(8,i,1)
g(0,i,0)=g(0,i,1)
end do
return
end
subroutine tcalcu(g,th,n,m)
real g(0:8,0:n,0:m),th(0:n,0:m)
do j=0,m
do i=0,n
ssumt=0.0
do k=0,8
ssumt=ssumt+g(k,i,j)
end do
```

```
th(i,j)=ssumt
end do
end do
return
end
subroutine rhouv(f,rho,u,v,cx,cy,n,m)
real f(0:8,0:n,0:m),rho(0:n,0:m),u(0:n,0:m),v(0:n,0:m),cx(0:8),cy(0:8)
do j=0,m
do i=0,n
ssum=0.0
do k=0,8
ssum=ssum+f(k,i,j)
end do
rho(i,j)=ssum
end do
end do
DO i=0,n
DO j=0,m
usum=0.0
vsum=0.0
DO k=0,8
usum=usum+f(k,i,j)*cx(k)
vsum=vsum+f(k,i,j)*cy(k)
END DO
u(i,j)=usum/rho(i,j)
v(i,j)=vsum/rho(i,j)
END DO
END DO
return
end
subroutine result(u,v,rho,th,uo,n,m,ra)
real u(0:n,0:m),v(0:n,0:m),th(0:n,0:m)
```

```
real strf(0:n,0:m),rho(-1:n,0:m)
open(5,file='streamt.txt')
open(2,file='uvfield.txt')
! Stream function Calculations
strf(0,0)=0.
do i=0,n
rhoav=0.5*(rho(i-1,0)+rho(i,0))
if(i.ne.0) strf(i,0)=strf(i-1,0)-rhoav*0.5*(v(i-1,0)+v(i,0))
do j=1,m
rhom=0.5*(rho(i,j)+rho(i,j-1))
strf(i,j)=strf(i,j-1)+rhom*0.5*(u(i,j-1)+u(i,j))
end do
end do
! ___________________
write(2,*)"VARIABLES=X,Y,U,V"
write(2,*)"ZONE ","I=",n+1,"J=",m+1,",","F=BLOCK"
do j=0,m
write(2,*)(i,i=0,n)
end do
do j=0,m
write(2,*)(j,i=0,n)
end do
do j=0,m
write(2,*)(u(i,j),i=0,n)
end do
do j=0,m
write(2,*)(v(i,j),i=0,n)
end do
write(5,*)"VARIABLES=X, Y, S, T"
write(5,*)"ZONE ","I=",n+1,"J=",m+1,",","F=BLOCK"
do j=0,m
```

```
write(5,*)(i,i=0,n)
end do
do j=0,m
write(5,*)(j,i=0,n)
end do
do j=0,m
write(5,*)(strf(i,j),i=0,n)
end do
do j=0,m
write(5,*)(th(i,j),i=0,n)
end do
return
end
```

Index

D

E

F

G

H

N

U

W

Kolibri 360 GmbH
Ferdinand-Jühlke-Straße 7
99095 Erfurt, Deutschland
produktsicherheit@kolibri360.de